STORYING MULTIPOLAR CLIMES OF THE HIMALAYA, ANDES AND ARCTIC

This book initiates multipolar climate/clime studies of the world's altitudinal and latitudinal highlands with terrestrial, experiential, and affective approaches. Framed in the environmental humanities, it is an interdisciplinary, comparative study of the mutually-embodied relations of climate, nature, culture, and place in the Himalaya, Andes, and Arctic.

Innovation-driven, the book offers multipolar clime case studies through the contributors' historical findings, ethnographic documentations, and diverse conceptualizations and applications of clime, an overlooked but returning notion of place embodied with climate history, pattern, and changes. The multipolar clime case studies in the book are geared toward deeper, lively explorations and demonstrations of the translatability, interchangeability, and complementarity between the notions of clime and climate. "Multipolar" or "multipolarity" in this book connotes not only the two polar regions and the tectonically shaped highlands of the earth but also diversely debated perspectives of climate studies in the broadest sense. Contributors across the twelve chapters come from diverse fields of social and natural sciences and humanities, and geographically specialize, respectively, in the Himalayan, Andean, and Arctic regions.

The first comparative study of climate change in altitudinal and latitudinal highlands, this will be an important read for students, academics, and researchers in environmental humanities, anthropology, climate science, indigenous studies, and ecology.

Dan Smyer Yü is Kuige Professor of Ethnology at Yunnan University, China, a Global Faculty Member of the University of Cologne, Germany, and the co-lead of Himalayan University Consortium Thematic Working Group on Environmental Humanities.

Jelle J.P. Wouters is Associate Professor of Anthropology and Sociology at Royal Thimphu College (RTC), Bhutan, and the chair of the Himalayan Centre for Environmental Humanities (HCEH).

"Just as the summit of a mountain gives a new viewpoint, this important collection gives a fresh and critically needed perspective in the environmental humanities. Edited by leading scholars in the field, the chapters collectively and convincingly make a case for 'multipolar clime studies' as a key area of research. Grounded in rich case studies and with reflective, careful, and engaging storytelling, this is environmental humanities at its best."

Emily O'Gorman, *Associate Professor, Discipline of Geography and Planning, Macquarie School of Social Sciences, Macquarie University, Australia*

"This important book aims at revolutionizing climate science. It proffers the concept of 'clime' to connect technical insight with human experience. As ecological beings, we are embedded in more-than-human geographies and histories that constantly shape our relationships with water, earth, and air. This book shows that we can face planetary climate crises only by taking such diverse local comprehensions and practices very seriously. A resounding call to both climatology and the environmental humanities, it is a truly pioneering work."

Willem van Schendel, *Professor of Modern Asian History, Amsterdam University, The Netherlands*

"This is a remarkable multidisciplinary book that initiates new climate studies in multipolar regions. The authors are at once rigorous and imaginative, comprehensive and creative. This book will speak to those interested in the climate challenges and solutions that are both local and global, affective and discursive, culturally aware and practically relevant. An indispensable and path breaking book for specialists and non-specialists alike."

Mary Evelyn Tucker, *Yale Forum on Religion and Ecology, USA; Co-author, Journey of the Universe*

Routledge Environmental Humanities

Series editors: Scott Slovic (University of Idaho, USA), **Joni Adamson** (Arizona State University, USA), and **Yuki Masami** (Aoyama Gakuin University, Japan)

The *Routledge Environmental Humanities* series is an original and inspiring venture recognising that today's world agricultural and water crises, ocean pollution and resource depletion, global warming from greenhouse gases, urban sprawl, overpopulation, food insecurity and environmental justice are all *crises of culture*.

The reality of understanding and finding adaptive solutions to our present and future environmental challenges has shifted the epicenter of environmental studies away from an exclusively scientific and technological framework to one that depends on the human-focused disciplines and ideas of the humanities and allied social sciences.

We thus welcome book proposals from all humanities and social sciences disciplines for an inclusive and interdisciplinary series. We favour manuscripts aimed at an international readership and written in a lively and accessible style. The readership comprises scholars and students from the humanities and social sciences and thoughtful readers concerned about the human dimensions of environmental change.

For more information about this series, please visit: www.routledge.com/Routledge-Environmental-Humanities/book-series/REH

STORYING MULTIPOLAR CLIMES OF THE HIMALAYA, ANDES AND ARCTIC

Anthropocenic Climate and Shapeshifting Watery Lifeworlds

Edited by
Dan Smyer Yü and Jelle J.P. Wouters

Routledge
Taylor & Francis Group
LONDON AND NEW YORK

earthscan
from Routledge

Designed cover image: Astrid Oberborbeck Andersen - Kangerlussuaq fjord and settlement, Greenland

First published 2023
by Routledge
4 Park Square, Milton Park, Abingdon, Oxon OX14 4RN

and by Routledge
605 Third Avenue, New York, NY 10158

Routledge is an imprint of the Taylor & Francis Group, an informa business

British Library Cataloguing-in-Publication Data
A catalogue record for this book is available from the British Library

ISBN: 978-1-032-38826-7 (hbk)
ISBN: 978-1-032-38835-9 (pbk)
ISBN: 978-1-003-34702-6 (ebk)

DOI: 10.4324/9781003347026

Typeset in Bembo
by codeMantra

CONTENTS

FIGURES

TABLES

ACKNOWLEDGMENTS

This book is the latest outcome of the generous support of the Himalayan University Consortium (HUC) led by Dr. Chi Huyen Truong (a.k.a Shachi). Housed at the International Centre for Integrated Mountain Development (ICIMOD), HUC, a network of ninety universities and research institutions based in the Himalayan region and around the world, has emerged as a strong supporter and practitioner of environmental humanities in Asia's Global Souths over the last half decade. We, editors and authors, are exceedingly grateful to HUC's global vision, interregional network, interdisciplinary wisdom, and policy engagements with the pressing issues of our time under the directorship of David Molden, Eklabya Sharma and Philippus Wester. Without its generous support of our publishing workshop, this book would not have been possible. We would like to give our heartfelt thanks to HUC's dedication to building bridges between the worlds of policy, social/natural sciences and humanities, and the public, especially shown in its outreach to scientists and scholars specializing in Andean and Arctic climate studies for this book project. We are truly indebted to Shachi and her colleagues Achala Sharma, Bhawana Syangden, Kritika Sharma, and Sanhita Sahasrabudhe. As HUC's secretariat lead, Shachi is an exemplary leader in the multifaceted field of sustainable development. This book is one of the newest examples of her tireless endeavors to synergize the collective wisdom of diverse cultural and national origins for the common good of the Himalayan region and the world.

Initiated by the HUC Thematic Working Group on Environmental Humanities co-founded by Dan Smyer Yü and Erik de Maaker, this book project was fortunate to partner with My Climate Risk, a Lighthouse Activity of World Climate Research Programme (WCRP) co-chaired by Theodore Shepherd and Regina Rodrigues, The Global South Studies Center at the University of Cologne led by Michael Bollig, The National Centre for Borderlands Ethnic

Studies in Southwest China (NaCBES) at Yunnan University led by Guan Kai, and the Himalayan Centre for Environmental Humanities (HCEH) led by Jelle J.P. Wouters at Royal Thimphu College. The innovative perspectives and advice from the leading scientists and scholars of our partnering institutions effectively co-shaped the concept note and the interdisciplinary orientation of this book project. This is another example of the innovative strengths empowered by collective wisdom situated in the HUC global network. We owe our profuse thanks to these partners.

Royal Thimphu College (RTC), the co-host of our book project, was the backbone of our week-long online workshop held in late 2021. The new norm brought by the Covid-19 pandemic presented challenging situations as well as giving us opportunities to test our resilience and adaptive creativity. Throughout the preparation and the completion of our workshop, RTC proved itself as a resourceful institution and a regional base of cutting-edge environmental studies. President Tshewang Tandin's warm welcoming speech, Dean Shivaraj Bhattarai's academic support, and Dean Samir Patel's administering of the workshop organizing team, the preparation of the online conference venue, and the globally accommodating schedules for individual presentations from Asia, Australia, Europe, North America, and South America all seamlessly ensured the success of our publishing workshop. We can't thank RTC enough for the fruitful outcome of our presentations, discussions, and peer thought exchanges.

For weaving the book together with the individual chapters and their intertextual conversations, we owe much debt to Sunil Amrith, Karsten Paerregaard, and Astrid Oberborbeck Andersen—the three keynote speakers of our publishing workshop. Their keynote addresses set the tone of the book and laid the foundation of the inter-polar conversations between our authors, respectively, specialized in the Andes, Arctic, and the Himalaya. Drawing insights from their historical and ethnographic research, Sunil, Karsten, and Astrid opened the path for their peer authors to visualize how water, in its liquid, solid, and vaporous states, moves altitudinally and latitudinally to shape climate patterns and facilitate climate changes in diverse lifeworlds. We are truly thankful for their inspiring contributions to the book.

Our indebtedness equally goes to Arupjyoti Saikia and Willem van Schendel. As a co-organizer of the publishing workshop, Arupjyoti devoted time to our organizing meetings and offering ideas for thematizing our different sessions. Due to time constraints, he was regrettably not able to join our post-workshop editorial work, but his inspirational input and support to the organization of our workshop sessions nevertheless found their fruitful metamorphosis in this book.

Willem's work is internationally celebrated. In the Himalayan worlds, Willem is widely known for his tireless, inspiring mentorship, and brilliant advice for both early career and well-published scholars. Without exception, the HUC Thematic Working Group on Environmental Humanities has been nourished by Willem's participating mentorship since its inception in 2017. Again, this time, his charismatic presence naturally drew crowds around him, revealing how his comments and suggestions point to the glimmers of light at the end of the tunnel for many of our authors. We are deeply indebted to his attentive mentorship and environmental vision of riverine plains as highlands flattened.

We would also like to extend our deepest thanks to our peer scholars around the world who supported our book project in various capacities. They are Eduardo Erazo Acosta, Elizabeth Allison, Milinda Banerjee, Evan Berry, Michael Bollig, Susanne Brandtstädter, Gunnel Cederlöf, Mareile Flitsch, Rajkumar Gade, Clemens Greiner, John Grim, Stéphane Gros, Guan Kai, Narayan Gyawali, He Ming, Michael T. Heneise, S. Mohammmed Irshad, Bengt G. Karlsson, Kinley Dorji, Robert LaFleur, Christof Lamer, Marco Lazzarotti, Li Zhinong, Lü Zhi, David Ludden, Elizabeth McAnally, Jean Michaud, Emily O'Gorman, Philipp Öhlmann, Joy Pachuau, Jean-Baptiste Pettier, Almut-Barbara Renger, Kate Rigby, Mandy Sadan, Sara Shneiderman, Juliane Stork, Tsebhe, Tsering Shakya, Guido Sprenger, Tanka B. Subba, Thinley Dema, Mary Evelyn Tucker, Mark Turin, Sarah Turner, Gustavo Valdivia, Thom van Dooren, Ute Wallenböck, Deki Yangzom, Li Zhang, Zhao Chunsheng, and Zhu Min.

Dan Smyer Yü and Jelle J.P. Wouters

CONTRIBUTORS

Ambika Aiyadurai is an Assistant Professor at the Indian Institute of Technology – Gandhinagar. Her research examines the social dimensions of wildlife conservation with a special interest in human-animal relations and community-based conservation projects. Her ongoing and long-term research aims are to understand how local and global forces shape human-animal relations. She has a PhD in Anthropology at the National University of Singapore. She is trained in both natural and social sciences with masters' degrees in Wildlife Sciences from the Wildlife Institute of India (Dehradun) and Anthropology, Environment and Development from University College London. She has published articles in *Conservation & Society* and co-edited a special issue in *Interdisciplinary Studies in Literature and Environment*. Her monograph *Tigers Are Our Brothers: Anthropology of Wildlife Conservation Northeast India* was published by OUP (2021).

Sunil Amrith is the Renu and Anand Dhawan Professor of History at Yale University, and current chair of the South Asian Studies Council. His research focuses on the movements of people and the ecological processes that have connected South and Southeast Asia. He is a 2017 MacArthur Fellow and recipient of the 2016 Infosys Prize in Humanities. He is the author of *Unruly Waters: How Rains, Rivers, Coasts, and Seas Have Shaped Asia's History* (Basic Books 2018), *Crossing the Bay of Bengal: The Furies of Nature and the Fortunes of Migrants* (Harvard University Press, 2013), *Migration and Diaspora in Modern Asia* (Cambridge University Press, 2011), and *Decolonizing International Health: South and Southeast Asia, 1930–1965* (Palgrave, 2006).

Astrid Oberborbeck Andersen holds a PhD in Social Anthropology from the University of Copenhagen and is Associate Professor of Techno-Anthropology at Aalborg University. Her research centers on human-environment relations and on how to make anthropological perspectives matter in interdisciplinary research on ecosystems, environmental relations, and climatic crises, as well as in public life. Astrid practices detailed and critical ethnography, and she has extensive fieldwork experience in Peru and Greenland. She has published on themes such as water politics, wildlife and environmental management, and the entanglement of digital and environmental worlds. She is a coeditor of the book *Anthropology Inside Out: Fieldworkers Taking Note* (Sean Kingston Publishing 2020) and the anthology *Rubber Boots Methods for the Anthropocene: Doing Fieldwork in Multispecies Worlds* (University of Minnesota Press, 2023).

Kalzang Dorjee Bhutia is a Visiting Scholar in the Asian Studies Program at the University of California Riverside. He is originally from western Sikkim, India, and completed his PhD in Buddhist Studies at the University of Delhi. He is currently completing a monograph on the environmental history of Sikkimese Buddhism. Publications related to this project include "Trees as Village Protectors, Guru Rinpoche's Wayfinders, and Adopted Family Members: Arboreal Imagination, Agency, and Relationality in Sikkim," in *Worldviews* 25.2 (2021); "Purifying Multispecies Relations in the Valley of Abundance: The Riwo Sangchö Ritual as Environmental History and Ethics in Sikkim," in *MAVCOR* 2 (2021); and "Foxes, Yetis, and Bulls as Lamas: Human-Animal Interactions as Resource for Exploring Buddhist Ethics in Sikkim," in *Journal of Buddhist Ethics* 25 (2018).

Malene K. Brandshaug holds a PhD in Social Anthropology from the University of Gothenburg (2020) and is a postdoctoral fellow at Oslo Metropolitan University. Her doctoral dissertation *Liquid Landscapes: Human-Water Interactions and Water Scarcity in Yanque, Peru* is based on in-depth ethnographic fieldwork in the Peruvian Andes and concerns more-than-human relations, water practices, and questions of decoloniality among small-scale farmers. Malene is currently working on her postdoctoral project Rights of Waters: An Ethnographic Study of Human-Environment Relations and Local Mobilization in Peru and Norway. She is a member of the research project Riverine Rights: Exploring the Currents and Consequences of Legal Innovations on the Rights of Rivers. Her publications cover topics such as human-environment relationships, water politics, infrastructure, climate change, environmental management, and indigeneity.

Anders Burman holds a PhD in Social Anthropology from the University of Gothenburg (2009) and is currently Associate Professor in Human Ecology at the School of Global Studies, University of Gothenburg where he teaches Political Ecology. He has published on issues concerning indigenous peoples and movements, activism and activist research, ritual practice, gender politics, climate justice, political ontology, decolonization, and knowledge production with a geographical focus on the Bolivian Andes. He is the author of *Indigeneity and Decolonization in the Bolivian Andes: Ritual Practice and Activism*, published 2016 by Lexington Books/Rowman & Littlefield.

Razzeko Delley holds a PhD in Tribal Studies and works as an Assistant Professor at Jomin Tayeng Government Model Degree College Roing. He is the author of two books *Idu Mishmi Shamanic Funeral Ritual Ya* (Bookwell Publications, 2018) and *IGU-Study on Four Idu Mishmi Shamanic Rituals* (Bookwell Publications, 2021). He has also contributed chapters to edited books, namely "Narrating the Tribe: Colonial Constructs and Voicing the Savages: A Post-Colonial Study of the Colonial Writings on the Mishmis of Arunachal Pradesh" and "Sineru Monomyth: The Idu Mishmi Funeral Mythography" in Sarit Chaudhury and Tarun Mene ed., *Change and Continuity among Tribes: The Idu Mishmis of Arunachal Pradesh* (Rawat Publication, 2019), and "Shamanic Performance Art During the Death Ritual among the Idu Mishmis of Arunachal Pradesh" in Eli Doye ed., *Cultural Heritage of North East India* (Aayu Publications, 2019).

Ruth Gamble is a Senior Lecturer in History at La Trobe University in Melbourne, Australia. She is a historian of Tibet and the Himalaya, with a particular interest in this region's rapidly changing environment. Her publications include *Reincarnation in Tibetan Buddhism: The Third Karmapa and the Invention of a Tradition* (Oxford University Press, 2018), and she is currently writing a history of the Yarlung Tsangpo (Brahmaputra) River. She has also published articles on the region's ecological politics, literature, and histories. She was a post-doctorate fellow at Ludwig Maximilians University in Munich, and a fellow at Yale University's School of Forestry & Environmental Studies. Before coming to La Trobe University, she taught Tibetan language studies and Asian Religions at the Australian National University. She was recently awarded an Australian Research Council DECRA Fellowship.

Iftekhar Iqbal is Associate Professor of History and International Studies at the Universiti Brunei Darussalam. He primarily works in environmental history with a focus on waterspaces of Asia. Before joining UBD, he held appointments at the University of Dhaka, King's College, London and Humboldt University, Berlin and was a Fellow with the Humboldt Foundation, British Academy and Aga Khan University, London. Iqbal's publications include *The Bengal Delta: Ecology, State and Social Change 1840–1943* (Palgrave 2010). His journal articles have appeared in the *Journal of Asian Studies*, *Environment and History*, and *Modern Asian Studies*. Iqbal's current research explores the connected histories southern China, eastern South Asia, and mainland Southeast Asia through the lens of the Tibetan-Himalayan rivers.

Karsten Paerregaard is Professor Emeritus of Anthropology at School of Global Studies, Gothenburg University. He has been research fellow at the Woodrow Wilson Center (2010–2011) and Swedish Collegium for Advanced Study (2020–2021) and Visiting Professor at the University of Florida (1994–1995, 2018). His research topics are ecology, water, livelihoods, power, migration, and indigenous cosmologies. His current research is focused on climate-driven migration.

His books include *Linking Separate Worlds: Urban Migrants and Rural Lives in Peru* (Berg, 1997), *Peruvians Dispersed: A Global Ethnography of Migration* (Lexington, 2008), and *Return to Sender. The Moral Economy of Peru's Migrant Remittances* (University of California Press, 2015). His forthcoming book is titled *Andean Meltdown. A Climate Ethnography of Water, Power, and Culture in Peru* (University of California Press, 2023). His journal articles have appeared in *HAU. Journal of Ethnographic Research* (2020), *Environmental Communication* (2020), *Water International* (2020), *Water Alternatives* (2019), *Climate and Development* (2018), and *WIRE's Water* (2018).

Theodore G. Shepherd is a dynamical meteorologist whose interests range from fluid dynamical fundamentals to climate modeling and data analysis, with a focus on atmospheric circulation. Ted worked at the University of Toronto before moving to the University of Reading in 2012 as the Grantham Chair in Climate Science. His research since then has highlighted the important role of atmospheric circulation in climate change, which has implications for regional adaptation and societal risk, and has led to the development of the storyline approach to climate risk and uncertainty. Ted has held leadership roles in scientific assessments of both climate (IPCC) and stratospheric ozone (WMO/UNEP), and in the World Climate Research Programme (WCRP), and is a Fellow of the Royal Society. He is increasingly involved in interdisciplinary collaborations, and co-chairs WCRP's "My Climate Risk" Lighthouse Activity. Since March 2022, Ted has a part-time appointment at the Jülich Supercomputing Centre in Germany.

Vandana Singh is Professor of Physics at Framingham State University in Massachusetts. For over a decade, she has been working on a trans-disciplinary, justice-centered pedagogy of climate change, at the intersection of science, social sciences, and the humanities. She has authored an educational case study on Arctic climate change and its impact on the Iñupiaq communities of the Alaskan North Shore, has co-conceptualized and helped run an interdisciplinary summer workshop for school teachers on climate pedagogy, and has contributed to the US National Academies deliberations on interdisciplinarity in the sciences. Vandana was born and raised in India, and is also a writer of speculative fiction, with two short story collections to her credit. She is a Climate Imagination Fellow (2021). Recent academic publications include two book chapters (one co-authored) in *Curriculum and Learning for Climate Action: Toward an SDG 4.7 Roadmap for Systems Change* (UNESCO Brill Publications Fall, 2021).

Dan Smyer Yü is Kuige Professor of Ethnology at Yunnan University and a Global Faculty Member of University of Cologne, Germany. He received his PhD in anthropology from the University of California at Davis in 2006. Currently he is the co-lead of HUC's Thematic Working Group on Himalayan Environmental Humanities, an elected board member of International Society for the Study of Religion, Nature, and Culture (ISSRNC) and a member of the Advisory Board of Yale Forum on Religion and Ecology. He is the author and editor of numerous publications, such as *Mindscaping the Landscape of Tibet: Place, Memorability, Eco-aesthetics* (De Gruyter, 2015), *Trans-Himalayan Borderlands: Livelihoods, Territorialities, Modernities* (Amsterdam University Press, 2017), and *Environmental Humanities in the New Himalayas: Symbiotic Indigeneity, Commoning, Sustainability* (Routledge, 2021). The Swedish Research Council, the British Academy, the Himalayan University Consortium, the China Social Sciences Foundation, and German Ministry of Education have funded his research.

Chi Huyen Truong uses her anthropological training (PhD from University of Toronto, 2001) to create impact by building genuine and productive interdisciplinary partnerships in research and higher education of and for sustainability. She currently leads the Secretariat of the Himalayan University Consortium, a thriving network of ninety university members in eight Hindu Kush Himalayan countries, hosted by the International Centre for Integrated Mountain Development (Kathmandu, Nepal). Her earlier research interests include social memory, education for indigenous children, child poverty, and post-disaster livelihood recovery and resilience in Cambodia, the Philippines, and Nepal. She taught at Vietnam National University (2001–2004) and served as the Lead Qualitative Researcher for Young Lives – an International Study of Childhood Poverty in Vietnam (Oxford University, 2008–2010).

Jelle J.P. Wouters is a social anthropologist and teaches in the Department of Social Sciences at Royal Thimphu College (RTC), Bhutan. He holds an MPhil (Distinction) in social anthropology from the University of Oxford and a PhD in anthropology from North-Eastern Hill University, Shillong, where he was also a Wenner-Gren grantee. Prior to joining RTC in 2015, he taught at Sikkim Central University, India, and was visiting faculty at Eberhard Karls University of Tubingen, Germany, under the "Excellence Initiative" of the German Research Foundation. He is the author of *In the Shadows of Naga Insurgency* (OUP 2018) and *Nagas as a Society Against Voting and Other Essays* (Highlander Press, 2020), the co-author of *Subaltern Studies 2.0: Being Against the Capitalocene* (Chicago/Prickly Paradigm Press, 2022), the editor of *Vernacular Politics in Northeast India* (OUP, 2022), and the co-editor of *The Routledge Handbook of Highland Asia* (2022), and the *Routledge Companion to Northeast India* (2022).

MULTIPOLAR CLIME STUDIES OF THE ANTHROPOCENIC HIMALAYA, ANDES AND ARCTIC

An Introduction

Dan Smyer Yü

Marcel Kurz, a Swiss topographer and mountaineer, was the first person to speak of Mt. Everest (Chomolungma) as "the Globe's Third Pole" after his 1930 expedition (Kurz 1933). Twenty-five years later, Günter Dyhrenfurth's (1955) *The Third Pole: History of the High Himalaya* extended the coverage of Kurz's metaphoric toponym from the highest mountain to the entirety of the Himalaya. Thus, the Himalaya became formally, though metaphorically, associated with the Arctic and the Antarctic. Not too long after, the "Third Pole" was not merely a phrase used by mountain climbers to express their impression of the Himalaya, but had entered the public's environmental consciousness and compelled scientists and scholars to connect the Himalaya with the Earth's two poles for a range of natural scientific and cultural understandings of global water, climate change, and ultimately connecting it with the fate of the Earth. Thanks to Kurze and Dyhrenfurth, what is obvious is that our perception of the Earth has evolved from a bipolar to a tri-polar vision. However, from the geological perspective, the Earth is a *multipolar* planet. The Himalaya is not the only altitudinal highland on Earth. The Andes, the Southern Great Escarpment, and the Rocky Mountains, to name a few such regions, are also parts of the Earth critical for the survival of humans and nonhumans. Additionally, the Arctic and the Antarctic are also home to altitudinal highlands, some of which are formed by ice, such as Greenland's North and South Domes, the Arctic Cordillera, and the Transantarctic Mountains. In the fields of climate and environmental studies, the geological vision of a multipolar Earth compels us to formally initiate, variously, "multipolar climate/clime studies," "multipolar environmental studies," or "multipolar environmental humanities."

Situated in environmental humanities, this book initiates *multipolar clime studies* of the world's altitudinal and latitudinal highlands with terrestrial, experiential, and affective approaches complementary to climate science's numerical

DOI: 10.4324/9781003347026-1

models, projections, and abstract discourses of climate change. The polar regions it covers are the Himalaya, the Andes, and the Arctic. As we are well aware of the ongoing discordant climate debates in both academic and public arenas (Hulme 2009; Bristow and Ford 2016), our initiation of multipolar clime studies is geared toward deeper, lively explorations and demonstrations of the translatability, interchangeability, and complementarity between the notions of clime and climate. "Multipolar" or "multipolarity" here connotes not only the two polar regions and the tectonically shaped highlands of the Earth but also diverse perspectives of climate studies in the broadest sense. In many ways, it is a type of geopolitics that takes the prefix "geo-" as the gravity of our clime studies. Throughout this book, *geo*, as its Greek root—*gaîa*—suggests, signifies both the physical Earth and the mother of all life conceived of as an Earth goddess, and is the integral union of the physical and the spiritual worlds. The indigenous Greek concept of *geo* resonates with its indigenous counterparts in the Himalaya, Andes, and Arctic. Thus, geopolitics in our multipolar clime studies are the interwoven politics of the Earth, humans, and nonhumans inclusive of animals, plants, and local deities, all of whom holistically help us shape our multipolar perspectives on clime/climate change. Given the Anthropocene context of our research, multipolarity also suggests the multiple polarizations of humans with the Earth and nonhuman beings evidenced by accelerated, human-induced global warming and other environmental transformations, and the mass extinctions of nonhuman species. All these human and nonhuman forces of change prompt us to initiate meaningful multipolar clime studies.

Appearing marginally in many climate studies' texts, the word "clime" was equated or juxtaposed with climate until James Rogers Fleming (2010, 7, 8) and Mark Carey and Philipp Garone (2014, 284, 285) invoked it in their call for linking together the local and the global, weather and place, nature and culture, and science and lived experiences to nurture sensible, meaningful understandings of climate change. Clime in their texts is not interchangeably used with climate in specific case studies or in general conceptual discussions because its meaning is largely etymologically associated with local places (Latin and Greek *clima/klima*). Clime thus connotes local not global climate in their works. When we formally use the word in our multipolar clime studies-in-the-making, the meaning of clime varies from local to global contexts and finds innovative applications among our authors. In our individual chapters, readers will find that clime change, either local or global, pertains to a multidimensional, perceivable, and observable transformation of weather conditions, physical environments, human societies, and nonhuman habitats.

Braided together by nine anthropologists, four historians, one climate scientist, one physicist, one literary scholar, and one Buddhologist, this book offers terrestrial understandings of climate knowledge and climate change articulated through our diverse conceptualizations and applications of clime in both global and regional contexts of the Himalaya, Andes, and Arctic. Situated in the greater interscientific and interdisciplinary discourse on climate change, it is intended to be a response to the calls for bridging climate change with daily lives (Dilling and Moser 2007, 15), closing the gap between climate science and the existential meaning of climate (Shepherd and Lloyd 2021, 17, 18), re-establishing the connection "between what we know and how we feel, between our affective, discursive and epistemological selves" (Bristow and Ford 2016, 6), and enacting "a deeper cultural and historical reading of climate and its meaning" (Hulme 2008, 6).

Demonstrated in many of our individual case studies, this book reconnects the disconnected knowing and feeling of climate by starting a critical rethink of humanity as an ecological species (Rose et al. 2012). Being ecological in our chapters means relational entanglements with many other species, rootedness in places, capability of migrating with the terrestrially shifting life-giving forces of the Earth when the fecundity of ancestral lands declines due to extreme weather conditions, and our ability to innovate new modes of human-earth interactions for survival because of environmental crises created by ourselves. When climate changes, it concurrently works through air, water, and land in both creative and destructive manners. Like many of our peer scholars and scientists, we also see climate change both as a proxy of the dual power of the Earth as the life-giver and the life-receiver, and as a consequence of human-induced environmental changes. With the geological agency of humankind in the Anthropocene considered (Chakrabarty 2021, 14, 15; Clark and Yusoff 2017, 5), the Earth appears to be more destructive than life-sustaining with instances of global warming, extreme weather events, environmental degradation, and the frequently involuntary migrations of humans and nonhumans. While climate models inform us of the current and projected average temperature change of the planet, climate change itself or, more accurately, clime change occurs not uniformly but variously in geographically-specific places in meteorological, ecological, and political terms. Rethinking ourselves as ecological beings affords us multiple terrestrial perspectives to recognize the omnipresent kinesis of climate embodied in the physical Earth and in the health, social behaviors, and psyches of humans, animals, and spiritual entities.

This book does not offer ready-made multipolar clime studies, but invites our readers to co-create it. It takes the initiative to expand and innovate the existing idea of clime in the altitudinally and latitudinally understood multipolar contexts. Written in first-person pronouns "we" and "I," this introduction showcases the common approaches to and conceptualizations of clime from the unique case studies of the contributing authors. Its modes of speech shift between "we" and "I" when I, the author, as the book project-maker and co-editor, speak of the common themes across the chapters and when I narrate my own research findings to strengthen my peer authors' common endeavors to deepen clime studies. This introduction thus informs our readers of how we, the contributing authors, climatically interlink the Himalaya, the Andes, and the Arctic in both historical and ethnographic terms, how each of us contributes new perspectives and usages to the powerful but not yet fully enacted notion of clime, and how each of us connects clime with indigenous environmental knowledge and experiences of climate change in a multipolar fashion in the Anthropocene, a scientific hypothesis commonly accepted as a new geological and climatic norm of our time. We hope our findings and conceptual discussions of *clime change* in this geologically and politically multipolar world will yield positive interdisciplinary and policy implications for building a sustainable future.

Interlacing the Anthropocenic Himalaya, Andes, and Arctic

The thread that interlaces the multipolar climes of the Himalaya, the Andes, and the Arctic in our book is water in its solid, liquid, vaporous, saline, and fresh states. Water's omnipresence is integral to shaping the habitability of places as well as individual ecosystems, multispecies worlds, cultures, civilizations, and human affective and spiritual relations with the creativity of the physical environment. Critical to all life communities on Earth, common environmental knowledge informs us that 2.53% of the water content in the hydrosphere is fresh water and 68.7% of this small portion is stored in the glaciers and permanent snow cover of the world's high mountains, and in the Arctic, and Antarctica (Babkin 2003, 13). While the two (traditional) polar regions host over 70% of the world's glaciers and snow cover, the Himalaya hosts 70% of the glaciers outside the polar regions with the remaining 30% mostly found in the Andes (Pant et al. 2018, 41). Himalaya-sourced water, food, and energy support three billion people in a dozen Asian countries (Amrith 2018, 299), and meltwater from the Andes supports ninety-five million people spread out in ten countries in South America (Gonda 2019, 14).

The global significance of the Himalaya, Andes, and Arctic does not stop at their capacities for storing fresh water for humans alone. The fate of the Earth lies in their human-induced climatic role, too, besides their inherent role in the global hydrological cycle (Pant et al. 2018; Pandit 2017; Serreze and Barry 2005; Gonda 2019). As amply documented by natural scientists, the three world regions continue to be severely affected by global climate change, locked inextricably into a positive feedback loop of global warming, accelerated melting of glaciers, and extreme weather events. In order for our readers to have an informed and engaged reading experience, this introduction provides a climatic contextualization of our thematically connected chapters regarding what we mean by the Anthropocenic Himalaya, Andes, and Arctic, and how we connect the three multipolar regions with a historical depth on a global scale.

While most of our authors focus on modern and contemporary climatic, environmental, and social issues between the nineteenth and the twenty-first centuries, the historical periods of the chapters by Gamble and Smyer Yü go back further, to as early as the 1200s, to the Little Ice Age. Although the Little Ice Age remains on the textual margins of this book, it nevertheless holds causal relations with the Anthropocene. To note, we recognize the Anthropocene as a hypothesis in the natural sciences; however, we are equally aware of the growing social scientific, humanistic, and public perceptions of it as if it were a proven reality.

Its starting point debatably varies from the global diffusion of humankind one hundred thousand years ago and the worldwide spread of agriculture 10,000 years ago to the highpoint of the Industrial Revolution nearly two centuries ago and the advent of the explosive increase in the human population half a century ago (Chakrabarty 2018, 30). Here, we lean toward the Industrial Revolution as the starting point or at the least the accelerating point of the Anthropocene (Crutzen and Stoermer 2000).

It was not an accident that the Little Ice Age's end in the mid-1800s marks the start of the Anthropocene. Both are embedded with natural and human forces of change. In particular, the environmental impacts of the Little Ice Age pushed many human societies, especially those from Eurasia, into a seven-century-long period of transregional and cross-continental migrations and propelled us toward the Industrial Revolution. Migration in this historical context entails water movements, weather conditions, individual persons and families looking for environmental refuge beyond their ancestral lands, as well as imperial expansions, colonial conquests, and international commerce. The multipolar worlds of the Himalaya, the Andes, and the Arctic were tightly looped into the impacts of the Little Ice Age and human migrations over time.

The Little Ice Age, a climate event under-researched in the field of environmental humanities, is well remembered in the arts, literature, and histories of Europe and Asia (Lamb 1982; Fagan 2019; Mann 2002; Zhu 1973). In climate science, it is understood as a climate event taking place mostly in the northern hemisphere between the 1200s and 1800s. Its periodization is not clear-cut as the meteorological changes it engendered varied from region to region. The starting point of the Little Ice Age in the case studies of Gamble and Smyer Yü aligns with Brian Fagan's assessment that the cooling began in the Arctic and Greenland in the 1200s and spread to the rest of Europe, Asia, and North America, onward until the mid-1800s (Fagan 2019, 18, 76). If we take the late meteorologist Zhu Kezhen's phenological approach to the historical cooling into consideration, we could push the dates in Asia to the 1100s, the century during which the agricultural productivity and the economic center of China (Song Dynasty) shifted from the north to the southeast coast (Zhu 1972, 175).

The natural scientific storyline of the Little Ice Age is that of an average 2 °C temperature drop (Lamb 1982, 209; Mann 2002, 3). Translated into environmental, livelihood, and political terms, it means latitudinal shifts in the movement of water, decreasing fecundity of pastoral and farm lands, hunger, famine, social unrest, regime changes, and outbound migrations. Longer winters and more snowfall led to the expansion of glaciers and permafrost. As more water shapeshifted from its liquid to its solid state, the temperature decreased and the Earth's surface became drier in many parts of the northern hemisphere.

In Gamble's historical account, the peopling of the Yarlung Tsangpo Gorge (the upper reach of the Brahmaputra) was a multispecies migration process of seeking refuge for survival while at the same time, the Himalaya range was undergoing the last glaciation before the glaciers began to recede in the mid-1800s

(Rowan 2017, 293, 294) when "the Modern Warm Period" began (Fagan 2019, 28). In Smyer Yü's storying of Asian elephants' shrinking habitats, the latitudinal lowering of the Little Ice Age's cold from the Arctic and the Inner Asian steppe to the Tibetan Plateau, the eastern Himalaya, and East Asia is traced with the southward migration of the elephants, as well as that of the Mongols, the Han, and the Manchus. Both the elephants and humans were climatic refugees. Among these multispecies refugees, the Mongols were also building a continental empire succeeded by the Ming and the Qing Dynasties, and the modern Chinese republics. Recent research outcomes based on court records suggest that the Mongols started out not as imperial conquerors, but as climatic refugees, fleeing deteriorating grasslands and subsequent famine. Their military invasion of China, Tibet, and part of Southeast Asia paralleled the trek of millions of Mongolian civilian refugees seeking food and shelter from their agricultural neighbors in the south (Li 2020, 155; Smyer Yü 2021a, 12). After the short-lived Mongol Yuan Dynasty perished, its inter-Asian imperial and territorial legacy continued to be embodied in the polities of its successors. Systematic mining, agricultural expansion, deforestation, and transitioning toward the use of fossil fuel have added to what we now call the anthropogenic forces of climate change (Ruddiman 2007, 2).

In Europe, hunger, famine, political instability, and migration were also the tangible impacts of the Little Ice Age. The Great Famine of the fourteenth century wiped out 10% of the northern European population; and the 1800s saw the famine-driven mass migration of the Irish (Collet and Schuh 2018, 4; Camenisch 2018, 69; Brooke 2014, 332, 374; Fagan 2019, 20). The Irish were not the only Europeans seeking refuge outside their homelands. The entire period of Europeans' "discovery" of the "new world" and building of new countries in non-European homelands on other continents coincided with the Little Ice Age. If we incorporate the climatic forcing of the Little Ice Age into our world historiographies, the "Age of Discovery" and the "Age of Exploration," marked by Christopher Columbus's landing in America in 1492, could also be written as an "age of climatic migration." Seen from this perspective, Europeans migrating out of their homelands could be likened to the Mongols' concurrently looking for refuge in and conquering of eastern Europe, Central Asia, East Asia, and Southeast Asia (Smyer Yü 2021a, 10–14). Similarly, Europeans were both climatic refugees and new world discoverers during the Little Ice Age.

As climatic refugees, building a new life on a new continent was, for poverty-ridden people, a treacherous choice to make. It could mean losing one's freedom or life. For instance, in the 1700s, over half of the European immigrants to the British colonies in the Americas were indentured laborers who sold their freedom for four to seven years. In Virginia, indentured Europeans composed up to 75% of the settler population. In the latter half of the 1800s, 75% of the indentured Europeans were under the age of twenty-five (Galenson 1984, 3). Among the young population, "94 percent of the men were under thirty years old, as were 98 percent of the women. About two-thirds of the men and four-fifths of the women were minors" (Galenson 1978, 44). Indentured servitude and immigration to other continents in the 1700s and 1800s, were contemporaneous with the inception and the advancement of the Industrial Revolution, which brought an increase in productivity and population, but also engendered high rates of unemployment, poverty, and social unrest.

As New World "discoverers," Europeans who settled in the Americas, Australia, and Indo-Pacific islands built "Neo-Europes" on the lands of aboriginal peoples and commenced the "Columbian exchange" (Crosby 2003, 2004). In Alfred Crosby's work, "Neo-Europes" refers to the New World countries with at least 80% of their populations being of European descent; these include Canada, the United States, Argentina, and Australia. Crosby's delineation of Neo-Europe is based on a demographic majority of Europeans. From the postcolonial perspective, however, many large, pre-independence, former European colonies can also be seen as Neo-Europes irrespective of their demographic makeup. For example, colonized South Asia, once ruled by a British demographic minority, was an indispensable resource base for the birth of the Industrial Revolution in the British Isles.

The Columbian exchange, then, was a series of environmental, colonial, political, industrial, technological, and human-induced climatic outcomes resulting from the historical intertwinement of Europe and Neo-Europes during the Little Ice Age, and from the continuing colonial approaches to environmental and population management in postcolonial, non-European nations (Amrith and Smyer Yü, this book). While New World crops diversified human food sources, the spread of Old World (Eurasian) diseases, for example, smallpox, measles, and influenza, as well as colonial violence decimated many indigenous populations (Crosby 2003). Immediately relevant to our book, the Columbian exchange laid the material ground for the worldwide transition from non-fossil to fossil fuels (Cho 2014, 7, 8). The latter was the foundational energy that made Industrial Revolution possible and marked the dawn of the Anthropocene.

So, the historical logic of the consecutive occurrences of the Little Ice Age and the Anthropocene in the mid-1800s are highly relevant for our individual case studies of the anthropogenic changes in the multipolar lifeworlds of the Himalaya, Andes, and Arctic. To be noted, while climate scientists correlate the inception of the Anthropocene with the Industrial Revolution, and the Modern Warm Period is understood as the period in which the global surface temperature rose 0.4 °C in the mid-1800s and a further 1 °C today with the increasing rate of 0.2 °C per decade (Fagan 2004, 250; NASA n.d.), the Little Ice Age rarely appears in the picture of the Anthropocene. In our book, it is an imperative contextual thread tying together the climatic histories of the "three poles" on the global scale.

Although many of us do not bring it up directly in our chapters, the consequences of the Columbian exchange, the Industrial Revolution, and their subsequent anthropogenic imprints are abundantly found in the environmental and social realities of our historical cases and ethnographic sites. Environmental flows and climate change recognize no human ethnic or national boundaries and, thus, leave no one untouched. Their transregional movements and global presence in the Anthropocene present common challenges to life as well as opportunities for adaptive innovations. When water's moving and shapeshifting nature is carefully examined from our comparative lenses, its travel in the sky with winds, flow in the form of rivers through life communities, and embodiment as local deities are increasingly intercepted by anthropogenic forces. At the same time, it is expressing forceful resistance to human-induced changes in the manner of natural disasters and a readjusting of the proportion of its solid, liquid, and vaporous states unfavorable to modern modes of energy use and resulting lifestyles. This is what we mean in brief by the Anthropocenic state of the Himalaya, Andes, and Arctic. Our individual chapters tell unique stories of water and climate change in multipolar climes from the Little Ice Age to the Anthropocene.

Revitalizing Clime in Climate Studies

Throughout the book, we use the word "clime" as a noun, a verb, and a gerund to capture the diverse environmental embodiments of global climate change in the mountains, plateaus, and floodplains of the Himalaya and the Andes, and in the Arctic tundra. By "embodiment," we mean physical place as both recipient and maker of climate change. We thus draw direct equations between climate, clime, and place for the purpose of identifying the planetary forces of change moving between terrestrial worlds and of returning the global to the local and vice versa. So, when place is seen as clime, it is place imbued with climate agency, history, pattern, and change. It may appear to be a passive recipient of climate change understood as numerical global averages in climate science; however, in the sensible world, we recognize clime as an agent of climate change, too, an inherently living dynamic of the Earth. In other words, a clime is both a place-based embodiment and an agent of climate change. As the gravity of this book is found in the concrete, terrestrial side of changing climate conditions over time and across the multipolar planet, it, therefore, matters a great deal when we see place as clime and extend this perceptual and conceptual synonymity to landscape, habitat, country, territory, region, and the planet.

Lexically, clime is only known as a noun that can be etymologically traced to Latin and Greek roots meaning "zone" or "region." When we use it as a noun, we expand its etymological meaning and coin new phrases with it in our individual studies of specific landforms, climatic patterns, ecosystems, environmental changes, and regional environmental histories:

- "Monsoon clime," "anthropocenic clime," "mountain clime," "riverine clime," "oceanic clime," and "clime perspective" (Sunil Amrith and Dan Smyer Yü);
- "Clime universe," "deep clime," "social climing" (Karsten Paerregaard);
- "Geological clime" and "tundra clime (Astrid Oberborbeck Andersen);
- "Clime maker," "ecospiritual climes," and "riverine clime" (Razzeko Delley and Ambika Aiyadurai);

- "Andean climes" and "water clime" (Malene K. Brandshaug);
- "Cryospheric climes" (Vandana Singh);
- "Storylining climes" and "storying clime" (Theodore Shepherd and Chi Huyen Truong);
- "Local clime" (Anders Burman);
- "Elephant clime," "more-than-human clime," "Zomian clime," "collapsing clime," and "clime change" (Dan Smyer Yü);
- "Clime diversity" and "assemblage of climes" (Ruth Gamble);
- "Deep ecological clime" and "terrestrial clime" (Iftekhar Iqbal);
- "Multilateral climes" (Jelle Wouters).

When these coinages are correlated here, they show the meteorological, ecological, and spiritual liveliness of the Earth as well as the intertwinement of nature and culture, and human and nonhuman. The multipolar climes we write about are diverse terrestrial worlds situated in their local and regional ecosystems, shaped by the geological and climatic forces of the Earth, and affected by modern anthropogenic forces. The efficacy of seeing place as clime is shown in the environmentally and socially observable presence of climate forces in our individual chapters. When climate is understood as clime, or concrete places viewed as diverse lifeworlds, climate change is then no longer encoded in the abstractly constructed numerical models but marks itself deeply in the physical environment, moves in the environmental flows, and enters the social behaviors of humans and nonhumans.

The conceptual outcomes we yield from applying clime as a noun afford us integral responses to what Mike Hulme calls "our discordant conversations" (2009, 4) or what Karsten Paerregaard refers to as "climate discord" (2020) concerning the debates on the persistently dichotomized conceptual pairings, namely human versus Earth, culture versus nature, climate science versus local knowledge, and global versus local. These dichotomies have been critiqued myriad times in multiple disciplines, many of which focus truistically on the philosophical and scientific origins of the dichotomies without offering feasible ways out of the expanding maze of the debates from both sides of the fulcrum. Like Chakrabarty, we believe that climate science is a multidiscipline and that those of us situated in the social sciences and humanities need to come to terms with the basics of climatological findings, arguments, and projections (Chakrabarty 2016, xvii). Of course, we expect the same from natural scientists regarding our qualitative approaches and outcomes. The current state of the interdisciplinary critiques of climate science/studies is lively but has, indeed, reached an impasse, meaning that we have not yet been able to move beyond the recurrent realization of the decoupled scientific knowledge of climate and lived experiences of climate change (Bristow and Ford 2016, 6; Jasanoff 2010, 249). Through this book, we are finding solutions in the idea of clime and in our inclusion of the work of natural scientists like Shepherd and Singh.

Situating our book in the field of environmental humanities, we find that our conceptual utilization of clime resonates with the recent critiques of the nature-culture and human-earth hyperseparations in the works of the late Val

Plumwood (2009, 116, 117) and Deborah Bird Rose (2013, 96, 97). Both environmental thinkers did not merely dwell on the epistemological traces of these hyperseparations in the modern West, but offered actionable suggestions "to resituate the human within the environment, and to resituate nonhumans within cultural and ethical domains" (Rose et al. 2012, 3). At the same time, the outcomes from our ecologically-engaged multipolar clime studies are also seeing successes in reversing the detachment of climatological knowledge "from meaning, [and the] distancing [of] received scientific facts from the thick textures of lived experience" (Bristow and Ford 2016, 6) and in recognizing "climate change in the sequences of weather experienced in given places" (Hulme 2009, 3).

Let me elaborate how our integral responses circumvent the dichotomized debates. Originating from Fleming, and Carey and Garone, but further applied in our individual case studies in the lifeworlds of the three multipolar regions, clime in our conceptual discussions throughout this book is an integral concept of climate grounded in specific places and interwoven in the regional and global histories of climate change. It inherently integrates the scientifically compartmentalized atmosphere, hydrosphere, cryosphere, lithosphere, and biosphere—the five components of the Earth's climate system.

Furthermore, given its terrestrial emphasis, the notion of clime in this book comfortably communicates with the concept of the Critical Zone from earth science, referring to the continuous Earth surface "between the top of the tree canopy and the base of weathered rock" (Dybas 2013, 1)—the Earth's most biologically and culturally active sphere (Smyer Yü 2020). Contingent upon the geological and geographical scale of one's individual research, clime can be either local or global or both. To think globally, the entire Critical Zone can be seen as a planetary clime. If we equate clime with the Critical Zone on the global scale, we may well revise the existing definition of the latter as the Earth surface "between *the mountain tops protruding into the troposphere* (the lowest layer of the atmosphere) and the base of weathered rock." The added emphasis is meant to point out that the high mountain tops are much more ecologically and climatically critical than tree tops and that clime, as a terrestrial concept, is not a passive recipient of climate change but shapes and changes climate with its geology, geomorphology, and ecology. The meeting of the mountain tops and the troposphere solidifies the climatic and ecological significance of the altitudinal polar regions. The chapters by Paerregaard, Burman, Gamble, Bhutia, Brandshaug, and Amrith and Smyer Yü in particular show how mountains are simultaneously susceptible to global climate change and are makers of local weather, transregional environmental flows, and of global climate change itself with their high, tectonically folded bodies. The idea of clime, further developed in our case studies, thus affords us an interdisciplinary niche in which natural and social scientists and environmental humanists communicate with one another effectively outside our disciplinary comfort zones.

One common arena of our individual research and conceptual discussions is worth highlighting: this is our historical and ethnographic articulations and discussions of varied local knowledge and experiences of climate change. To take a step back, in comparison with the short history of the word "climate," clime, etymologically indigenous to Europe, is an ancient, meteorologically and environmentally grounded word. In the course of its meaning's evolution, it finds a group of synonymous and associative words, such as climate, acclimate, place, air, atmosphere, ambience, and aroma, to name a few (Boardman 2016). The modern use of climate began in the mid-1800s in England (Möller 2010, 324; Miller 2018, 539). As the quantitatively-oriented modeling of climate change research has expanded, the modern connotation of climate observably reveals experiential detachment from clime. This is the aforementioned gap disconnecting climate science from the experiential knowledge and meanings of climate change. Clime, admittedly, thus has little place in climate science.

However, clime has now begun to find ample room in interdisciplinary fields of climate studies. In our chapters, it comfortably connects with non-European indigenous words for climate and associated environmental knowledge. In her chapter concerning melting ice and environmental transformation in Kangerlussuaq, Greenland, Andersen builds the conceptual connection between clime and *sila*, a pan-Inuit word for weather. While both words connote weathered and storied place, *sila* obviously holds more affectively, cosmologically, and spiritually entwined meanings of weather, place, and consciousness, as it means a relationally-bound physical world on the one hand and the spirit or the breath of the world on the other. While the modern sense of climate refers to the weather conditions over time in the external environment, *sila* is understood as the spiritual force that makes weather and alerts the human and nonhuman residents of the appropriate times for waiting, traveling, and hunting. It is understood as the sentient force making the world alive. Andersen thus adds more conceptual meanings to clime during her discussion of an indigenous concept of weather and climate. Like Andersen, Delley and Aiyadurai build up the interchangeability of clime with *eno*, an Idu Mishmi word for ecospiritual space, in their case study of what they call "riverine climes" in the eastern Himalaya. While *eno* in the physical sense is a habitat of humans, animals, and plants, its mountains, rivers, and streams are revered as the embodiments of spiritual guardians. Like *sila* as the sentient essence of the Inuit homeland, *eno*, as Delley and Aiyadurai ascertain, spiritualizes the dwelling places of the Idu Mishmi, too. Likewise, the chapters by Brandshaug, Iqbal, Bhutia, Paerregaard, and Burman connect local terms and concepts of weather and climate with clime.

When clime is used as a noun in our book, it signifies the embodiment of climate and climate change in place, and, thus, is synonymous with climate except that it is physically implaced, thereby marking and remembering changing weather conditions and human interventions over time. In addition to its physical attributes, clime resonates naturally with the ongoing environmental humanist discourse on the livingness, the intelligence, and the agency of the Earth. Given the demographic fact that 80% of the world's humans who are situated as the constituencies of world religions and indigenous spiritual traditions hold diverse knowledge of the Earth as an animate and spiritual planetary being, our authors, especially those working with indigenous communities in both altitudinal and latitudinal polar regions, are inevitably compelled to contribute their findings of the affective, spiritual, and moral meanings of clime incarnated as mountains, glaciers, lakes, rivers, and forests. These intangible meanings are invaluable for current global endeavors to build new environmental ethics for sustainable living. Based on our research, clime is thus geologically and ecologically concrete, terrestrially sensible, multiculturally inclusive, interdisciplinarily vital, scientifically viable, methodologically malleable, and fully open to affective and spiritual approaches to climate change and sustainable living.

Climing

This part of the introduction is intended to share with our readers how we conceptualize the terrestrial kinesis of climate change and how we convey the dynamic meanings of clime not only as place embodied with climate history and changes but also as an assemblage of entangled agencies, movements, creations, destructions, adaptations, and renewals of ecosystems, humans, animals, and plants. Multipolar climes are particularly packed with the actions of the physical Earth, the environmental flows, and humans and nonhumans.

Among our authors, Karsten Paerregaard innovatively takes the lead to utilize clime as a verb and a gerund. We believe this innovation is groundbreaking, bringing forth a set of new perspectives in the interdisciplinary field of climate studies. Thanks to Paerregaard's innovative "verbing" of clime, climate change is understood also as climing processes. Such newly coined phrases in our chapters are meant for case-specific use as well as for interdisciplinary discussions on the terrestrial meanings of climate change in kinetic and agentive terms:

- "Deep climing," "social climing," "global climing," "climing practices," and correlations of "to clime" and "to climb" (Paerregaard; Gamble);
- "Water climing" and "climing as more-than-human practices" (Brandshaug);
- "To clime with water" (Bhutia);
- "Climing as ontological disobedience" (Burman);
- "Climing the Great Gorge," "more-than-human rock clim(b)ers," "more-than-human climing," "climing for refuge," and "re-climing" (Gamble);
- "Climing elephants" and "inter-climing experience" (Smyer Yü).

As one can see, the use of clime as a verb is concentrated in our case studies of Andean and Himalayan highlands. It is correlated with another verb "to climb." If we look at the physical movements of climbing and climing between different elevations, the two words convey similar meanings. When climate change is factored in, climing can mean two intertwined but different types of kinesis. One refers to the adaption-compelled movements of life communities from deteriorating dwelling places to more habitable places due to climate change.

Another signifies the shifts of the life-giving and/or life-taking forces of the Earth moving across the Critical Zone or the planetary clime. The former pertains to the migration of human, animal, and vegetational residents of the Earth under the pressure of climate change, while the latter expresses the climatic agency of the Earth, entailing either its own acts or those induced by humans but enacted through the meteorological flows and the fluctuating temperature of the Earth. Additionally, regarding the vegetational residents, if we look at single trees or single patches of grassland, our environmental perception most likely tells us that plants are immobile; however, when we situate a forest in its greater environment, it responds to climate change by latitudinally or altitudinally advancing or retreating. For instance, due to the current global warming, the taiga (coniferous forest of high northern latitudes) advances into the tundra (treeless Arctic region) because of the melting of the permafrost. Likewise, the Gobi Desert expanded into the adjacent grasslands of Mongolia due to the dry and cold weather brought by the Little Ice Age. So, like animals and humans, plants also clime but in their own ways.

Undoubtedly, the habitability of the Earth's surface shifts with changing weather conditions and the availability of water over time. Likewise, all living beings reflexively orient themselves to habitats that are fertile. Each species, dead or alive, is also an archive of climate change, e.g. human oral and written histories, tree rings, and mammoth fossils. The physical environment and neighboring species remember the traces of other species' climing. Paerregaard's coinage of deep climing and social climing is exemplary in translating the climing processes of the southern Peruvian highlands and the community of Mauricio, his indigenous ethnographic collaborator, into an interdisciplinarily-intelligible description. While deep climing shows us the three-layered environmental reality of Mauricio's mountainous dwelling place—the geological, biological, and hydrological—social climing pertains to the ritualized interactions of the community with the deep climing of the highlands by giving offerings to mountain deities. This affective and spiritual interaction with the Earth via mountain deities is also found in the chapters of Brandshaug, Burman, Bhutia, Gamble, Delley and Aiyadurai. Deep climing and social climing are thus inextricably entwined with each other. The latter, as Paerregaard puts it, "offers an interpretive opportunity to not only translate the nourishing but feared power of the Earth" (in this book).

When we look at the climing of the Earth and living beings from the perspective of Deep Time, climate change, or rather clime change as understood in our book, is a long process of what Ron Nixon calls "slow violence," which is "neither spectacular nor instantaneous, but rather incremental and accretive, its calamitous repercussions playing out across a range of temporal scales" (Nixon 2011, 2). Although Nixon's use of slow violence is contextualized by humans' releases of toxic materials such as radioactive wastes and microplastics which are persisting into the "deep futures" (Farrier 2019, 6) of a ruined Earth, I nevertheless find it equally efficacious when we take a deep reading of past and present climate changes.

The notion of slow violence repurposed in this introduction retains Nixon's initial definition, but starts with the understanding of clime changes, which are as old as the Earth itself. It conveys the senses of Paerregaard's deep climing and social climing as well as of more-than-human climing discussed in the multispecies and the multidimensional contexts of the chapters by Bhutia, Brandshaug, and Gamble. Clime changes in the Himalayan, Andean, and Arctic cases are undoubtedly processes of slow violence, tectonically, meteorologically, ecologically, socially, and politically. If we look at slow violence from an Earth-centered perspective, it can mean the concurrent creative and destructive forces of the Earth. Without the slow but violent tectonic collisions, the Himalaya and the Andes would not have risen to be the elevationally-niched homes of their indigenous populations of humans, animals, and plants. They would not have their awe-inspiring sublime magnificence. As Donna Haraway (2016), Deborah Bird Rose (1996), and other environmental humanists point out, Mother Earth is nourishing but not always cuddly. The Earth's creativity comes with destruction and vice versa. With anthropogenic forces on a global scale added, the Earth may in some ways offer more destruction than life-sustaining provision to the contemporary world.

Multispecies Indigeneity and Indigenous Climes

All our chapters touch on or offer in-depth studies of indigenous environmental knowledge and adaptive responses to climate change. Some use the word "indigenous" or "indigeneity" in upper case, while others prefer lower-case or the use of local ethnonyms. Some, instead, use words like "local" and "traditional" but convey ecological, experiential, and affective meanings historically rooted to the geographical locations of their case studies. Some prefer to address the meanings of indigeneity in universal terms encompassing indigenous experiences across the board. Some focus on ethnolinguistically and eco-geologically specific places and people. These diverse approaches and perspectives reflect our authors' varied disciplinary backgrounds, historical and ethnographic research results, lived experiences, interpretive approaches, and sociopolitical assessments. They also reflect the diversity and the unsettledness of indigenous studies and debates situated in different disciplines and public arenas. The coinages of clime-related phrases in our chapters can be summarized as what could be called "indigenous climes," whether pre-colonial and premodern or externally and anthropogenically altered or ruined. Fuller delineation will be given toward the end of this section.

Intertextually combing through our chapters, it is safe for me to categorize three strands of discussion on indigenous knowledge and experiences from our individual research. They can be phrased in these terms – "indigenous Earth," "human indigeneity," and "multispecies indigeneity." Indigenous Earth is synonymous to the pre-human Earth in the sense of Deep Time as shown in the chapters by Paerregaard, Andersen, Amrith and Smyer Yü, Iqbal, Delley and Aiyadurai, and Smyer Yü. Indigenous Earth in these chapters is a pre-human

Earth with a planetary biography of its own. To be specific, it is a Deep Time Earth with a 4.5 billion-year history in comparison with the three hundred thousand-year history of *Homo sapiens*. Although we are currently making deep geological impacts on the Earth, we are nevertheless a shallow species when contrasted with the cosmologically understood indigeneity of the planet itself. Paerregaard devotes a significant portion of his chapter to tell a story of what he calls "deep climing" revealing the interactive relations between the shallowness of humans and the timelessness of the Andean mountains, glaciers, and rocks. In his ethnographic narration, the indigenous Andeans express their deep reverence for and negotiated relations with the geological indigeneity of the Earth in the midst of overwhelming changes of external origins, e.g. global warming, mining and international tourism.

Likewise, the chapters by Iqbal, and Amrith and Smyer Yü give full recognition of the geological and ecological deepness of the wetlands and riverine habitats in Bengal before the arrival of the colonial and the postcolonial eras. The entwinement of the indigenous Earth and pre-colonial and premodern humans in their case studies show human-earth relations in a negotiated equilibrium in terms of humans' reliance on the natural provision of the Earth and respect for the innate freedom of the Earth's environmental flows facilitated by the monsoon and Himalayan waters. In the chapters by Andersen, and Delley and Aiyadurai, the Deep Time Earth speaks respectively through the words of indigenous humans—Inuits' *sila* and Idu Mishmis' *eno* as aforementioned. The animate, spiritual, and affective human-earth relations expressed in the meanings of these words show the Earth as a planetary union of material and spiritual worlds.

The indigenous Earth understood in the sense of Deep Time does not refer to the Earth in the past tense but rather shows the presence of Deep Time in the past, the present, and the future. Regardless of how much our scientific hypothesis of the Anthropocene compels us to see the current phase of the Earth as a human Earth, Deep Time Earth reacts robustly to our anthropogenic disruptions and interventions. It calls us "to 'thicken' the present with an awareness of the other times and places" and to renew our planetary imagination of the geologic sublime by looking deeply into "the abyss of time," a notion brought forward by John Hutton and John Playfair, two proponents of the Deep Time concept over two centuries ago (Farrier 2019, 9). Deep Time in our texts does not require geological dating of rocks, mountains, water bodies, and icesheets, but couples the geological and climatological findings with our environmental humanist approaches to demonstrating the livingness of the Earth in physical, social, affective, and spiritual terms under multiple anthropogenic threats and devastations in this multipolar world.

The indigenous Earth, as an embodiment of Deep Time, undoubtedly speaks through the geologically shallow human indigeneity in its variety. Human indigeneity is currently one of the most debated, contentious topics of research and public discourse. The output of its human politics considerably exceeds that of its geological and ecological dynamics. This is where, I believe, we authors are divergent on how we use the words "indigenous" and "indigeneity" in upper- or lower-case, when we tell our stories and offer our thoughts on indigenous environmental and climatic issues. Covering three world regions, our book affords us and our readers to rethink the modern delineations of indigeneity on the global scale and from comparative perspectives.

Not representing our authors but showing a greater context of how we use the words and associated concepts, my own assessment is that the politics of indigeneity is a legacy of Neo-Europes primarily in the Americas and Australia. It can also be understood as a consequence of the Columbian exchange in the matter of inter-human encounters and subsequent deadly physical conflicts over lands, resources, and political power. The five-centuries-long period, overlapping the Little Ice Age, was a time in which Europe was being replicated or reinvented as Neo-Europes, or rather, as the transformations of the ancient lands into what is now called the New World. The loss of native populations and their cultural heritages in Neo-Europes are remembered as a trauma of humankind.

This historical trauma echoes in the United Nation's definition of indigenous people as "those which, having a historical continuity with pre-invasion and pre-colonial societies that developed on their territories, consider themselves distinct from other sectors of the societies now prevailing on those territories, or parts of them" (UNHRC 2013). To unpack this carefully worded definition provided by the largest intergovernmental organization on Earth, being indigenous often means living in poverty, being discriminated against, being treated as second-class citizens, and being stigmatized as backward and primitive though some indigenous communities in North America are economically better off now because of their willingness to open casinos on their ancestral lands (de la Cadena and Starn 2007, 2). This understanding of being indigenous is what I call the "political indigeneity" rightfully intended for "the protection of indigenous peoples' rights and justice" (Smyer Yü 2021b, 244). On this front, I share with Marisol de la Cadena and Orin Starn the historical assessment of political indigeneity as being a "European invention" (de la Cadena and Starn 2007, 4).

A side effect of this politically-oriented understanding of indigeneity in the academic world is that the gravity of indigenous studies is found more in the issues of colonial encounters, identity, and rights but less in the social and spiritual ethos and ecologically-grounded lifeworlds of indigenous peoples, past and present. Burman's chapter here is most critical of the concurrent subjugation of indigenous lifeworlds and the romanticization of indigenous people in Bolivia as "guardians of nature." It offers an integral understanding of the ecospiritual role of humans in the more-than-human world. It gives us hope that a political ecology of an indigenous society does not have to highlight politics while marginalizing ecology.

Perhaps we can consider expanding the human-centered politics into Earth-centered politics or a type of geopolitics "not limited to humans but [one that] includes all the elements or entities deemed part of the composition of the common world" (Latour 2014, 1). In fact, the diversely narrated human indigeneity in our chapters speaks of the Earth as a planet consisting of many storied and weathered worlds (Ingold 2011, 141; Andersen, this book). It does not stand alone but is deeply entangled in the web of life in the high mountains, plateaus, and riverine plains that are home to animals and plants, too.

This is where I also see the multispecies indigeneity shown in our chapters, particularly in those by Gamble, Smyer Yü, Delley and Aiyadurai, Burman, Andersen, and Iqbal. By "multispecies indigeneity" I mean that being indigenous connotes more-than-human experiences in geographically and ecologically specific places, in which human and nonhumans spin together their shared but negotiated and contended lifeworlds. In addition to animals and plants, nonhumans in the Himalaya and the Andes also include locally-cherished sentient gods, goddesses, spirits, and supernatural beings, many of whom are incarnated in mountain- and water bodies. Grounded in the maternal principle of the Earth (Berry 2009, 75), human-nonhuman relations weave together their co-habited worlds—or a pluriverse understood as a world of many worlds (Escobar 2018; de la Cadena and Mario Blaser 2018; Andersen, and Iqbal, this book). From this multispecies perspective, the Himalaya and the Tibetan Plateau can be seen, for example, as the nation of snow leopards (Gamble, this book); the Peruvian Andes as a nation of llamas, the Arctic as intertwined nations of polar bears, penguins, and harp seals. We can continue with Yunnan as a surviving Asian-elephant nation (Smyer Yu 2021a, 15), and many interconnected parts of India, Bangladesh, Nepal, and Bhutan as a nation of Bengal tigers (Aiyadurai 2021). Examples are ample. The point here is that the geographies of multispecies climes disrupt human cartographies.

When indigeneity is seen from this multispecies perspective, it offers an ecologically and biologically more enlightened picture of the Earth. Adding the Deep Time perspective, many nonhuman species existed long before the arrival of humans on Earth. With the unique inclusion of Earth-based deities as sentient and ecological beings in many of our chapters, the lifeworlds of multispecies indigeneity also encompass a multitude of spiritual worlds in the Himalaya, the Andes, and the Arctic. Climate change is inevitable; however, our ethics for sustainable human-nonhuman relations always have the option of being renewed as multispecies collaboration for collective survival and flourishing.

Returning to our multipolar clime studies, when we connect the indigeneity of the Earth, and humans and nonhumans in geological, ecological, and climatic terms, we find the historical and contemporary Himalaya, Andes, and Arctic are undoubtedly multispecies indigenous climes, in which being indigenous entails horizontally shared, trans-species experiences of common climate histories and changes since pre-human times. Specific environmental wisdoms accumulated over time are embodied and instrumentalized in living beings' unique bodily forms that are adapted to particular ecological niches and climatic patterns. Each species has its unique relation with the Earth and with other species. In this sense, indigenous climes are symbiotic climes, meaning that everything and everyone is entwined in physical contact (Margulis 1998, 5). However, symbiosis does not mean only harmonious and peaceful interactions, entanglements, and intertwinements. Animal-human relations are both violent and affectionate. So are plant-human and human-to-human relations.

The slow violence of climate change has its quick moments or tipping points as seen in accelerated global warming, melting glaciers, extreme weather patterns, devastating environmental changes, and radical behavioral changes of animals and humans. Polar bears and grizzly bears are meeting in the Canadian Arctic and reproducing polar-grizzly bears or "pizzly bears" (Pongracz et al. 2017; Macdonald and Wester 2019). Lowland leopards from western China are entering the habitats of snow leopards in the Tibetan Plateau (Li 2021). Elephants prefer to eat human-farmed crops and drink humans' home-brewed rice wine in Yunnan, while humans, due to population increases and global market demands, are replacing ancient forests with cash crops, e.g. rubber and tea plants, so that vast areas become "green deserts." Many indigenous communities in the Andes were converted to Catholicism but syncretized it with their native spiritual beliefs (de la Cadena 2015; Paerregaard; Brandshaug; Burman, this book). Non-Arctic nations like China and India are claiming exploratory rights for natural resources in the Arctic Ocean. Everything moves in an accelerated and precarious fashion in the Anthropocene (Tsing 2015, 27–36; Andersen, this book). It is not surprising that the contemporary Arctic and Himalaya are being referred to as "the New Arctic" and "the New Himalaya" due to modern anthropogenic changes (Evengård, Larsen and Paasche 2015, 1; Smyer Yü 2021c). The Columbian exchange continues beyond the Americas. The anthropogenically conceived new worlds are admittedly increasing in the once sparsely populated Himalaya, Andes, and Arctic.

In sum, indigenous climes in this book are multispecies, symbiotic climes. With the cosmological and geological indigeneity of the Earth included, they are simply diverse climes of the Earth that are terrestrially connected with each other via ecotones and that are entwined in the global climate system concurrently as recipients and agents of climate change. Situated in the Anthropocene, indigenous climes are anthropogenic climes encountering environmental and social challenges from both national and geo-economic developments. Given our focus on altitudinal and latitudinal highlands, our chapters are the inaugural outcomes of the multipolar clime study proposed in the opening of this introduction. They demonstrate that high and long mountain ranges and the two polar regions influence the fate of the Earth under the current circumstance of global warming. They help us connect dots and draw lines between climatic events, changing weather patterns, shapeshifting water bodies, changing environments and livelihoods, and human and nonhuman migrations on a global scale. The multipolarity of the climate system and climate change becomes obvious in concrete, multispecies climes when our individual chapters are read intertextually.

Thematics

Given the diverse findings and conceptual discussions in our chapters, the book is thematized into four sections: Climing Earth Summits, Water Climes, Bridging Disciplines and Teaching Clime Changes, and Multispecies Clime Change from the Little Ice Age to the Anthropocene. By no means is this book an all-encompassing survey of climatic, environmental, cultural, and political aspects of the Himalayan, Andean, and Arctic regions; however, it is sincerely intended as a multipolar clime study for the purpose of establishing more legible, meaningful, and complementary alliances between climate science and earth science. All chapters offer engaging ethnographic and/or conceptual understandings of clime as place and as the agentive movements and acts of the Earth, humans, animals, plants, and their spiritual companions embodied in different landforms. Abundant throughout the book, the topics of indigeneity and water are the main threads weaving together our geographically and ethnolinguistically different case studies. We hope the thematization of our chapters below helps our readers to geographically and topically navigate the book.

Climing Earth Summits features the contributions of Sunil Amrith and Dan Smyer Yü, Karsten Paerregaard, and Astrid Oberborbeck Andersen. Amrith, Paerregaard, and Andersen were the keynote speakers for the generative publishing workshop. To be noted, my co-authorship with Amrith was a post-workshop development out of our mutually sought out alliance for co-building a case of the climatic nexus of the mountainous, riverine, and oceanic worlds; however, the initiatives and the foundational strengths all come from Amrith's keynote presentation. The three chapters show how climate change speaks through places and how places change with changing weather conditions and the shapeshifting states of water. While humans, animals, and plants make place, place nevertheless has a life of its own that is animate, sentient, and agentive; this is particularly shown in the chapters by Paerregaard and Andersen. While Amrith and Smyer Yü focus on how the monsoon animates the transregional meteorological flows and changing climate patterns in the Himalaya-Bengal-Indian Ocean nexus, Paerregaard and Andersen ethnographically capture how Peruvian and Arctic indigenous communities adaptively interact with climate change through their ritualized communications with mountain deities (Earth beings) and through their affective consciousness of the Earth synchronized with the weather-incarnated spirit (*sila*) holding the universe. The potential take-away for our readers from these three chapters can be found in the authors' unique conceptualizations of clime as a noun and a verb as mentioned. To reiterate, Amrith and Smyer Yü regard the mountain climes of the Himalaya and the riverine climes of Bengal as Anthropocenic climes when they correlate multiple human and climatic events in the mid-1800s, such as the ending of the Little Ice Age, the beginning of the Anthropocene, and the heyday of British colonial agro-economic development. Paerregaard's innovative use of clime as a verb effectively captures how the changing surface of Peruvian mountain bodies, melting glaciers, and human moral sensitivity make the invisibility and slow violence of climate change visible. Likewise, Andersen's Greenlandic case study attests to the mutual legibility, intelligibility, and translatability of clime and *sila* when they speak through each other in her ethnographic accounts and conceptual discussions.

Water Climes groups together the chapters by Razzeko Delley and Ambika Aiyadurai, Malene K. Brandshaug, Kalzang Dorjee Bhutia, and Iftekhar Iqbal. The commonality of the four chapters is a set of human-earth spiritual and affective relations mediated through the indigenous perceptions and experiences of the Earth as sentient, with place-embodied local deities, and with ecologically-grounded human bonding with nonhuman beings. Delley and Aiyadurai's ethnographic depiction of the Idu Mishmi's ecospiritual space—*eno*—shows us how water climes in the eastern Himalaya are not merely composed of habitats nourished by streams, rivers, and lakes, but are also cosmological worlds where supernatural spirits find their presence in households, farm fields, mountains, and water bodies. These beings do not always act as guardians of human well-being. When that is the case, it often means that humans are not giving them ritual offerings of wine and domestic animals as the sign of the respect they are due, or that we are harming forests and defiling rivers and lakes. In relating the indigenous physical-spiritual environments to climate studies, Delley and Aiyadurai's ethnographic work emphatically demonstrate that the indigenous sense of climate and climate change is found in actual weather, seasons, rainfall, snowfall, and in the negotiated human-spirit relations.

Delley and Aiyadurai's cosmological perspective highlights the more-than-human sociality of the Indu Mishmi's water climes, which conceptually resonates with Brandshaug's storying of the cosmpolitics of land, humans, and "earth beings" in the southern Peruvian Andes. In many ways, Brandshaug's chapter is a deepened study of the cosmopolitics of humans and Earth beings that de la Cadena writes of in her *Earth Beings* (2015), specifically, in the climatic reading of the shared lifeworld of Yanqueños and their mountain gods (the eponymous Earth beings). The climatic dynamic among them is expressed in what Brandshaug terms "water climing," a signifier of the entangled agencies of water, humans, and Earth beings in the context of global climate change. Water not only threads together the mountains, humans, and Earth beings as environmental flows, but is considered a sentient being itself. The multifaceted meaning of climing in Brandshaug's chapter demonstrates that climate encounters entail a series of more-than-human communications, negotiations, and moral responses.

Similarly, Bhutia's chapter historically and ethnographically depicts the animatedness of water in the shapes of mist and fog climbing/climing up and down the Sikkimese Himalayan mountain bodies. Like Delley, Bhutia, who is writing from his indigenous experience and disciplinary background as a scholar of religious studies, offers a vivid account of water moving elevationally between the Himalayan highlands and the Bengali plain. This account speaks descriptively of mountain bodies as both interceptors of water traveling in vaporous forms and storing the captured water. The Sikkimese Himalayan mountains have their own versions of Earth beings, but Bhutia specifically highlights *lu* (a type of water deity) who animate water to clime up and down from the plains and the foothills to the high mountains into the shared habitats of humans, animals, and plants.

Climing water creates an ethos of places, aside from its climatic role. Iqbal's chapter features a case of the enmeshed state of water and land known in the foothills of the Meghalaya mountain range of the far-eastern Himalayan Massif as the *haor*. As they receive the most rainwater, the *haor* are the wettest wetlands in the region. Iqbal depicts the entire region as an ecological commons hosting a multispecies world with its unique nature-culture intertwined ethos. He characterizes this more-than-human ethos as "a felt habitat" that bonds human, nonhuman, and spiritual worlds together. Through conversing with Paerregaard's application of the posthumanist idea of multinaturalism, Iqbal navigates readers from the meteorologically and ecologically shaped *haor* to their combined affective state of being. Native human societies' bonding with the *haor* rests upon their gratitude for the fertile wetlands for sustenance and agro-economic gains as well as upon their felt spirituality of the living water saturating the land. In other words, the exuberant rainwater not only creates local ecosystems but is also an inherent foundation of the region's human cultures and languages. However, as the *haor* entered the colonial and the postcolonial eras, it shared the same anthropogenic fate with the rest of Bengal in which water and land were artificially and rigidly separated for agricultural development.

Bridging Disciplines and Teaching Clime Changes hosts chapters by Theodore Shepherd and Chi Huyen Truong, and Vanda Singh. In many ways, their chapters showcase how bridges are being built between the natural and social sciences and the humanities in ethnographic, conceptual, and pedagogical terms. Both chapters give social scientists and humanists an opportunity to respond to Chakrabarty's aforementioned call to come to terms with the basics of climatological knowledge and approaches. The widely critiqued disconnection of climate science and experiential knowledge of climate change is, in fact, an artificial remnant of disciplinary orientations and boundaries that have been shaped over a long period of time; however, it is also up to us, scientists, humanists, and the members of the public, to narrow the gap and eventually close it up with our consistent collaborative research outcomes and shared experiences, truly making climate knowledge democratically available and intelligibly accessible for everyone. These chapters tell us that qualitative and quantitative approaches can find a home in the same textual body.

Shepherd and Truong's chapter is a fine example, showing the storylining approach found in climate science and its resonance with the storying approach found in environmental humanities. With the authors' geographical specializations in the Himalaya and in the Arctic, the chapter offers the stories of the Melamchi flood disaster in Nepal and the catastrophic landform collapses in the Yup'ik country of Alaska. The authors adopt Robin Wall Kimmerer's "braiding sweetgrass" metaphor to show that the climatological storylining approach to climate change and the environmental humanist storying approach to clime change can work together seamlessly in both empirical and conceptual terms.

Emanating the same interscientific spirit, Singh's chapter offers a socially and politically engaged pedagogy of climate change and innovatively proposes that the climate crisis itself is a teacher of climate knowledge. This is where she promotes what she calls a "transdisciplinary, justice-centered pedagogy" grounded in three "interconnected metaconcepts," namely balance-imbalance, limits and thresholds, and complex entanglements. Her pedagogical innovation is intended to transcend the local-global, human-nonhuman, and social-ecological dichotomies.

Multispecies Clime Change from the Little Ice Age to the Anthropocene comprises chapters by Ruth Gamble, Anders Burman, and Dan Smyer Yü. Their historical periods primarily cover the Little Ice Age (1200s–1850s) and the Anthropocene (1850s–present), two consecutive climatic cycles. In climate science, both are currently subsumed under the Holocene. Gamble's chapter tells a story of "more-than-human rock clim(b)ers" in the Yarlung Tsangpo Gorge in the eastern Himalaya. It is a story of multispecies climing in the mountain folds of the gorge. In it, climing is a long migration process in which plants and animals moved into the diverse niches in the mountain folds of the gorge first and humans followed suit. It can be seen as a process of "discovery"; however, given the modern human histories in the region, the chapter problematizes the idea of "discovery" by examining the modern imperial encounters with the British and the Qing Dynasty as well as the ongoing hydropower development projects initiated by the Chinese and Indian governments. It makes a compelling argument that human discovery of the gorge is "a form of colonization and erasure" of the local nonhuman and human histories and knowledge of the gorge.

The anthropogenic changes of the Anthropocene are not limited to the climate and the physical environments of the Earth. They also saturate human social spheres and, thus engender changed political climates within human societies. Burman's chapter shows a case of the intertwinement of Anthropocenic climate change and the political climate in which Bolivian government officials and environmental activists romanticize and politically represent indigenous environmental knowledge. In his findings, while the government presents itself as an indigenous government and the environmental movements promote indigenous people as stewards of nature, they nevertheless sustain the public presence of the indigenous population as an Other. Burman deems this political climate the "coloniality of reality" in which the indigenous constituencies are ontologically marginalized and made subaltern. Like Paerregaard, Brandshaug, Gamble, and Bhutia, Burman also uses "clime" as a verb; however, climing, evidenced with his ethnographic presence with indigenous shamans, is a political act of what he calls "ontological disobedience" that is intended to let suppressed voices and modes of being reemerge in the public through interstices, cracks, and nooks. His ethnographic account recalls the shamans' meeting and communion with the *achachilas* (mountain beings), other-than-human persons who know what human-induced climate change means to the mountains and their indigenous humans.

Lastly, Smyer Yü's chapter also tells a story of other-than-human persons—in this case, Asian elephants in Yunnan, a Southwestern Chinese province on the eastern end of the Himalaya. Beginning with the recent year-long excursion of fifteen elephants from the National Nature Reserve in Yunnan to more distant farm fields, towns, and cities, the chapter delves into a climatic account of how the Little Ice Age and the Mongol Yuan Dynasty and its successors from the 1200s to the modern republics have led to Asian elephants in China becoming an endangered species. By correlating climatological findings and historical materials, the chapter traces the southward migrations of elephants and humans from higher latitudes to lower latitudes. It argues that the humans—Mongols, Han, and Manchus—from the north, were both climatic refugees and imperial conquerors; whereas the elephants were refugees fleeing climate change and both human imperialists and refugees. This elephant story is thus a story of the shrinking elephant clime. Understood as place and region, clime in the chapter undergoes the slow violence of the Little Ice Age and the fast human-led transformation of elephant habitats and the elephants' neighboring indigenous human societies. In facilitating his argument, Smyer Yü utilizes perspectives from new animal geographies to convey the environmental and political meanings of the climatic entanglements of elephants and humans and to show the elephants' current mode of being in Yunnan as an "elephant modernity" in which the lifestyle of the wild elephants is being humanized.

References

Aiyadurai, Ambika. 2021. *Tigers Are Our Brothers: Anthropology of Wildlife Conservation in Northeast India*. Delhi: Oxford University Press India.

Amrith, Sunil. 2018. *Unruly Waters: How Rains, Rivers, Coasts, and Seas Have Shaped Asia's History*. New York: Basic Books.

Babkin, V.I. 2003 "The Earth and Its Physical Features." In *World Water Resources at the Beginning of the Twenty-First Century*, edited by I.A. Shiklomanov and John C. Rodda. Cambridge: Cambridge University Press, 1–18.

Berry, Thomas. 2009. *The Sacred Universe: Earth, Spirituality, and Religion in the Twenty-first Century*. New York: Columbia University Press.

Boardman, Miriam. 2016. "An Introduction to the Word Climate." https://www.science.smith.edu/climatelit/intro-to-the-word-climate/.

Bristow, Tom and Thomas H. Ford. 2016. "Climates of History, Cultures of Climate." In *A Cultural History of Climate Change*, edited by Tom Bristow and Thomas H. Ford. London: Routledge, 1–14.

Brooke, John L. 2014. *Climate Change and the Course of Global History: A Rough Journey*. Cambridge: Cambridge University Press.

Camenisch, Chantal. 2018. "Two Decades of Crisis: Famine and Dearth During the 1480s and 1490s in Western and Central Europe." In *Famines During the 'Little Ice Age' (1300–1800): Socionatural Entanglements in Premodern Societies*, edited by Dominik Collet and Maximilian Schuh. Cham: Springer, 69–90.

Carey, Mark and Philip Garone. 2014. "Forum Introduction," 284–295. In "Forum: Climate Change and Environmental History," edited by Mark Carey, Philip Garone, Adrian Howkins, Georgina Endfield, Lawrence Culver, Sherry Johnson, Sam White, and James Roger Fleming, *Environmental History*: 281–364.

Chakrabarty, Dipesh. 2016. "Forward." In *A Cultural History of Climate Change*, edited by Tom Bristow and Thomas H. Ford. London: Routledge, xvii–xix.

Chakrabarty, Dipesh. 2018. "Anthropocene Time." *History and Theory* 57(1) (March): 5–32.
Chakrabarty, Dipesh. 2021. *The Climate of History in A Planetary Age*. Chicago: University of Chicago Press.
Cho, Ji-Hyung. 2014. "The Little Ice Age and the Coming of the Anthropocene." *Asian Review of World Histories* 2(1) (January): 1–16.
Clark, Nigel and Kathryn Yusoff. 2017. "Geosocial Formations and the Anthropocene." *Theory Culture & Society* 34(203): 3–23. Special Issue: Geosocial Formations and the Anthropocene.
Collet, Dominik and Maximilian Schuh. 2018. "Famines: At the Interface of Nature and Society." In *Famines During the 'Little Ice Age' (1300–1800): Socionatural Entanglements in Premodern Societies*, edited by Dominik Collet and Maximilian Schuh. Cham: Springer, 3–16.
Crosby, Alfred W. 2003. *The Columbian Exchange: Biological and Cultural Consequences of 1492*. Westport, CT: Praeger.
Crosby, Alfred W. 2004. *Ecological Imperialism: The Biological Expansion of Europe, 900–1900*. Cambridge: Cambridge University Press.
Crutzen, Paul J. and Eugene F. Stoermer. 2000. "The 'Anthropocene'." *Global Change Newsletter* 41 (May): 17, 18.
De la Cadena, Marisol. 2015. *Earth Beings: Ecologies of Practice Across Andean Worlds*. Durham: Duke University Press.
De la Cadena, Marisol and Mario Blaser, eds. 2018. *A World of Many Worlds*. Durham: Duke University Press.
De la Cadena, Marisol and Orin Starn. 2007. "Introduction." In *Indigenous Experience Today*, edited by Marisol de la Cadena and Orin Starn. New York: Berg.
Dilling, Lisa and Sussanne C. Moser. 2007. "Introduction." In *Creating a Climate for Change: Communicating Climate Change and Facilitating Social Change*, edited by Susanne C. Moser and Lisa Dilling. Cambridge: Cambridge University Press, 1–27.
Dyhrenfurth, Günter O. 1955. *To the Third Pole: The History of the High Himalaya*. Translated from the German by Hugh Merrick. London: Werner Laurie.
Dybas, Cheryl Lyn. 2013. *Discoveries in the Critical Zone: There Life Meets Rock*. Alexandria, VA: National Science Foundation.
Escobar, Arturo. 2018. *Designs for the Pluriverse: Radical Interdependence, Autonomy, and the Making of Worlds*. Durham: Duke University Press.
Evengård, Birgitta, Joan Nymand Larsen, and Øyvind Paasche, eds. 2015. *The New Arctic*. Cham: Springer.
Fagan, Brian. 2004. *The Long Summer: How Climate Changed Civilization*. London: Granta Books.
Fagan, Brian. 2019. *The Little Ice Age: How Climate Made History 1300–1850*. New York: Basic Books.
Farrier, David. 2019. *Anthropocene Poetics: Deep Time, Sacrifice Zones, and Extinction*. Minneapolis: University of Minnesota Press.
Fleming, James Roger. 2010. *Fixing the Sky: The Checkered History of Weather and Climate Control*. New York: Columbia University Press.
Galenson, David W. 1978. "British Servants and the Colonial Indenture System in the Eighteenth Century." *The Journal of Southern History* 44(1) (February): 41–66.
Galenson, David W. 1984. "The Rise and Fall of Indentured Servitude in the Americas: An Economic Analysis." *The Journal of Economic History* 44(1) (March): 1–26.
Gonda, Catalina María. 2019. *El Cambio Climático y La Criósfera Andina* (Climate Change and the Andean Cryosphere). Translatated by Elisa Predassi Bianchi. Fundación Ambiente y Recursos Naturales (FARN).

Haraway, Donna J. 2016. *Staying with the Trouble: Making Kin in the Chthulucene*. Durham: Duke University Press.
Hulme, Mike. 2008. "The Conquering of Climate: Discourses of Fear and their Dissolution." *The Geographical Journal* 174(1) (March): 5–16.
Hulme, Mike. 2009. *Why We Disagree about Climate Change: Understanding Controversy, Inaction and Opportunity*. Cambridge: Cambridge University Press.
Ingold, Tim. 2011. *Being Alive: Essays on Movement, Knowledge and Description*. London: Routledge.
Jasanoff, S. 2010. "A New Climate for Society." *Theory, Culture & Society* 27: 233–253.
Kurz, Marcel. 1933. "Die Erschliessung des Himalaya." *S.A.C. Die Alpen*: 9, 7–11.
Lamb, H. H. 1982. *Climate, History and the Modern World*. London: Routledge.
Latour, Bruno. 2014. "Some Advantages of the Notion of 'Critical Zone' for Geopolitics." *ScienceDirect*, www.sciencedirect.com.
Li, Tana. 2020. "The Mongol Yuan Dynasty and the Climate, 1260–1360." In *The Crisis of the 14th Century: Teleconnections between Environmental and Societal Change?* edited by Martin Bauch and Gerrit Jasper Schenk. Berlin: De Gruyter, 153–168.
Li, Xueyang. 2021. "Where Golden Wind and Jade Dew Meet: The Presence of Lowland Leopards and Snow Leopards in Yushu" (金风玉露，何处相逢——金钱豹雪豹同现玉树). *Papers*, https://www.thepaper.cn/newsDetail_forward_14805712.
Macdonald, Catherine and Julia Wester. 2019. "Public Perceptions of the Hybridization of Polar (*Ursus maritimus*) and Grizzly Bears (*Ursus arctos horribilis*)." *Human Dimensions of Wildlife*. 2019, https://doi.org/10.1080/10871209.2019.1581859.
Mann, Michael. 2002. "The Little Ice Age." In Michael MacCracken and John Perry, eds. *Encyclopedia of Global Environmental Change*, vol. I. Chichester: John Wiley & Sons, Ltd.
Margulis, Lynn. 1998. *Symbiotic Planet*. Amherst, MA: Basic Books.
Miller, Elizabeth Carolyn. 2018. "Climate Change and Victorian Studies: Introduction." *Victorian Studies* .60(4): 537–542.
Möller, Detlev. 2010. *Chemistry of the Climate*. Berlin: De Gruyter.
Nixon, Rob. 2011. *Slow Violence and the Environmentalism of the Poor*. Cambridge: Harvard University Press.
NASA. n.d. "Overview: Weather, Global Warming and Climate Change." https://climate.nasa.gov/resources/global-warming-vs-climate-change/#:~:text=Global%20warming%20is%20the%20long, gas%20levels%20in%20Earth's%20atmosphere.
Paerregaard, Karsten. 2020 "Communicating the Inevitable: Climate Awareness, Climate Discord, and Climate Research in Peru's Highland Communities." *Environmental Communication* 14(1): 112–125.
Pandit, Maharaj K. 2017. *Life in the Himalaya: An Ecosystem at Risk*. Cambridge: Harvard University Press.
Pant, Govind Ballabh, P. Pradeep Kumar, Jayashree V. Revadekar, and Narendra Singh. 2018. *Climate Change in the Himalayas*. Cham: Springer.
Plumwood, Val. 2009. "Nature in the Active Voice," *Australian Humanities Review* 46: 113.
Pongracz, Jodie D., David Paetkau, Marsha Branigan, and Evan Richardson. 2017. "Recent Hybridization between a Polar Bear and Grizzly Bears in the Canadian Arctic." *Arctic* 70(2) (June): 151–160. https://doi.org/10.14430/arctic4643.
Rose, Deborah Bird. 1996. *Nourishing Terrains: Australian Aboriginal Views of Landscape and Wilderness*. Canberra: Australian Heritage Commission.
Rose, Deborah Bird. 2013. "Val Plumwood's Philosophical Animism: Attentive Interactions in the Sentient World." *Environmental Humanities* 3: 93–109.
Rose, Deborah Bird, Thom van Dooren, Matthew Chrulew, Stuart Cooke, Matthew Kearnes, and Emily O'Gormand. 2012. "Thinking Through the Environment, Unsettling the Humanities." *Environmental Humanities* 1: 1–5.

Rowan, Ann V. 2017. "The 'Little Ice Age' in the Himalaya: A Review of Glacier Advance Driven by Northern Hemisphere Temperature Change." *The Holocene* 27(2): 292–308.

Ruddiman, William. 2007. "The Early Anthropogenic Hypothesis: Challenges and Responses." *Reviews of Geophysics* 45, RG4001/2007: 1–37.

Serreze, Mark C. and Roger G. Barry. 2005. *The Arctic Climate System*. Cambridge: Cambridge University Press.

Shepherd, Theodore G. and Elisabeth A. Lloyd. 2021. "Meaningful Climate Science." *Climatic Change* 169: 17. http://doi.org/10.1007/s10584-021-03246-2.

Smyer Yü, Dan. 2020. "The Critical Zone as a Planetary Animist Sphere: Ethographing an Affective Consciousness of the Earth." *JSRNC* 14(2): 271–290. http://doi.org/10.1558/jsrnc.39680.

Smyer Yü, Dan. 2021a. "Perpendicular Geospatiality of Corridors and Borderlands: An Introduction." In *Yunnan-Burma-Bengal Corridor Geographies: Protean Edging of Habitats and Empires*, edited by Dan Smyer Yü and Karin Dean. London: Routledge.

Smyer Yü, Dan. 2021b. "Symbiotic Indigeneity and Commoning in the Anthropogenic Himalayas." In *Environmental Humanities in the New Himalayas Symbiotic Indigeneity, Commoning, Sustainability*, edited by Dan Smyer Yü and Erik de Maaker. London: Routledge, 239–261.

Smyer Yü, Dan. 2021c. "Situating Environmental Humanities in the New Himalayas: An Introduction." In *Environmental Humanities in the New Himalayas Symbiotic Indigeneity, Commoning, Sustainability*, edited by Dan Smyer Yü and Erik de Maaker. London: Routledge, 1–24.

Tsing, Anna Lowenhaupt. 2015. *The Mushroom at the End of the World*. Princeton: Princeton University Press.

United Nations Human Rights Council (UNHRC). 2013. *The United Nations Declaration on the Rights of Indigenous Peoples: A Manual for National Human Rights Institutions*. Sydney, Australia: Asia Pacific Forum of National Human Rights Institutions.

Zhu, Kezhen. 1972. "A Preliminary Research on Climate Chang of China in the Last Five Thousand Years." *Scientia Sinica* 3(2): 168–189.

PART I

Climing Earth Summits

1

THE HIMALAYA AND MONSOON ASIA

Anthropocenic Climes Since the 1800s

Sunil Amrith and Dan Smyer Yü

As a historian of the Indian Ocean and an environmental anthropologist of Himalayan highlands, we team up in this chapter to address the intimate connection between the oceanic and mountainous worlds as symbiotic weather makers and climate changers. Our focus is on the 1800s to the present, a period that coincides with the Anthropocene or the human-induced geological epoch of the earth (Crutzen and Stoermer 2000). The anthropocenic contextualization of this chapter is meant to underscore a commonly recognized period during which climate change has entailed a series of intertwined natural and human events with transformative outcomes in the Himalaya, South Asia, and the Indian Ocean.

The mid-1800s in particular were a time of globally and locally concurrent changes at multiple scales. To count a few, it was the ending point of the Little Ice Age, a seven-century-long global cooling period that mostly took place in the northern hemisphere from the 1200s to the 1800s (Matthews and Briffa 2005, 17; Fagan 2000, 18; Smyer Yü, in this book). Among its many effects, the climatic disruptions of the Little Ice Age were one trigger of conflict-laden, global migration of people seeking cross-continental refuges for survival, building new empires, innovating new modes of production, or inventing new nations (Crosby 2004). The mid-1800s were also the starting point of "the Modern Warm Period" (Fagan 2000, 28), which is contemporaneous with the rapidly advancing Industrial Revolution and the recently hypothesized Anthropocene. It was simultaneously the economic heyday of British India, a time of rigorous alteration of the physical environments of South Asia on an industrial scale, and an era of hunger and famine. While the mid-1800s serve as the historical focus of this chapter, it builds out from there toward postcolonial lessons and perspectives on the climatic meaning of being humans in the Anthropocene.

Like our peer authors in this book, we interchangeably use the words "climate" and "clime" but give more precedence to the latter when we address

DOI: 10.4324/9781003347026-3

geographically and ecologically-specific, human-induced climate and environmental changes from the Himalayan highlands to the floodplains of Bengal. While we recognize the point made by Smyer Yü in the introductory chapter about the local-global relationality and the conceptual complementarity of clime and climate in natural and social scientific fields, we see the terrestrial efficacies of clime in helping us understand local experiences of how global climate change takes place locally and regionally. The initial proponents of the notion of climes—James Roger Fleming (2010), Mark Carey and Philip Garone (2014)—emphatically draw our attentiveness to local environmental changes as proxies of global climate change—and to this we would add that local environmental changes are important in their own right and not only as proxies for larger transformations. Like many of our peer authors, we regard clime as climate concretely embodied in physical habitats, seasons, weather, lives, livelihoods, and polities. Making climes is the work of the earth, of humankind, and of other species in the Anthropocene. The clime perspective thus allows us to see *place*, specific places in all their particularity, as embodiments of climate as well as agents of climate change in both natural and anthropogenic senses.

Specific to the Himalaya, South Asia, and the Indian Ocean, clime in this chapter is overall understood as a monsoon clime or a series of monsoon-engendered environments in which water, in the form of ice, snow, rainfall, vapor, streams, rivers, and wetlands, is the primary thread connecting mountains with ocean via the floodplains of Bengal. To facilitate our conversation with our readers, we then geologically and ecologically break down this monsoon clime into a "mountain clime," a "riverine clime," and an "oceanic clime." In essence, these are all interlinked by water cyclically moving between the atmosphere, the hydrosphere, and the lithosphere in different seasons. As both an agent of climate making and a target of human economic development, monsoonal water cycles between mountains, plains, and the ocean in different elevations and landforms. By adding the anthropogenic forces of change, we also consider the evolution of the monsoon climes of mountainous and riverine regions from the colonial era to their postcolonial and current states, which we call the "anthropocenic clime," which will be elaborated below.

We begin with the connection between the Himalaya and the Indian Ocean through the modeling perspective of climate science. Needless to say, it is from the work of climate scientists that we have apprehended the sheer scale of the threats that confront the region in an era of climate change (Pant et al. 2018). Meanwhile, we offer thoughts on the ongoing critique of the recognized epistemological gaps between climate science and regionally-specific experiential knowledge of climate and climate change (Hulme 2015, 176; Shepherd and Truong in this book). We believe the value of the interdisciplinarity of natural sciences, social sciences, and humanities lies in their mutually beneficial complementarity. This leads us to the gravity of our co-authorship in connecting the floodplains, the coastal riverine frontier, and the high mountains in more intimate, terrestrial terms, using a nested series of local experiences and knowledge embedded in both landscapes and mindscapes: categories of thought and understanding that are less easy to model, and even harder to constrain within the boundaries of nation-states. This is where we start visualizing climate change in the above-proposed imageries of clime change in different elevations and habitats based on local knowledge and local understandings of modern anthropogenic forces of change. Having tried, in earlier works, to see climate change in the Bay of Bengal (Amrith 2013, 2018) and the Himalayan highlands (Smyer Yü 2021a, Smyer Yü and Michaud 2017) through local eyes, our hope is that through conversations and collaborations like this, we will find new ways to link the mountains, the plains, and the ocean, and to imagine sustainable futures founded upon a set of new, locally-inspired planetary ethics.

Monsoon Climes of the Himalaya and Bengal

The Himalaya, inclusive of its plate-tectonic extension into the Tibetan Plateau, is often represented through metaphors such as the "roof of the world," the "third pole" of the earth, and the "water tower" of Asia, all of which evoke its global geological magnificence and environmental significance. Truly, its water sources are an inherent part of the nourishing earth sustaining the habitats of over a billion human- and countless nonhuman lives in Asia. In depicting Bangladesh as a land of water and silt expanding into the Indian Ocean, Willem van Schendel writes:

> In a sense, Bangladesh *is* the Himalayas, flattened out. Every spring the mountain snow melts and the icy water sweeps along particles of soil, forming into rivers that rush to the sea. As these rivers reach the lowlands, they slow down and deposit those particles, building up a delta. This age-old process has created the territory that we now know as Bangladesh—a territory that pushes back the sea a little further with every annual deposit of new silt.
> *(Van Schendel 2020, 3)*

Expanding Van Schendel's formulation beyond Bangladesh, we could say that large parts of South Asia are "the Himalayas, flattened out": that flattening, in the form of water and silt, is a critical source of natural nourishment to human civilizations and other beings in the plains of South Asia. The pantheons recognized by the region's humans find their abodes in the vertically awe-inspiring geological magnificence of the earth (Pandit 2017, 5, 16). Human reverence for the natural and spiritual wonders of the high mountains is found in all mountainous parts of the world. While "about half of all humankind directly depends on mountain resources, primarily water," Eklabya Sharma and his colleagues write, "mountains support 25% of world's terrestrial biodiversity and include nearly half of the world's biodiversity 'hotspots'" (Sharma et al. 2019, 2). High mountains and mountain ranges in the human spiritual perceptual field are revered as theophanies in which humanity and divinity find their communion (Smyer Yü 2011, 56). The Himalaya has thus nourished and inspired multiple human cultures and civilizations in the lowlands of Asia. The ecological experience and affective perception both recurrently affirm the life-supporting power of the Himalayan mountains and water.

While the Himalayan mountains move unnoticeably to human eyes in the geological pace of ancient but ongoing tectonic forces, their relative immobility in the human perceptual field creates an impression of the water they hold as the starting point of a downpour into the plains of South Asia and the Indian Ocean. However, climatically speaking, water moves on its own accord, alternatingly high and low, thick and thin, visible and invisible. It makes seasons, weather and climate patterns, and climate changes. Ultimately the flow of water forms climes or places and regions with pronounced geomorphological and ecological variations due to geological and climate forcing. The Himalaya is a pivotal station in the cyclical journey of water between the atmosphere and the Indian Ocean. As part of our effort to germinate conversations between natural/social scientists and environmental humanists, we find climate science helpful for us to visualize the water flow between the mountain and oceanic worlds, particularly in respect to how the monsoon propels the movement of water and threads together highlands, lowlands, and the edging frontiers of land, ocean, and atmosphere.

The South Asian monsoon is one of the six global monsoon systems affecting 70% of the world's human populations in both the Southern and the Northern Hemispheres (An et al. 2015, 30). The other five are the African, East Asian, Australian, North American, and South American monsoons. Etymologically traced to the Arabic word *mawsim* or season, monsoon in both climate sciences and daily use refers to "a seasonal prevailing wind," which is "driven by the annual cycle of solar insolation and the differential heating of the land and ocean, which lead to the seasonal reversal of surface winds as well as the contrast between wet summers and dry winters" (Endo and Kitoh 2016, 48). The precipitation the South Asian monsoon brings to the Himalaya is not uniformly dispersed. To be precise, the South Asian monsoon is most responsible for 80% of the annual precipitation in the central and eastern Himalaya (Shroder 2016, 21), with the western Himalaya not receiving as much monsoonal rain during the summer due to its longer geographical distance from the Indian Ocean. Exclusive of the Indian Ocean monsoon, most of precipitation in the Himalaya occurs between November and May, borne on the westerly, katabatic (downslope), and anabatic (upslope) winds, respectively, from the Atlantic and Mediterranean, the Tibetan Plateau, and the lowland deserts of South and Central Asia (Shroder 2016, 9). Unlike the monsoonal rainfall in its central and eastern regions, the form the precipitation in the western Himalaya takes is mostly snowfall, which is the source of the ancient, spectacular glaciers in the Pakistani and Afghan Himalaya as well as in their neighboring mountain ranges in Central Asia. The uneven precipitation across the eastern, central, and western Himalaya characteristically shapes the regions' respective climatic and meteorological patterns. While the east receives more monsoonal rain than it does snowfall, the west is the opposite, explaining why the western Himalaya has more glaciers than the eastern part of the range—a feature that now has planetary implications in an era of anthropogenic climate change.

As this chapter focuses mostly on the nexus of the Indian Ocean and the central-eastern Himalaya (see Figure 1.1), the ongoing research outcomes from us and our peer scholars all affirm that the South Asian monsoon is the annual mover of water between the ocean, the mountains, and the floodplains of Bengal. Smyer Yü (2022, 48) connects the monsoon with the Tibetan Plateau: "Thanks to the geological uplift, the southern and eastern edges of the Tibetan Plateau all possess high altitude-induced steep ecological gradients and are wide open to annual monsoon from South Asia." Arupjyoti Saikia (2019, 2) traces the force of the monsoon from the Brahmaputra to the Bay of Bengal: "During the monsoon, the Brahmaputra's swelling waters turn violent and unpredictable; its waters spread as far as the eye can see…the Brahmaputra supplies water at the rate of approximately 19,800 m^3/second to the Bay of Bengal."

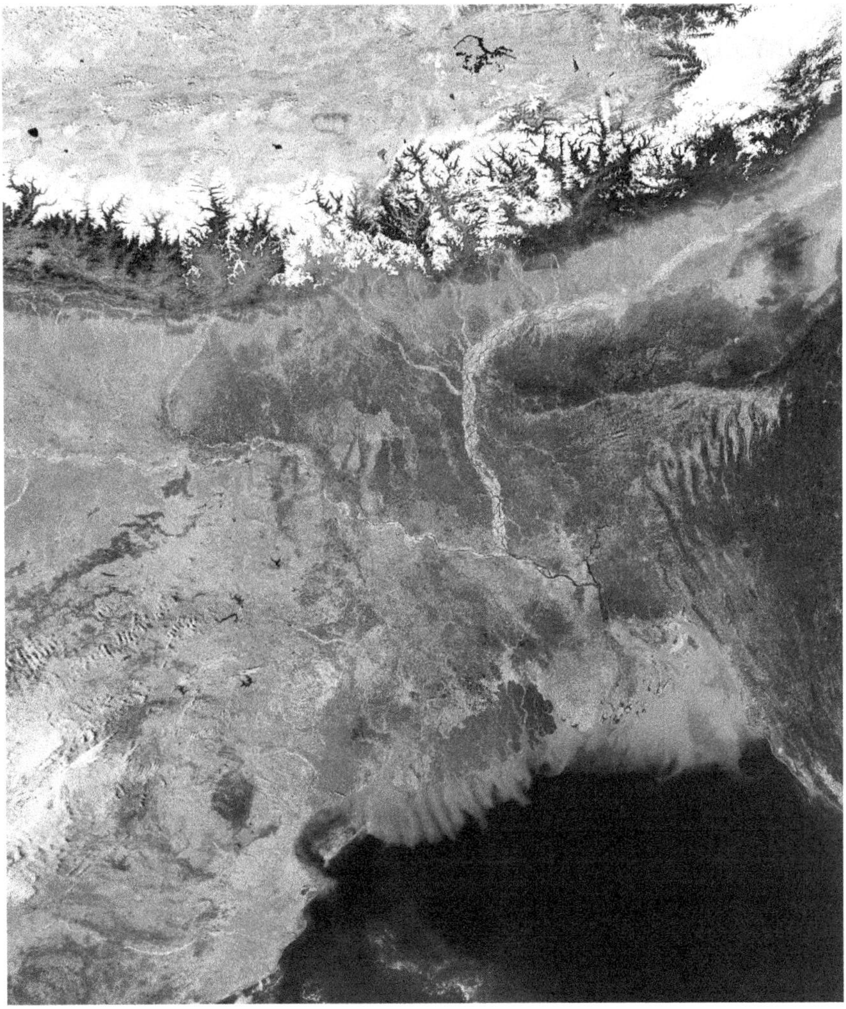

FIGURE 1.1 Satellite view from the Himalaya to the Bay of Bengal.
Source: ESA.

By connecting the findings from climate and social sciences, and the humanities, we see climate not merely expressed in abstract, statistical models and charts, but also expressed in concrete habitats and environmental flows. By pushing for deeper interdisciplinary conversations and perspectival exchanges, we recognize climate as clime, and climate change as clime change, as evidenced in our experiential worlds. While climate is often understood as prevailing meteorological conditions over a long period of time, the idea of clime affords us experientially-integral perspectives on how the five components of the climate system—atmosphere, oceans, cryosphere, lithosphere, and biosphere—interact and intertwine with each other as concrete living environments. Clime convinces us that the integrated understanding of the five components can help us see the omnipresence of climate in air, water, and land since prehuman times. This naturally leads us to interlace together the Himalayan mountains, the Tibetan Plateau, the riverine landscapes of Bengal, and the Indian Ocean as a series of monsoon (induced) climes. We thus further identify monsoon climes as a set of interconnected physical environments that are characteristically shaped by different forms of moving water, such as vapor, rain, hail, and snow, patterned in the annual monsoonal climatic cycle. With the range of elevations, landforms, ecosystems, and ecotones considered—alongside diverse modes of human life, patterns of settlement, and degrees of economic inequality—the monsoon climes can be further parsed into "mountain clime," "riverine clime," and "oceanic clime" leading from the Himalayan highlands to the Bengali floodplains and the Indian Ocean. We are interested both in their interconnections and in their singularity.

By interlacing the highlands, the lowlands, and the land-oceanic frontiers together, we see the role of the monsoon in shapeshifting water between liquid, solid, and vaporous forms in its circular journey between the mountain and oceanic worlds. From our climatic perspective, it is hardly possible to pinpoint the starting point of this moving water because it travels in cycles. It thus makes perfect sense when Van Schendel (2020, 3) sees the flattened Himalaya in Bangladesh, when Smyer Yü (2021b, 9, 10) regards the Himalaya and the Tibetan Plateau as a "terrestrial ocean," and when Amrith (2018, 92) captures how water has shaped the climate of South Asia as "the descent of water from the Himalayan rivers and the ascent of water vapor from the Bay of Bengal and the winds stirring the ocean surface and transporting clouds to shore." Either the mountains or the oceans can be the starting points of the environmental flows between them thanks to the omnipresence and climate-generating force of water.

Colonial Visions of the Himalayan-Bengal Climes

We reconstruct the monsoon climes mentioned above without much of a human presence because it is our intention to converse with natural scientists, to point out that the Anthropocene is a shallow, human-centered history of the earth since the Industrial Revolution, and to reacquaint ourselves with the deep history of the earth known among scientists and now the public as Deep Time. The earth has a life of its own (Rose 1996) and its geological and ecological forces shaped the landforms and ecosystems of the Himalaya and South Asia before the emergence of *Homo sapiens*. Our species' two-hundred-thousand-year-history is incredibly short when contrasted with the 4.5 billion years of the earth's history (Black 2021). However, the stories of the Anthropocene tell us that since the 1800s, humans' destructively-creative activities have been physically transformative enough to alter the earth's geological structure and climate patterns. For some, the Anthropocene is celebrated as the scientific triumph of humankind; for others, it is the time in which current global warming and interlocking environmental crises are largely the proxies of our unsustainable use of land and resources. Holders of both opinions recognize the intelligence and the technological power of humankind but with divergent teleological and ethical orientations. We thus see the Anthropocene as a critical time to re-educate ourselves about the earth's own planetary biography and to rethink what sustainability means in more-than-human terms.

For this purpose, we now tell a story of humans' envisioned transformation of the Himalaya and the Bay of Bengal since the advent of the Industrial Revolution and the heyday of the colonial economy in South Asia. It is a story of the colonial cartographical envisioning of the vast region to be a source of natural and human resources and a nodal point of international trade, with climatic and environmental implications continuing to be present in contemporary South Asia.

In the 1860s, a view produced in distant London aimed to present an encompassing picture of South Asia when, in 1868, Trelawney Saunders took over the collection of maps and charts at the India Office, the British government department responsible for the administration of colonial India. In 1870, Saunders published his *Sketch of the Mountains and Rivers of India*. It is not clear whether Saunders ever visited India himself; he pieced his map together from sketches and route plans and travel accounts accumulated over a century and lying unsorted in India Office vaults. He synthesized the expansive exercise in mapping that was the Great Trigonometrical Survey of India, which first measured the height of the Himalaya through a process of triangulation. Saunders wanted the reader's eye to move in two directions—up "from the sea to the summits of the land," and down "from the sources of the rivers to their outfalls." In both maps, he argued, "the Himalaya mountains will be found in a new form" that would change the viewer's perception of India and its frontiers. He wrote of the Tibetan Plateau, source of Asia's great rivers, that "all detailed knowledge of the interior of this extraordinary country is wanting." But by 1870 Saunders still knew enough to redraw the map.

Saunders's map was far from static—taking a bird's eye view of the rivers' descent drew attention to the constant movement of people and goods along those routes (see Figure 1.2). By following their path he could see that "the people of south-eastern Asia are naturally connected with the Himalayan system." He observed that trade from India to Central Asia "passes from the Indus over this Afghan plateau to Persia and Turkistan," and from there fed into both Russian

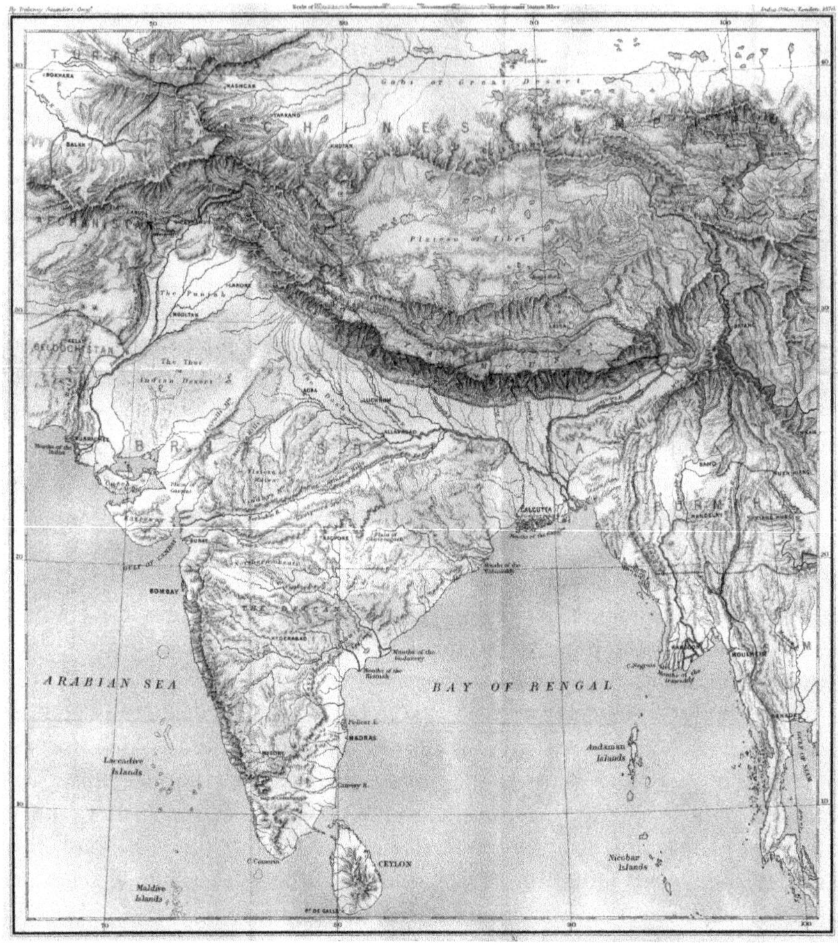

FIGURE 1.2 Source: Saunders, *A Sketch of the Mountains and River Basins of India, in Two Maps, with Explanatory Memoirs* (London: India Office, 1870).

and Chinese circuits. The "enfrachisement of trade and intercourse along the vast frontier," he concluded, would lead to "a trade of unparalleled magnitude with countries abounding in raw products, and populations devoid of machinery and manufactories on a large scale" (Saunders 1870, 9). Saunders's maps of India's rivers and mountains presented a set of propositions about how places were connected to one another—and which places mattered. They contained an argument about India's internal and external frontiers, and about India's place in Asia. Saunders wanted his map to draw attention to an untapped "frontier" for British commerce. It was a view of the Himalaya suffused by colonial power.

A decade later, the official compendium of India's history and geography, the *Imperial Gazetteer*—published in eight volumes in 1881 and edited by William Wilson Hunter, a keen meteorologist as well as an ethnographer—insisted that "by India we now imply not merely the wide continent which stretches southward from the Himalayas to Cape Comorin, but also the vast entourage of mountainous plateaux and lofty ranges." India, Hunter insisted, "can no longer be considered apart from the wide hinterland of uplands." This was a geopolitical as much as a geo-economic imperative. "India," he wrote, "must be held to include those outlying territories over which the Indian administration extends its control, even to the eastern and southern limits of Persia, Russia, Tibet, and China." He was envisioning "a future of railways developments" with a "rush of motor traffic" and predicting "the land approaches to India" would "rival those of the sea" by saying "we shall take out tickets in London for Herat, and change at Kandahar for Kabul or Karachi." Among these frontiers, the least explored, but one "potentially destined to play an important part in Indian history," was "the great highway of the Brahmaputra valley from the plateau of Tibet to the plains of Assam" (Hunter 1901, 5, 6).

Hunter drew two conclusions. The first was that "the material wealth of India largely depends" on its "capacity for the storage of that water supply which carried fertility to its broad plains." The other was that British India's security depended on "guarding the gateways and portals of the hills," thus preventing "those landward irruptions" that had reshaped Indian history on many occasions in the past. These propositions would endure; they would outlast the British Empire that Hunter and his contemporaries wanted above all to preserve. And here is the contradiction at their heart: envisioning India through its rivers encouraged a far-reaching consideration of its connections with distant places, of "world highways" (or potential "highways") that situated India in relation to the flow of goods, people, money, and water—to and from China, Tibet, and the expanse of Central Asia. By contrast, to see the Himalaya as a natural barrier, one always under threat of breach, was to advocate for a vision of India as a bounded place: an "amphitheater" sealed off from the rest of Asia. Natural frontiers became metonymous with the security of the realm. Hunter's concern with the economic value of water to Indian agriculture reinforced this bounded view. Water was a resource to be stored, possessed, harnessed, and put to work: it was the essence of *India's* "material wealth."

With the advancement of storm science came a different way of thinking about space—and about India's place in the world. In the same decades, British explorers and scientists began to study the Himalaya. From there, botanist Joseph Hooker observed the monsoon from the mountain peaks. In Sikkim, Hooker witnessed a watery realm reaching from the ocean to the atmosphere; the sky was a mirror to the sea. He described a climatic system where "vapours raised from an ocean whose nearest route is 600 km distant are safely transported without the loss of one drop of water to support the rank luxuriance of this far distant region." And then, "the waste waters are returned by the rivers to the oceans, and again exhaled, exported, recollected and returned" (Hooker 1854, 97).

Two features of this imperial vision of the Himalaya are of lasting importance: it was an *extractive* vision, and a *bounded* one. The Himalaya was thus envisioned as a form of territory to provide "water resources" to the colony, and in due course, a naturally-formed security barrier to the nation-state of India.

This dual-purposed vision found new expression in postcolonial India, where dams were declared by Prime Minister Jawaharlal Nehru as the "temples of new India" and where water resources were seen as the key to planned national development. Dreams of hydraulic engineering were inseparable from dreams of national freedom. Kanwar Sain, the second head of India's water authority, wrote that "the river valley projects constitute the biggest single effort since independence to meet the material wants of the people, for from irrigation springs ultimately the sinews of man, from power the sinews of industry." He voiced the hopes of many of India's planners and architects when he declared that the dams "are indeed the symbols of the aspirations of new India, and the blessings that stream forth from them are the enduring gifts of this generation to posterity" (India 1957). Note here, the fusion of the language of science and faith, reason and rapture.

Toward the end of the twentieth century, those ambitions to secure the "sinews" of power led the Indian state, engineers, and industrialists to eye the upper reaches of the Himalayan rivers as an untapped frontier—this took place, as Pomeranz (2009) shows, in parallel and potentially in conflict with Chinese efforts to construct large dams on the Tibetan Plateau. Faced with a raft of water-related conflicts and a fast dwindling aquifer in the agrarian heartland of the northwest, the Indian approach has been to entrench further a top-down, engineer's vision of water. We can see a direct line between the territorializing vision of the colonial map, depicting the Himalaya in terms of the water security of the colony, and this map below (see Figure 1.3), illustrating India's mammoth water-linking project. Notice, here, the bounded map—as if the rivers, the glaciers, and the clouds stop at the borders of the nation-state; as if India's sovereign power to engineer water would have no consequences downstream (Khera 2020).

Advent of Anthropocenic Changes in the Himalaya and Bengal

Mapping, in the colonial history of South Asia, was a teleological act for acquiring natural and human resources, expanding colonies, building global markets, and providing future prospects of the empire. One hundred years before Saunders's mapmaking endeavor in London, the East India Company already had a fairly clear picture of what it had accomplished and what it would achieve in Bengal for the coming years. In essence, Saunders's compilation of the existing maps was a showcase of the British Empire's existing accomplishments in South Asia. In 1780, James Rennell, a surveyor for the East India Company, published his *A Bengal Atlas* (Rennell 1780) serving the commercial, military, and administrative purposes of the Company. From our current environmental perspective, it was a harbinger of the anthropocenic changes in Bengal in the coming two centuries. Its cartographical information provided colonial administrators with detailed navigations for the forthcoming expansion of the colonial economy and

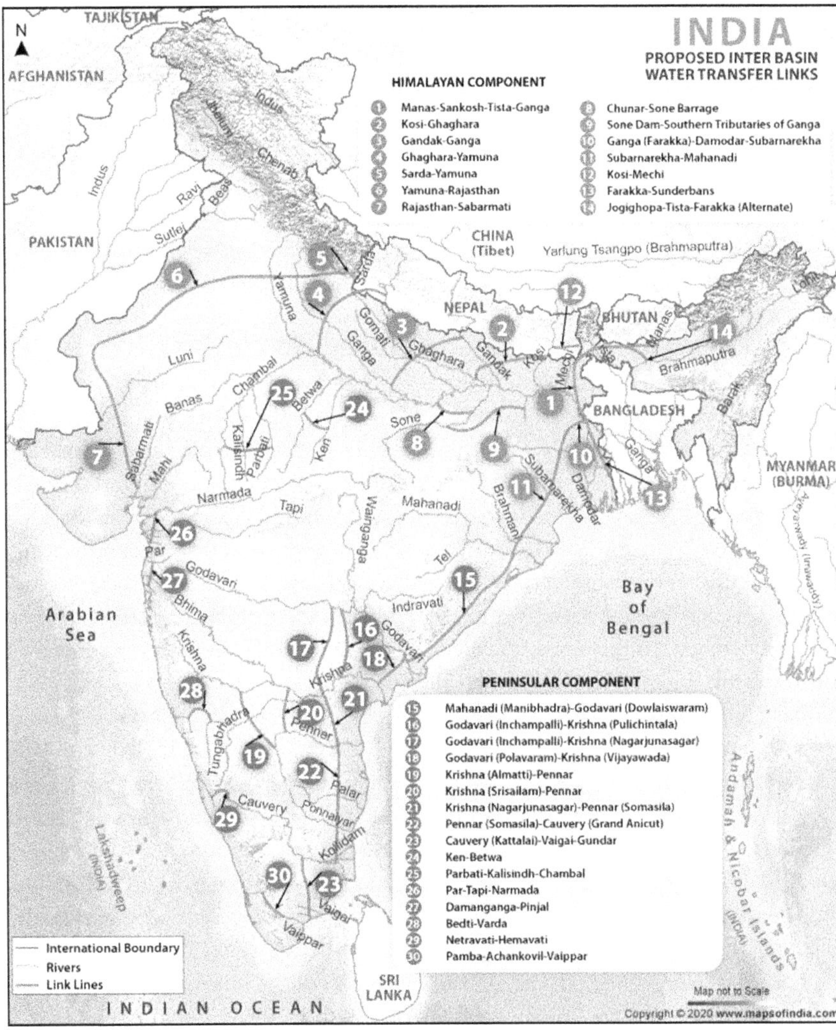

FIGURE 1.3 India's Water-Linking Project.
Source: Srishti Khera 2020.

subsequent environmental transformations. Before we delve further into the discussion of anthropocenic changes, we would like to open up the idea of a riverine clime, to understand better the precolonial landscape that Rennell and his successors sought to make more legible and to harness to the British imperial project.

Lying between the Himalayan Mountains and the Indian Ocean, Bengal, aided by the power of the monsoon, has annually received the "flattened Himalaya" in the form of torrential rivers and megatons of silt, and has relayed much of this matter into the Indian Ocean since the Indo-Australian plate collided into the

Eurasian plate over fifty million years ago, before the emergence of humankind. Because its water and land have been so entwined with each other, this vast region is characterized by scholars as "monsoon landscapes" (Cederlöf 2014a, 29) an "aqueous land" (Lahiri-Dutt 2014b, 508), "a braided landscape of land and water" (Amrith 2018, 182), "riverine communities" (Saikia 2019, 5), and "a fluvial Asia" (Iqbal 2021, 105). All these water-saturated phrases encourage us to see Bengal as a riverine clime that sustained native human societies since ancient times, but that has incessantly undergone modern anthropogenic alterations and human-induced clime changes starting in the colonial era and continuing to the present day.

By a riverine clime in the case of Bengal, we mean a vast habitat that is geologically contiguous to the Himalaya and the Indian Ocean and that is braided together by Himalaya-sourced rivers, silt, and the monsoon-driven, transregional environmental flows. To be noted, we use the word "braided" in reference to both rivers and lands, although we see its predominant use favoring rivers, in phrases like "braided rivers" (Saikia 2019; Richardson and Thorne 2001), "braided river systems" (Lahiri-Dutt and Samanta 2013, 35), and "braided river channels" (Chalov 2001). While we recognize these environmentally accurate depictions of the braided rivers in the landscape of Bengal, we also see terrestrial braids interspersed with river braids. Without the former, the latter could be hardly identified. In this sense, the riverine clime signifies the intertwinement of shifting lands and rivers, which then shapes the ecosystems, ecotones, environmental conditions, weather patterns, and ecologically-afforded habitable lands for both humans and nonhumans.

As we treat Bengal geologically and climatically as an inherent part of the Himalaya-Indian Ocean nexus, the locally-understood riverine clime is indeed a braided riverine landscape, while the regionally-understood one offers a larger picture of how the active geological and meteorological forces of the Himalaya and the monsoon—legacies of deep, geological time—precondition the annually shapeshifting activities of both water and land. Moving rivers shape and reshape the surface of land, while land, understood as a part of the lithosphere, moves with the ancient plate-tectonic forces of the earth and shifts the courses of large rivers, such as the Brahmaputra and the Ganges, often at an unnoticeable speed until the perceivable environmental tipping points signal us of their new courses induced by extreme weather and/or seismic activities. Both the locally- and regionally-understood riverine climes compel us to address the question of climate change, or rather more aptly, clime change, as both natural and human-induced occurrences.

To break down the human implications of the monsoon and the moving Himalayan water and silt, let us take a look at how the Brahmaputra, with the aid of the Himalayan elevation and the monsoon, moves both water and land and shapes the human world of Bengal. In his biography of this ancient river, Arupjyoti Saikia writes:

> During the monsoon, the Brahmaputra's swelling waters turn violent and unpredictable; its waters spread as far as the eye can see. The Ganga and

> the Brahmaputra together discharge the highest volume of water of the world's rivers; the Brahmaputra supplies water at the rate of approximately 19,800 m^3/second to the Bay of Bengal. In comparison, the Indus supplies only one-third of this to the Arabian Sea. The Brahmaputra, however, can be lethargic in the dry winter months. Only sixteen rivers on earth carry more than hundred million tons of sediment to the sea and the Brahmaputra ranks as one of the highest amongst them.
>
> *(Saikia 2019, 2)*

The geologically, ecologically, and climatically forceful Brahmaputra, along with other Himalaya-sourced rivers and their tributaries, shaped precolonial human livelihoods into "a highly sophisticated water-based artisan economy—the blessing of rivers—irrigated naturally by the monsoon rains and annual flooding," as Kuntala Lahiri-Dutt recounts:

> In this water-based artisan economy, the countryside was dotted with dispersed local markets attesting to a highly commercialized economy that boasted impressive industries. The rivers were not just channels of water, they carried a thriving trade, transporting people and goods from one part of the delta to another.
>
> *(Lahiri-Dutt 2014a, 17)*

While rivers, lakes, and streams, especially during the monsoon season, changed courses, inundated human dwelling places and farmlands, and overran roads, humans were adaptive, resilient, interactive, and innovative with the riverine clime that environed them. Their modes of subsistence, livelihoods, collective well-being, art, and literature, in fact, thrived with the ebb and flow of the rivers and with the changing landforms and water bodies. In his *The Changing Face of Bengal*, the philosopher and sociologist Radhakamal Mukerjee, writing in the 1920s, depicts the lifeworld at the confluence of the Brahmaputra and the Ganges:

> The waters here widen out into endless vistas, and the lands are in a process of perpetual building and unbuilding by a wide and interlaced network of canals and streams, while an endless procession of cocoanut-clad villages and itinerant hamlets, containing a swarming population, fringes the marshes or the river-banks, which are higher than the surrounding country, and are always accessible by water. Fishes are abundant here throughout the year, and the paddy-lands extend right up to the doors of the peasants' huts. Man here is essentially a child of the rivers. His crops and farming practices are closely adjusted to the timely inundations of red water; and indeed, if the rivers do not rise in flood, and submerge the country, he will be a fish out of water indeed.
>
> *(Mukerjee 1938, 19, 20)*

If the biblically-rendered idiom "the land of milk and honey" in Western cultures refers to a land of cornucopian abundance, precolonial Bengal could be seen as a land of abundant rice and fish, which were the material basis of its thriving artisan economy. Of course, "rice" here also symbolizes other crops such as cotton and sugarcane, among others. These nonstaple crops were traded in local as well as regional markets connected by waterways. Larger rivers were highways between towns and regions, while smaller ones served as "streets" linking together households and markets in riverine villages and towns. Boats, large and small, dotted the riverine clime of Bengal for transportation and fishing (Saikia 2019, 107). Emblematic of its artisan economy, Bengal's native textile industry spread out in its riverine clime, where "spinning, weaving, and trading were mainly rural occupation" (Van Schendel 1991, 41) and "it is difficult to find a village in which every man, woman, and child is not employed in making a piece of cloth" (Lahiri-Dutt and Samanta 2013, 53; Orme 1805). Before British rule, Bengal was not a perfect human world, but the riverine clime was able to nourish approximately twenty million people there (Van Schendel 1991, 38).

This riverine clime of precolonial Bengal, which combined pockets of self-reliance with significant commercial expansion during the Mughal era, would soon be repurposed as the agricultural basis of British India (Richards, 2003, 25–38). The riverine clime of Bengal was transformed into an anthropocenic clime through its physical restructuring and economic reorientation for the maximization of yield and market values. In our climate study of Bengal, we deem the anthropocenic clime a physical sign of the Anthropocene in its inchoate phase in Asia, before its effects were fully visible. In the global discourse of the Anthropocene, humans are frequently looked upon as a "geological agent" (Clark and Yusoff 2017, 5; Halperin 2015, 81). In the case of Bengal, the geological and ecological agency of British India expressed itself as a forceful anthropogenic conveyor of change, particularly in its rigid separation of land from water. In the process, the agricultural use of land was privileged over water. In the historical studies of Bengal's environmental and social changes, scientists and scholars commonly agree that the relentless hyperseparation of land from water under British rule permanently altered the geomorphology and ecology of Bengal. While embankments and barrages began to dominate the riverine clime, rivers were being disciplined to flow within human-perceived or -prescribed courses, the land was classified into different grades based on humans' assessment of their agricultural value and productive capacity. The introduction and implementation of private land ownership accompanied the physical separation of land from water (Lahiri-Dutt and Samanta 2013, 51–77). The common use of land among native Bengalis of the past became less and less possible. The predominantly colonial ownership and transformation of the land toward agricultural use soon eroded the native water-based artisan economy and made it impossible for millions of Bengalis to continue their age-old livelihoods relying on and synchronized with the ebb and flow of rivers and the commonly shared, shapeshifting lands.

Environmentally, the construction of higher embankments and larger barrages was, in principle, intended for irrigation and protecting humans from destructive floods during the monsoon season. However, the geological and ecological forces of the earth do not always agree with human economically-driven environmental acts undertaken on an industrial scale. The works of Cederlöf (2014a, 2014b), Iqbal (2010, 2014), Saikia (2019), Lahiri-Dutt (2014a, 2014b), Singh (2014), and others all recount extreme environmental consequences of British India's efforts to alter the riverine clime of Bengal. Among them, in his *An Imperial Disaster*, Benjamin Kingsbury elaborates his historical and environmental argument that the devastating cyclone of 1876 in Bengal, which killed over three-hundred thousand people, was "not simply a 'natural' disaster, but one shaped by all-too-human patterns of exploitation and inequality" and was thus a natural disaster imbued with "human elements of calamity" (Kingsbury 2018, xiii).

When the equilibrium of human-earth relations is maintained, it is often because humans are willing to negotiate with and proactively work with the earth's natural forces. In the colonial context, when the logic of human environmental intervention rests upon the intention of harnessing local resources and productivity for nonlocal use, the burdens inevitably stretch the carrying capacities of both the land and its native residents. The responses from the earth to these anthropogenic shocks are often meteorologically and ecologically forceful enough to engender natural disasters. This type of destructive response from the earth is now deemed anthropocenic due to humans' alteration of the hydrosphere, lithosphere, and biosphere that have consequentially induced clime change and extreme weather and seismic events. The environmental state of colonial Bengal was an example of the initial anthropocenic effects in the expression of natural disasters induced by modern, colonial humans as geological agents. In this sense, the colonial riverine clime was an anthropocenic clime in which land and water were artificially segregated for a new environmental order undertaken with a new teleology to serve the geo-economic interests of a modern empire.

Planetary Engagements with Anthropocenic Climes

One of the environmental and political ironies in modern Asian history is that the colonial anthropogenic practices continue in the postcolonial era in spite of regained national freedom from colonial rule. In other words, and in this case, the environmentality of the colonial and postcolonial times largely remain the same in former-Bengal now partitioned into the territories of free India and Bangladesh (former East Pakistan) in 1947. Understood as political governance of the environment by either colonial powers or postcolonial states, the environmentality in the South Asian context suggests an anthropocenic environmental consciousness that was molded and shaped in the colonial era but finds its longevity in the postcolonial governing bodies and their subjects. Apparently, the British agricultural management of the environment in Bengal "left an indelible mark on Indian political consciousness" (Agrawal 2005, 166; Appadurai 1996,

124). To reiterate, this political consciousness of the environment is centered on the maximization of agricultural yield through the rigid separation of land and water with the privilege given to land. Postcolonial states fully inherited this environmentality, scientifically justified, as an integral part of their new nation-building processes. The historical circumstance was that after the nation was depleted of its natural resources and its people became tired and wretched from serving the global demands of the empire, it desperately needed healing, replenishment, and a peaceful and abundant future without deprivation, hunger, or foreign invasion. With the same environmentality, the productivity of the land was then assigned with a new national imperative—to save the nation—with water playing a supporting role as the source of irrigation and hydropower. Madhav Gadgil and Ramachandra Ghuha give a vivid account of free India after the colonial plunder:

> As the natural world recedes, so shrink the capacities of local ecosystems to support these people. Dams and mines, for instance, have physically displaced millions of peasants and tribals in independent India. Others have fled as forests and, with them, springs have vanished. These people constitute the ecological refugees who live on the margins of islands of prosperity.
>
> *(Gadgil and Ghuha 1995, 3, 4)*

It was thus inevitable that Nehru cherished dams as the "temples of new India" seventy-five years ago. Dammed water gave him hope for the new India, as his administration repurposed the colonial management of land and water for India's own use. This environmentality is not unique in South Asia but is rather a pan-Asian environmental consciousness found among former colonies and semi-colonies, whose national sentiments are split between the painful memory of the colonial depletion of natural resources and the faith in modern science introduced from colonial countries as the modern method of saving the nation. China, Pakistan, Bangladesh, Nepal, and Bhutan are in the forefront of dam building for the energy needed for their day-to-day use, desired for increasing their GDPs by selling surplus hydro-electricity and wished for their national cornucopian futures. Each of these nations has its own development reasons to dam Himalaya-sourced rivers passing through its national territory. Bhutan prioritizes their dam building for exporting 75% of its hydro-electricity to India. Nepal plans to send electricity from its hydropower stations to the half of its rural population without electricity. Pakistan's dam building is intended to meet the goal of covering the needs of its entire agricultural sector which employs up to 66% of the nation's workers and supports 80% of its export economy. China and India, the two most populated Himalayan countries, are in a fierce competition over Himalayan transboundary hydraulic power to serve their growth-led economies (IR 2008; Gamble 2019). By 2019, there were over "550 hydropower projects in existence or under construction" (Xu et al. 2019, 144) in the Himalayan region. Obviously, these postcolonial nations are generating considerable stress on the Himalayan hydrosphere in pursuit of short-term developmental goals (Smyer Yü 2021a, 8).

At the same time, since the 1960s after the first astronauts sent images of the earth back from outer space, humans' consciousness of the earth on a planetary scale has expanded dramatically. These images afford us a new way and a new scale on which to recognize and imagine the environmental connectedness of the entire earth. Benefiting from anthropocenic "environing technologies" (Sörlin and Wormbs 2018), the planetary vision in this chapter combines our terrestrial perspectives with an altogether larger view of the Himalayan waters as expansive, borderless, and vast in its connections with the ocean and the atmosphere. It shows just how completely water crosses human borders in horizontal and vertical dimensions, and how Asia is shaped by a vast connected climate system. The Himalaya belongs to the world! Expressions that have now become commonplace—the "third pole" and the roof of the world"—gained currency only in the late twentieth century (see Figure 1.4, on the "third pole"), signifying that the fate of the world's flourishing depends on the health of Himalayan glaciers, just as it relies on the global importance of the Andean and Arctic cryospheres (Carey 2010; Sörlin 2017; Paerregaard 2020). The outer space perspective brings sharply into view the dependence of humanity on water and the ways that human activity is reshaping the earth. Our understanding of the role of the Himalayan region in shaping earth's climate is of pivotal importance to our planetary future.

But there is a striking absence from the outer space view, too: there are no people, no animals, and no plants in it—a similar picture also found in statistics- and graph-packed climate models. This is where we experiment with our clime study from the terrestrial perspective to fill in the details of life on earth. As clime is synonymous with place, the terrestrially-intended clime approach affords us to feel the pulsation of the earth in wind, fog, rain, snow, ice, moving water, and living beings. It offers to us empathetic understandings of place-based human life in terms of how water has entwined itself with the local environment and regional climate history since time immemorial, how modern human economic activities have forcefully altered physical landscapes/waterscapes and increased the occurrences of extreme weather and natural disasters, and how water is reaching a tipping point that prompt native and nonnative people to call for climate justice and re-appreciation of local traditional environmental knowledge in worldwide endeavors for sustainable living. The deeper meanings of being human as an ecological species are rooted in the earth. The earth surface is a planetary clime in its entirety that is pieced together with a diversity of bio-geologically specific climes.

In our case, we link the mountain clime, the riverine clime, and the oceanic clime together as an anthropocenic clime to address how climate change in the Himalaya-Indian-Ocean nexus expresses itself in the dynamic activities of water in both ancient and modern times. While the Himalayan rivers terrestrially interlace the high mountains, the floodplains, and the ocean together, the monsoon provides water through atmospheric airways, moving it between the ocean, the

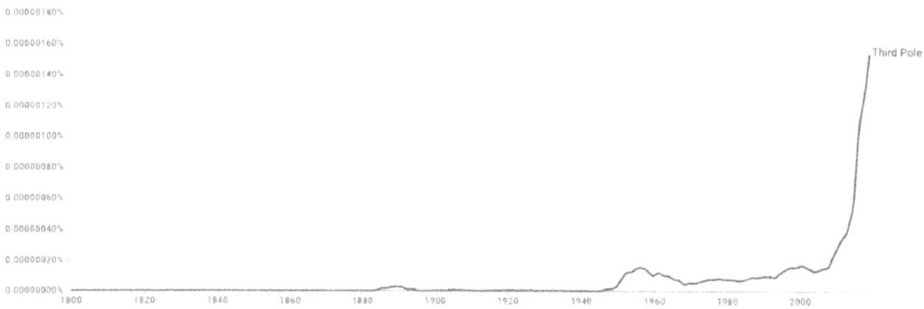

FIGURE 1.4 Google ngram Viewer showing the textual occurrence of the term, "Third Pole," by year.

mountain tops, and the surface of the land. By following the Himalayan rivers and the monsoonal airways, we have learned that people in the Himalaya and Bengal never had the luxury of the climate blindness as they have lived with the annual touchdown of the monsoon and the rivers that wildly dance with it. The anthropocenic clime of South Asia has taught us, as one of us has written elsewhere:

> The history of water shows that nature has never truly been conquered. Water has served as a material constraint on every Promethean plan of growth and plenty. The sheer ferocity of a wet climate—a climate of monsoons and cyclones—remains a source of fear, and no fear is as great as the fear of water's absence, in drought. The cultural history of water is one of reverence as much as hubris. And water has its own chronology—the chronology of the seasons; the episodic chronology of sudden, intense disasters; the imperceptible chronology of cumulative damage, as manifested in the effects of human activity on the oceans.
>
> *(Amrith 2018, 9)*

And on continents as well, to add. Water does not only work within the hydrosphere but also saturates the atmosphere, the biosphere, and the lithosphere in addition to its perceptible bodies, large and small, as glaciers, oceans, rivers, lakes, ponds, puddles, raindrops, hailstones, snowflakes, and vaporous droplets.

The coinage of the term anthropocenic clime in this chapter is thus intended to raise new awareness from correlations of relevant historical and current findings regarding the global climatic significance of the Himalaya and the monsoon, such as how Saikia's biographing of the Brahmaputra affords us to hear the voices of the fishes, the farmers, and the planners (2019), Lahiri-Dutt and Samanta's thick description of the colonial hyperseparation of land and water (2013), Iqbal's depiction of Bengal as a fluvial Asia (2021), and Smyer Yü's oceanic view of the Himalayan terrestrial ice (2021b). Looking again at the transregionality of the monsoon, we see an integral whole of the mountains, the plains, the ocean, and the sky and clouds above them. They afford us a polyvocal perspective on the region. Our coinage is thus meant to show there are many profound ways that local people across the whole region, stretching from the heights of the Himalayas to the coast, imagine their interconnectedness—with each other and with other species. Connections are felt and imagined through both natural events and cultural objects and symbols that move along the old trading routes, waterways, and shifting landscapes. The awareness also attaches to other forms of life: to the plants and animals that bind together high to low, and wet to dry climes. Irreversible changes that many of these ecologies of connection suffer bring an irreparable sense of loss. Thus, understanding the scale of the climate crisis in the Himalaya and riverine South Asia is essential to our planetary future; but so, too, are the intimate, local understandings of connection and loss on the part of those who feel them most directly.

Conclusion

Bracketing the complex legacies of the colonial past in these Himalayan nations, the current anthropogenic transformation of Himalayan rivers is environmentally continuous with colonial precedents, though now on an even greater scale as a result of new financial and technological capacities. The anthropocenic climes, which began to take shape in the colonial era, are expanding in the twenty-first century with alarming consequences for both human and nonhuman life. This compels us to rethink the meaning of sustainable development initiated by the United Nations in the postcolonial era (Brundtland 1987) and promoted by each of these Himalayan nations. What is being sustained? What is being developed? Does sustainable development mean to sustain unbridled economic development or to develop sustainable ways of using natural resources? What is the compass of our environmental ethics in the Anthropocene for sustaining our conflicting desires for unlimited material abundance on the one hand and the health of the earth on the other? Would the "sustainable" use of Himalayan hydrological resources likely happen if the Himalayan nations treated the natural integrity of rivers as an unowned transboundary commons rather than nationally-owned properties? One of the tragic legacies of the colonial past has been to obscure, or even to make deliberately invisible, these broader, transboundary commons.

While we do not yet have realistic answers to these apparently anthropocenic questions, we nevertheless feel the urgency to initiate our response with anthropocenic perspectives by interlacing the earth's deep history with an imagined sustainable future. Our emphasis on a deeper understanding of the Anthropocene is not tautological, but is an engaged stance to confront the ecological and ethical challenges engendered by the irreversible, climate-induced environmental degradations that have already taken place and to minimize the forthcoming detriments of more changes predicted by climate sciences. So, with our like-minded collaborators in this book and beyond, we are exploring new vistas and approaches to the same but larger-scale modern environmental challenges humans have created since the inception of the Anthropocene. Seeing climate change as terrestrially experienced clime change, our historical findings, on the one hand, resonate with the anthropocenic hypothesis of humans as a geological agent (Chakrabarty 2009, 206; Yusoff 2016, 5; Latour 2014, 15) in the sense of our technology-empowered ability to usher in a new, human-made geological era. When interlacing this young geological time with Deep Time, we, on the other hand, see the convincing vision of the late geoscientist Lynn Margulis, "Our tenacious illusion of special dispensation belies our true status as *upright mammalian weeds*" (Margulis 1998, 119; emphasis added).

With both anthropocenic perspectives combined, we see humankind as both a fragile and a technologically daring species. As we are ontologically entangled in the anthropocenic web of life, we join forces with those who are actively nurturing a new planetary ecological consciousness for re-embracing the maternal principle of the earth, restoring the health of the earth and building new environmental ethics for equitable, just, and sustainable living for human and nonhuman beings (Berry 2009; Grim and Tucker 2014; Rose et al. 2012). Planetary perspectives help us globally situate our case study of clime change in the Himalaya, Bengal, and the Indian Ocean and connect our environmental ethical concerns and approaches to sustainable living with the rest of the world.

References

Agrawal, Arun. 2005. "Environmentality: Community, Intimate Government, and the Making of Environmental Subjects in Kumaon, India." *Current Anthropology*, vol. 46, No.2 (April): 161–190.

Amrith, Sunil. 2013. *Crossing the Bay of Bengal: The Furies of Nature and the Fortunes of Migrants*. Cambridge: Harvard University Press.

Amrith, Sunil. 2018. *Unruly Waters: How Rains, Rivers, Coasts, and Seas Have Shaped Asia's History*. New York: Basic Books.

An, Zhisheng, Wu Guoxiong, Li Jianping, Sun Youbin, Liu Yimin, Zhou Weijian, Cai Yanjun, Duan Anmin, Li Li, Mao Jiangyu, Cheng Hai, Shi Zhengguo, Tan Liangcheng, Yan Hong, Ao Hong, Chang Hong, and Feng Juan. 2015. "Global Monsoon Dynamics." *The Annual Review of Earth and Planetary Sciences*, vol. 43: 29–77.

Appadurai, Arjun. 1996. *Modernity at Large: Cultural Dimensions of Globalization*. Minneapolis: University of Minnesota Press.

Berry, Thomas. 2009. *The Sacred Universe: Earth, Spirituality, and Religion in the Twenty-first Century*. New York: Columbia University Press.

Black, Riley. 2021. *Deep Time: A Journey through 4.5 Billion Years of Our Planet*. London: Welbeck.

Brundtland, Gro Harlem. 1987. *Our Common Future, From One Earth to One World: An Overview by the World Commission on Environment and Development* (also known as the Brundtland Report). United Nations General Assembly document A/42/427.

Carey, Mark. 2010. *In the Shadow of Melting Glaciers: Climate Change and Andean Society*. Oxford: Oxford University Press.

Carey, Mark and Philip Garone. 2014. "Forum Introduction." In *Forum: Climate Change and Environmental History*, edited by Mark Carey, Philip Garone, Adrian Howkins, Georgina Endfield, Lawrence Culver, Sherry Johnson, SamWhite, and James Roger Fleming. *Environmental History*, 281–364.

Cederlöf, Gunnel. 2014a. "Monsoon Landscapes: Spatial Politics and Mercantile Colonial Practice in India." In *Asian Environments: Connections across Borders, Landscapes, and Times*, edited by Ursula Münster, Shiho Satsuka, and Gunnel Cederlöf. Munich: RCC Perspectives, 29–35.

Cederlöf, Gunnel. 2014b. *Founding an Empire on India's North-Eastern Frontiers 1790–1840*. New Delhi: Oxford University Press.

Chakrabarty, Dipesh. 2009. "The Climate of History: Four Theses." *Critical Inquiry*, vol. 35, No.2 (Winter): 197–222.

Chalov, R. S. 2001. "Intricately Braided River Channels of Lowland Rivers: Formation Conditions, Morphology, and Deformation," *Water Resources*, vol. 28, No.2: 145–150

Clark, Nigel and Kathryn Yusoff. 2017. "Geosocial Formations and the Anthropocene." *Theory, Culture & Society*, vol. 34, No.2–3: 3–23.

Crosby, Alfred W. 2004. *Ecological Imperialism: The Biological Expansion of Europe, 900–1900*. Cambridge: Cambridge University Press.

Crutzen, Paul J. and Eugene F. Stoermer. 2000 "The Anthropocene." *Global Change Newsletter*, vol. 41: 17, 18.

Endo, Hirokazu and Akio Kitoh. 2016. "Projecting Changes of the Asian Summer Monsoon Through the Twenty-First Century." In *The Monsoons and Climate Change: Observations and Modeling*, edited by Leila Maria Véspoli de Carvalho and Charles Jones. New York: Springer, 47–66.

Fagan, Brian. 2000. *The Little Ice Age: How Climate Made History 1300–1850*. New York: Basic Books.

Fleming, James Roger. 2010. *Fixing the Sky: The Checkered History of Weather and Climate Control*. New York: Columbia University Press.

Gadgil, Madhav and Ramachandra Guha. 1995. *Ecology and Equity: The Use and Abuse of Nature in Contemporary India*. London: Routledge.

Gamble, Ruth. 2019. "How Dams Climb Mountains: China and India's State-Making Hydropower Contest in the Eastern-Himalaya Watershed." *Thesis Eleven*, vol. 150, No.1: 42–67.

Grim, John and Mary Evelyn Tucker. 2014. *Ecology and Religion*. Washington: Island Press.

Halperin, Ilana. 2015. "Physical Geology/The Library." In *Art in the Anthropocene: Encounters Among Aesthetics, Politics, Environments and Epistemologies*, edited by Heather Davis and Etienne Turpin. London: Open Humanities Press, 79–84.

Hooker, Joseph Dalton. 1854. *Himalayan Journals: Notes of a Naturalist in Bengal, the Sikkim and Nepal Himalayas, the Khasia Mountains &c.* London: J. Murray, vol. 1.

Hulme, Mike. 2015. "Climate: Living Lexicon for the Environmental Humanities." *Environmental Humanities*, vol. 6: 175–178.

Hunter, William Wilson. 1901. *The Imperial Gazetteer of India* (revised edition, Oxford: Clarendon Press).

India, Central Water & Power Commission. 1957. *Major Water & Power Projects of India* (Bhagirath Pamphlet 1, June 1957).

International Rivers. 2008. *Mountains of Concrete: Dam Building in the Himalayas*. Berkeley: International Rivers.

Iqbal, Iftekhar. 2010. *The Bengal Delta: Ecology, State and Social Change, 1840–1943*. New York: Palgrave.

Iqbal, Iftekhar. 2014. "Governing the 'Wasteland': Ecology and Shifting Political Subjectivities in Colonial Bengal." In *Asian Environments: Connections across Borders, Landscapes, and Times*, edited by Ursula Münster, Shiho Satsuka, and Gunnel Cederlöf. Munich: RCC Perspectives, 39–43.

Iqbal, Iftekhar. 2021. "Rivers of Mobility: Multi-ethnic Societies and Ecological Commons in a Fluvial Asia." In *Yunnan-Burma-Bengal Corridor Geographies: Protean Edging of Habitats and Empires*, edited by Dan Smyer Yü and Karin Dean. London: Routledge, 105–120.

Khera, Srishti "PM Modi Scheme of River-Linking Projects in India." *My India*, March 16, 2020, https://www.mapsofindia.com/my-india/government/pm-modi-scheme-of-river-linking-projects-in-india.

Kingsbury, Benjamin. 2018. *An Imperial Disaster: The Bengal Cyclone of 1876*. Oxford: Oxford University Press.

Lahiri-Dutt, Kuntala. 2014a. "Commodified Land, Dangerous Water: Colonial Perceptions of Riverine Bengal." In *Asian Environments: Connections across Borders, Landscapes, and Times*, edited by Ursula Münster, Shiho Satsuka, and Gunnel Cederlöf. Munich: RCC Perspectives, 17–22.

Lahiri-Dutt, Kuntala. 2014b. "Beyond the Water-land Binary in Geography: Water/lands of Bengal Re-visioning Hybridity." *ACME: An International E-Journal for Critical Geographies* vol. 13, No.3: 505–529.

Lahiri-Dutt, Kuntala and Gopa Samanta. 2013. *Dancing with the River: People and Life on the Chars of South Asia*. New Haven: Yale University Press.

Latour, Bruno. 2014. "Agency at the time of the Anthropocene." *New Literary History*, vol. 45: 1–18.

Margulis, Lynn. 1998. *Symbiotic Planet: A New Look at Evolution*. New York: Basic Books.

Matthews, John and Keith R. Briffa. 2005. "The Little Ice Age: Re-evaluation of an Evolving Concept." *Geografiska Annaler*, vol. 87 A, No.1: 17–36.

Mukerjee, Radhakamal. 1938. *The Changing Face of Bengal: A Study in Riverine Economy*. Calcutta: University of Calcutta.

Orme, Robert. 1805. *Historical Fragments of the Mogul Empire, of the Morattoes, and of the English Concerns in Indostan: From the Year MDCLIX. Origin of the English Establishment, and of the Company's Trade, at Broach and Surat; and a General Idea of the Government and People of Indostan*. London: Printed for F. Wingrave.

Paerregaard, Karsten. 2020. "Searching for the New Human: Glacier Melt, Anthropogenic Change, and Self-Reflection in Andean Pilgrimage." *HAU: Journal of Ethnographic Theory*, vol. 10, No.3: 844–859.

Pandit, Maharaj K. 2017. *Life in the Himalaya: An Ecosystem at Risk*. Cambridge: Harvard University Press.

Pant, G. B., P. Pradeep Kumar, Jayashree V. Revadekar, and Narendra Singh. 2018. *Climate Change in the Himalayas*. Cham, Switzerland: Springer.

Pomeranz, Kenneth. 2009. "The Great Himalayan Watershed." *New Left Review*, vol. 58: 5–39.

Rennell, James. 1780. *A Bengal Atlas Containing Maps of the Theatre of War and Commerce on That Side of Hindoostan*. London: East India Company.

Richards, John F. 2003. *The Unending Frontier: An Environmental History of the Modern World*. Berkeley: University of California Press.

Richardson, W. Roy and Colin R. Thorne. 2001. "Multiple Thread Flow and Channel Bifurcation in a Braided River: Brahmaputra–Jamuna River, Bangladesh." *Geomorphology* 38 (3):1 85–96.

Rose, Deborah Bird. 1996. *Nourishing Terrains: Australian Aboriginal Views of Landscape and Wilderness*. Canberra: Australian Heritage Commission.

Rose, Deborah Bird, Thom van Dooren, Matthew Chrulew, Stuart Cooke, Matthew Kearnes, and Emily O'Gorman. 2012. "Thinking Through the Environment, Unsettling the Humanities." *Environmental Humanities*, vol. 1: 1–5.

Saikia, Arupjyoti. 2019. *The Unquiet River: A Biography of the Brahmaputra*. Oxford: Oxford University Press.
Saunders, Trelawney. 1870. *A Sketch of the Mountains and River Basins of India, in Two Maps, with Explanatory Memoirs*. London: India Office.
Sharma, Eklabya, David Molden, Atiq Rahman, Yuba Raj Khatiwada, Linxiu Zhang, Surendra Pratap Singh, Tandong Yao, and Philippus Wester. 2019. "Introduction to the Hindu Kush Himalaya Assessment." In *The Hindu Kush Himalaya Assessment: Mountains, Climate Change, Sustainability and People*, edited by Philippus Wester, Arabinda Mishra, Aditi Mukherji, and Arun Bhakta Shrestha. Cham, Switzerland: Springer.
Shroder, John. 2016. "Introduction to Physical Characteristics of Water." In *Transboundary Water Resources in Afghanistan: Climate Change and Land-Use Implications*, edited by Shroder, John and Sher Jan Ahmadzai. Amsterdam: Elsevier, 1–22.
Singh, Vipul. 2014. "Gangetic Floods: Landscape Transformation, Embankments, and Clay Brick-Making." *Asian Environments: Connections across Borders, Landscapes, and Times*, edited by Ursula Münster, Shiho Satsuka, and Gunnel Cederlöf. Munich: RCC Perspectives, 23–28.
Smyer Yü, Dan. 2011. *The Spread of Tibetan Buddhism in China: Charisma, Money, Enlightenment*. London: Routledge.
Smyer Yü, Dan. 2021a. "Situating Environmental Humanities in the New Himalayas: An Introduction." In *Environmental Humanities in the New Himalayas: Symbiotic Indigeneity, Commoning, Sustainability*, edited by Dan Smyer Yü and Erik de Maaker. London: Routledge, 1–24.
Smyer Yü, Dan. 2021b. "Symbiotic Indigeneity and Commoning in the Anthropogenic Himalayas." In *Environmental Humanities in the New Himalayas: Symbiotic Indigeneity, Commoning, Sustainability*, edited by Dan Smyer Yü and Erik de Maaker. London: Routledge, 239–260.
Smyer Yü, Dan. 2022. "The Middle Highlands of Modern China as a Historical Inter-Asian Zomia: Human-Nature Diversity in the Hengduan Mountains." In *Routledge Handbook of Highland Asia*, edited by Jelle J.P. Wouters and Michael T. Heneise. London: Routledge, 43–55.
Smyer Yü, Dan and Jean Michaud, eds. 2017. *Trans-Himalayan Borderlands: Livelihoods, Territorialities, Modernities*. Amsterdam: Amsterdam University Press.
Sörlin, Sverker. 2017. "Scaling the Planetary Humanities: Environmental Globalization and the Arctic." In *The Routledge Companion to the Environmental Humanities*, edited by Ursula Heise, Jon Christensen, and Michelle Niemann. London: Routledge, 433–442.
Sörlin, Sverker and Nina Wormbs. 2018. "Environing Technologies: A Theory of Making Environment." *History and Technology*, https://doi.org/10.1080/07341512.2018.1548066.
Van Schendel, Willem. 1991. *Three Deltas: Accumulation and Poverty in Rural Burma, Bengal and South India*. New Delhi: Sage Publications.
Van Schendel, Willem. 2020. *A History of Bangladesh*. Cambridge: Cambridge University Press.
Xu, Jianchu, Ram P. Chaudhary, and Pratikshya Kandel. 2019. "Sustaining Biodiversity and Ecosystem Services in the Hindu Kush Himalaya." In *The Hindu Kush Himalaya Assessment: Mountains, Climate Change, Sustainability and People*, edited by Philippus Wester, Arabinda Mishra, Aditi Mukherji, Arun Bhakta Shrestha. Cham: Switzerland: Springer, 127–165.
Yusoff, Kathryn. 2016. "Anthropogenesis: Origins and Endings in the Anthropocene." *Theory, Culture & Society*, vol. 33, No.2: 3–28.

2

CLIMING THE ANDES

Vertical Complementarity, Transhuman Reciprocity, and Climate Change in the Peruvian Highlands

Karsten Paerregaard

Let me take you on a walk in one of Peru's many mountain communities. Or rather a climb, which is synonymous with walking in Tapay, a community in Peru's southern highlands situated on a mountain slope that stretches from 2,250 m to 5,400 m (Paerregaard 1992, 1997). On our climb, we'll be guided by Mauricio, a sixty-five-year-old man who will introduce us to the many plots of land that he and his wife own in Tapay. Starting in a subtropical climate at 2,300 m, where Mauricio has several orchards that yield citrus and native fruits such *pacay* and *chirimoya*, we ascend to 2,600 m, where he owns a field that produces figs and pears. Further up at 2,800 m Mauricio has two fields where he has planted avocado trees and where he grows corn and a variety of vegetables. At 3,000 m he uses a couple of plots to grow apples and peaches and at 3,200 m he cultivates native grains such as *kinoa* and *kiwicha*, pumpkins, beans, and other legumes. Then, in the temperate climate at 3,400, he sows barley and potatoes. The slope between 3,500 and 4,200 m is too steep to cultivate, but on the *puna*, as the dry highland between 4,200 and 4,800 m is called, Mauricio owns an area of grassland where he keeps llamas that serve as beast of burden and alpacas that yield meat and wool.

I have known Mauricio since 1986 when I conducted my first ethnographic fieldwork in Tapay. My research was focused on the relation between Tapay's ecology and the villagers' social organization and cosmology and during the year I spent in the village I became particularly intrigued by the villagers' interaction with the more-than-human beings they believe control the physical elements that constitute the landscape and the atmosphere. By participating in both the villagers' daily chores and Tapay's public and ritual gatherings I learned that while offerings to these beings are part of even the most trivial activities such as irrigating and planting the fields, alcohol consumption, and even strong inebriation are common in the most solemn moment of the villagers' offering ceremonies. More bluntly, the sacred and the mundane are inextricably linked.

DOI: 10.4324/9781003347026-4

My current research in Tapay is focused on how climate change transforms the village's livelihoods and worldview, which has drawn my attention to the way rapid environmental change challenges the villagers' perception of the climate and the geophysical surroundings on the one hand, and, on the other, the way it challenges their offering rituals. Accordingly, this chapter has two aims. One is to explore what I call the villagers' climing practices which, I suggest, consist of two distinct but interwoven engagements: deep climing and social climing. The second is to discuss how climate change confounds the villagers' understanding of the two climing practices and urges them to engage in what I call global climing. Before presenting my field data, however, an introduction to the chapter's conceptual framework, argument, and organization is pertinent.

Concept, Argument, and Structure

To Tapay's villagers climbing is a daily routine that takes them to their fields and therefore is critical for their livelihoods which are agriculture, fruit growing, and cattle raising. But to Mauricio and the other villagers, climbing means more than mere physical mobility. It implies what I shall call climing; that is, observing, interpreting, and responding to microclimatic changes in the many ecological niches that define the vertical ecology of the Andes (Masuda et al. 1985; Murra et al. 1986) and other mountain regions such as the Himalayas (Gagné 2018; Pandit 2017; Smyer Yü and de Maaker 2021; Smyer Yü and Michaud 2017). The term "clime" is conventionally used as a substantive to scale climate change and capture the specific ways it interlaces with place and fashions people's climate perceptions (Carey and Garone 2014). Unlike the terms "place" or "landscape" that refer more broadly to the environment as an external setting and static scenery in people's lives, "clime" is immediately associated with climate and climate change, which is critical in a mountain region such as the Andes that is defined by its climatically and ecologically diverse climes. "Clime" also forefronts people's perception of the climate and frames the way they position themselves in and imagine the environment, which is essential to environmental humanities. Building on this understanding of clime, I employ it as a verb that enables me to explore how Andean people create ideas of climate change and engage in practices to manage it by not only registering microclimatic alterations but also socializing with the climate and engaging in personal contact with water, earth, mountains, and other physical elements that have an acknowledged liveliness, sentience, and agency. The empirical questions guiding my inquiry are, How do Andean people "clime" mountains? And how do they "re-clime" them in response to climate change and the water shortage it generates? My theoretical questions, on the other hand, are, How does a local, place-based approach to climate challenge established meanings of global climate change? What are the local proxies of global climate change? And what contributions can an environmental humanities approach to indigenous and marginalized people's climing make to climate change research?

My claim is that climing in the Andes involves two kinds of environmental engagements: deep climing and social climing. In "deep climing" Mauricio and other villagers read the environment as signatures of Earth's "deep time," which, according to anthropologist Richard Irvine, is "time as it manifests itself in the landscape" (Irvine 2020, 2) and which, in the words of geologist E-an Zen, "provides us with a framework by which to gauge rates of natural changes and the Earth's capacity to regenerate life-supporting resources" (Zen 2001, 8). Thus, when Mauricio engages in deep climing he finds himself in a seemingly "timeless" world of stones, rocks, and other material elements where anthropic activities are located "not only in social terms, but as part of a wider system of relations within a physical and biological environment" (Irvine 2014, 168). To capture the temporal as well as the spatial dimension of Mauricio's deep climing, I explore how this induces him to not only engage in the prehuman geological world that shapes Tapay's landscape, but also the contemporary climatic world that supplies Tapay's irrigation system with water and enables Mauricio's crops and trees to grow and ensures his animals' wellbeing. My proposal is that in deep climing Mauricio simultaneously reaches back in time (deep time) and out in space (deep space) which, I suggest, renders him a sense of being a "planetary agent" who co-inhabits and co-creates the Earth together with natural elements such as stones, plants, water, frost, heat, and wind.

By contrast, in "social climing" Mauricio inscribes Tapay's physiographic, climatic, and biological environment in a social universe that is fashioned by the human perception of nature and its anthropomorphic ideas of the landscape. In the Andes this is congruent with a cosmology that presents the environment as inhabited by what anthropologist Marisol de la Cadena (2015) labels "earth beings" who Andean people believe inhabit the environment and to whom they attribute the power to control access to natural resources such as water and soil, and to punish humans who fail to show them respect. Examples of such earth beings are *pachamama* (Mother Earth), the *gentile* (remote ancestor inhabiting the rocks), and the *apu* (mountain deity). An important activity in social climing is Andean people's ritual interaction with and offerings to earth beings which they regard essential to sustain a reciprocal relation with *pachamama*, the *gentile*, and the *apu*. In return for the gifts that Mauricio brings to the earth beings he expects them to release the mountains' meltwater, provide him with a good harvest, and keep his animals healthy. By drawing our attention to the social nature of Mauricio's engagement with Tapay's earth beings, "social climing" offers an interpretive opportunity to not only translate the nourishing but feared power of the Earth via the mountain deities and unseen spiritual forces into a set of environmental/ecological terms but also provide a clearer understanding of human gifting as a form of mutual exchange in recognition of the affordances provided by the Earth for the sustenance and sustainability of human life. As I discuss below, such a reciprocity is particularly relevant in regard to water, which, as pointed out by my peer authors (Amrith and Smyer Yü; Andersen) in this book, is important for human life not only for existential but also for political and legal reasons.

Deep and social climing are analytical tools I, as a researcher, employ to unpack the human-environment nexus in the Andes and understand the physical and metaphysical relations that shape Mauricio's worldview. But on his daily climbs in Tapay the two forms of engagements converge and while at one point Mauricio observes how a rock shields his field from the sun and slows down the evaporation of the water he uses to irrigate, at another point he offers the same rock a gift to appease its spirit. Similarly, even though Mauricio from time other spends a moment listening to the refreshing sound of water running in Tapay's canals, he is equally eager to know when it reaches his fields and when the water allocators' call on him to receive it, just as he never misses an opportunity to show the mountain deities and Mother Earth his gratitude for their favors by bringing them gifts. Thus, in Mauricio's world, geological time and climatic space are woven into social relations and configurations and it is only by recognizing the landscape and the climate as a universe of beings that are more-than-human and yet approachable to humans that he is cable of grasping the depth of deep time and scope of deep space. More recently, however, climate change has undercut the villagers' environmental engagements and laid bare the vulnerability of their livelihoods. By causing ice- and snowmelt, rising temperatures, irregular precipitation, and other forms of rapid environmental change, climate change has also led many to question the extra-human powers of the earth beings and the effect of the gifts villagers offer them. Furthermore, in response to the new opportunities state and private agents such as a mining company that has recently started to operate in Tapay are offering, the villagers are now engaging in what I call "global climing" which implies adopting the modern view of humans as the planet's dominating agent with the privileged right to manage its environment and exploit its resources.

I organize the chapter into two sections. I start by introducing climing practices in Tapay in two acts: the first describes how the villagers engage in "deep climing" by making a phenomenological reading of the climatic, hydrological, and geological environment and adjusting their daily activities to these observations, and the second considers how they practice "social climing" by interacting with Tapay's earth beings through making ceremonial offerings. In the second section I first discuss how climate change is affecting Tapay and how the villagers perceive the environmental changes they currently are experiencing. I then examine how recent changes such as the construction of a road to Tapay, the growing tourist industry, and the appearance of a gold mine have transformed the villagers' relation with the outside world and challenged their belief in earth beings and their idea of deep and social climing. Finally, I introduce the notion of "global climing," which I use to explain how the villagers envision the opportunities these changes are generating. In the conclusion, I return to the concept of clime and reflect on how it attains new meaning in a time of global climate change and on how this conceptual turn speaks to environmental humanities.

Deep Climing

So how exactly does Mauricio clime while he climbs? In Tapay, "deep climing" is like opening a Russian nested-doll. Each doll represents a distinct clime universe of knowledge, experiences, and practices, and every time you open one another appears. The first universe comprises the rocks, stones, and mountains that are formative of Tapay's landscape. As segments from an antediluvian time they represent nature per se and constitute the geophysical setting that makes possible but also constrains the villagers' lives. Used to construct houses, roads, fences, terraces, and canals, stones are essential building material in Tapay. Equally, rocks and mountains serve as important landmarks to navigate by in Tapay's landscape and are objects of adoration in Andean culture, as discussed below. But rocks and stones are not only a resource. They are also a hurdle which the villagers are reminded of when they clear their fields before planting the seed and when they experience landslides in the rainy season. Thus, in Tapay removing stones from the fields is a never-ending undertaking just as the risk of falling stones and rocks constitutes a constant worry. Finally, Tapay's many ecological niches are molded by its geology. Even though the sun shines from a blue sky almost daily in the dry season, it reaches Mauricio's fields at different intervals during the day because of the shadows the nearby mountains throw on Tapay's landscape. Knowing exactly at what time the mountains shade or yield to the sun's rays is therefore critical to Mauricio's deep climing and assessment of when to cultivate his fields.

The next nested-doll in Mauricio's deep climing is a world of soil, plants, and animals that are vital to the villagers' principal livelihoods: agriculture, fruit growing, and animal farming. The key to an efficient and sustainable economy in Tapay is to know the biodiversity that prevails in its ecological niches and how variations in temperature, moisture, and humidity affect such products as corn, grain, potatoes, apples, pears, peaches, meat, and wool. Thus, when Mauricio makes his daily climbs in Tapay he pays special attention to the microclimatic conditions that may vary considerably within the range of a few hundred meters and considers the options these offer him to cultivate the land and feed his animals. At the heart of Mauricio's deep climing is the idea of vertical complementarity, which is widespread in the Andes and other mountain regions, and which implies making maximum use of the ecological niches and spreading the risks by planting a variety of crops and fruit trees and moving the animals to best grasslands (Paerregaard 1992). And, as described above, when making use of Tapay's natural resources the villagers engage with not only the elements from Tapay's deep time but also their own social world. Just as the construction of houses and the maintenance of the infrastructure require the cooperation of Tapay's households, Mauricio must mobilize his social networks to plant, harvest, and manage his grasslands.

A third nested-doll in Mauricio's deep climing is Tapay's hydrology and the villagers' management of its water supply. In the western slopes of the Andes, rain is irregular, and it only falls from January to March which makes the villagers' livelihoods dependent on irrigation (Gelles 2000; Guillet 1992; Paerregaard 2018a; Rasmussen 2015, Travick 2003). Tapay's water supply includes fifty-three water sources of which twenty-one are springs and thirty-two are offtakes from the six rivers and streams that carry snowmelt from the surrounding mountain peaks (Paerregaard 1994a). A total of twenty-one reservoirs in Tapay's many settlements allow the water users to store the meltwater overnight. Capturing the water from the water sources, storing it overnight, and managing it during the day allow Mauricio and other villagers to extend the growing season and providing their animals with grass beyond the rainy season and in times of drought. Deep climing, therefore, means staying informed about the water flow in Tapay's infrastructure of canals and reservoirs that connects its rivers and springs with Mauricio's and other villagers' terraces, fields, orchards, and grasslands. To facilitate the water flow and ensure a fair use of its water resources Tapay is divided into irrigation clusters that are managed by local user committees and that appoint one of their members to allocate the water each year (Paerregaard 1997). During the months of irrigation, the water users' major challenge is to anticipate the allocators' time schedule and be present when they open a hole in the canal to let the water flow into the fields. Forecasting where and when the water flows in Tapay's irrigation infrastructures by anticipating the water allocators' moves are therefore important skills in Mauricio's ability to navigate in Tapay's water world. In other words, deep and social climing are closely linked in the management of the hydraulic resources.

Social Climing

The blurred line between the two forms of environmental engagement comes even more to the fore when directing our attention to the villagers' image of and interaction with Tapay's more-than-human world. Stones and rocks have always played a crucial role in the mythology of Andean people, who attribute to their Inca ancestors superhuman powers due to their ability to use them to build houses and large infrastructure works. The geological landscape is also the object of legendary tales about Tapay's pre-Inca inhabitants who, according to the villagers, went into hiding during an earthquake but who they believe live in the rocks and may cause disease and death. Another important geological element in the Andean cosmology and Mauricio's social climing are mountains and the *apu* that inhabits them. The villagers believe that these more-than-human agents control the water flow and that they punish humans who fail to show them respect and bring them offering gifts. Thus, making *tink'a*, that is engaging in what I label transhuman reciprocity by paying tribute to the earth beings, is considered fundamental to maintaining a symbiotic relation with the environment, preventing natural disasters, and engendering a friendly climate (Salomon 2018).

Of Tapay's many offering ceremonies the most important is the *tink'a* to Mt. Seprigina, the village's highest mountain (5,400 m) and its principal water source. The event is the culmination of a series of annual offerings to Tapay's water sources that begins on June 24 and ends on November 1 at Mt. Seprigina. The many ceremonies are performed at different places and times but even though they appear fragmented and unrelated they are linked in a cosmological order that becomes evident when viewed as a sequence (Paerregaard 2013). Time and place vary for all first offerings, while the second offerings partially coincide in time even though they take place at the same locations as the first. As for the third, place and time merge in a single event: November 1 on Mt. Seprigina when the water users act as one community paying respect to a single power believed to control Tapay's many water sources.

The *tink'a* to Mt. Seprigina not only suggests that the villagers understand that their water sources are controlled by the same power and that Tapay's springs, rivers, offtakes, canals, and reservoirs constitute one cohesive water body; it also shows that they perceive the water flow as a hydrological cycle which constantly renews Tapay's water supply, a notion that resonates with a more general Andean notion of the world as floating on a great sea which connects the freshwater of rivers, springs, and lakes with the salt water of the oceans (Bastien 1978; Urton 1981). Essential to this perception is the idea that the offerings appease the mountain deities and cultivate their readiness to release the water. The ingredients constituting the gifts are seeds and leaves of the coca plant, *konuqa* (an herb from the *puna*), *qorilibro* and *qolqelibro* (small objects of gold and silver), an alpaca fetus and *pichuhuira* (fat from the breast of llama), corn of three different colors, *qochayuyo* (seaweed), starfish, and seawater. As a collection of the key elements of Tapay's clime universe (plants, metals, animals, etc.) they represent not only its geographic and biological but also its hydraulic environment, of which three of the offering items are evidence: seaweed, starfish, and seawater (Paerregaard 2021a). Originating from the sea they point to a tie between the apparently independent springs, and echo the pan-Andean idea of the Earth as floating on a sea that unites the world—suggesting not only that the worlds' supplies of salt and freshwater are connected but also that the offering gift ties humans to the more-than-human beings that inhabit the mountains, rocks, stones, and other elements the landscape, and to the force that drives the hydrological cycle (Paerregaard 2018a).

In the minds of the villagers, then, the environment is constituted by more than stones, plants, substances, and living beings, and the climate by more than heat, frost, rain, wind, and other atmospheric forces which we usually associate with the two terms. To the villagers these phenomena are all instances of the power that *pachamama*, the *gentile*, and the *apu* command to control the resources that sustain their livelihoods. In short, when climbing Tapay's mountains Mauricio not only deep climes the landscape by observing geological, biological, hydraulic, and meteorological changes. He also social climes it by engaging in transhuman reciprocity with a universe of more-than-human beings that links these to the water flow and coproduces his deep climing.

Climate Change

The villagers have used the term *la clima* (rather than *el clima* which is grammatically correct in Spanish) to describe the weather since I first came to Tapay in 1986. But when I did my restudy in 2011 *cambio clímatico* (Sp. climate change) had been added to the villagers' glossary (Paerregaard 2013). And even though few had detailed knowledge about global climate change, many complained that

Mt. Seprigina and the nearby mountains were only snow covered during the rainy season and not the rest of year as they used to be. The villagers' claim resonated with my own observations and was supported by the household survey on climate perceptions I conducted in 2016 showing there is a broad consensus in Tapay that climate change represents a serious threat to the village. The most notable environmental changes reported by the interviewees were rising temperatures, new insects, and pests appearing in crops and fruits, irregular precipitation, and shrinking water supplies that in different ways damage Tapay's agriculture and fruit growing.

The fifteen villagers I interviewed in the household survey gave me a variety of reasons why the environment was changing, referring to anything from geological and meteorological to moral and teleological causes (Paerregaard 2020a). While several of them adeptly employed the glossary of global warming to account for climate change, they disagreed when I suggested that it was a problem caused by industrialized countries. Some of the interviewed argued that rising temperatures are due to the introduction of a modern lifestyle in Tapay and Peru's contaminating mining industry. Others suggested that climate change is a cyclical phenomenon that people in the Andes have known for centuries and that is related to natural disasters such as earthquakes, hunger, and plagues. A man who runs small hostel for backpackers and who take great interest in conversing with his visitors said: "In Europe you're worried about climate change, but we've known it for a long time. Today water is scarce, but things will change again." On a similar note, a woman said that even though rising temperatures are inevitable, they are only temporary and, like other natural disasters such as earthquakes, they will be followed by better times. Another woman told me that the villagers had experienced worse changes. She said: "My grandmother recalled a year of drought when the Colca River dried up and there was nothing to eat." Another villager pointed out to me that climate change is a phenomenon that occurs locally, not globally, and that global warming is caused by the encounter between warm water coming up from the earth and the cold water falling as rain from the sky. Yet others see climate change as sign that the world is growing old and morally corrupt. According to one villager, rising temperatures and lack of rain are omens of an apocalypse that will end the world. A similar interpretation was offered by a man who asserted that the sun is no longer protected by "the shirt" it used to wear and because it is now "naked" it burns more. Expressing the same fatalistic worldview as the other villagers I interviewed, the man said: "It doesn't even help to make offerings anymore. There is nothing we can do."

In fact, the offerings have been a contested issue for many years. During my field research in the early 1990s, a small group of Protestants called *hermanos* campaigned against not only the Catholic church but also Andean customs and ceremonial practices such as making offerings to the mountain deities and the ancestors, and they repeatedly tried to obstruct the villagers' ritualization of the annual cleaning of the water reservoirs and canals in Tapay (Paerregaard 1994b). Even so, the vast majority continued to participate in these events and in the late

1990s many of the *hermanos* gave up their struggle and reconverted to Catholicism. In 2011, I noticed that climate change and, particularly, Tapay's shrinking water supply had prompted some Catholics to question Tapay's offerings too. Nonetheless, they continued to participate in the offerings not because they believed the tributes would produce more water but because the rituals are a custom and, even more important, because omitting to make them could bring misfortune on the water allocator and his family (Paerregaard 2013).

Climate change is having a critical impact on Tapay; but even though the villagers concur that it has important consequences for its water supply and their livelihoods, they interpret its causes very differently. Some view global warming as a geophysical phenomenon explaining it as the effect of earthquakes while others attribute it to humans' moral decline and the anger of earth beings. Still others believe climate change is cyclical. Referring to the stories their grandparents told them about natural disasters they say that rising temperatures, irregular precipitation, and droughts are periodical occurrences that have happened before and that there will be better times again. The discrepancy that comes to the fore in the villagers' interpretations of the changing environment is evidence of the confusion and uncertainty that climate change is creating in Tapay. By confounding the villagers' reading of the landscape's deep time and the climate's deep space as well as the powers of the earth beings to control Tapay's water supply it undercuts their climing practices and urges them to re-clime their environmental engagement.

Other Changes

But climate change is not the only change in town. It is happening at a time when Tapay is undergoing a major transformation (Paerregaard 2023). Since the early 2000s, Tapay has become a haven for trekkers and backpackers who view the village as an exotic relic from premodern times. To attend the many visitors, more than a dozen villagers have put up hostels and restaurants that provide them with an extra income. Most of the tourists come from Europe and other Western countries and through the stories they tell about modernity and globalization the villagers have appropriated terms such as carbon-dioxide emissions, contamination, and global warming, and have been introduced to the modern notion of climate change. Even though many villagers disapprove of these ideas as I discussed above, the global vocabulary on climate change introduced by the tourists has provided them with a new discursive framework to review and articulate the environmental changes they are experiencing. And the tourists are not alone in instituting a new climate awareness in Tapay.

When I did fieldwork in 1986 the Peruvian state seemed utterly remote (Paerregaard 1997). The only visible proof of its existence was Tapay's schools and the identification papers some of the villagers had managed to obtain. Although Tapay celebrated Peru's day of independence (July 28) by hoisting the Peruvian flag and watching school children march while citing lyric homages to *la patria* (Sp. the fatherland), few villagers felt any personal connection to Peru or the Peruvian state. This has changed dramatically in the past two decades. Today multiple government institutions and programs are operating in Tapay, offering its water committees cement, pipes, and other material to improve their offtakes, canals, and reservoirs, which not only moderates the water flow but also eases the task of cleaning them. Using similar provisions from the state, Tapay's newly formed drinking-water committees have installed running water in all its households. Most of them have sewage service, too. The state has also installed electricity in Tapay, providing light around the clock both in private homes and on streets and squares. Other services include a health post and a host of social programs to support people below the poverty line, the elderly, single parents, and children. Perhaps the most powerful proof of the state's presence is the newly opened road to Tapay. Though bumpy and one-tracked, it allows the villagers to take the bus instead of walking to neighboring Cabanaconde to do their shopping, or travel to other places. At the same time private agencies provide such services as cable TV, and have put up a telephone antenna that enables the population to communicate with relatives in other parts of Peru and receive the daily news. And there is more to come!

In December 2016, a goldmine started to operate in Tapay, which has altered villagers' vision of the future (Paerregaard 2019). Initially, the mine caused much concern among the villagers who feared it would contaminate or even exhaust their freshwater supply. To object to its activities the villagers formed an association that organized two marches to the mine, which is situated above Tapay's main settlement at 4,800 m, not far from Mt. Seprigina; and to back the villagers' protests and affirm their concern for Tapay's environment, migrants arranged similar events in the city of Arequipa. The mining company, on the other hand, has tried to gain Tapay's support by offering employment to almost 200 villagers and migrants and sponsoring several public works in the village including the construction of new school buildings, new municipal offices, water pipes from some of Tapay's springs to enhance its water discharge, the purchase of a bulldozer to repair Tapay's new road and a bus (and a paid driver) to transport children from the village's smaller settlements to school. Finally, and most importantly, it has agreed to construct a channel to direct water to Tapay from a nearby river, a project that currently is in progress and that promises to enhance the village's water discharge significantly. The mine's investments in Tapay's infrastructure and, not the least, the employment of its villagers and migrants as workers have changed the image that particularly young villagers hold of its interests and activities. Its presence, however, has also created deep divides within Tapay's population and, although the prospect of a new channel generates hopes for the future among the young, many senior villagers still fear the mine will contaminate its water.

On my latest visit to Tapay in 2019, people told me that the water allocators were still making offerings to the village's many springs, offtakes, and reservoirs as well as to Mt. Seprigina. However, other very visible changes were going on in

Tapay: the mine had put up a compound housing several hundred workers who were working on the construction of the promised channel. Although the workers lived apart from the local population, the mine's engineers and technicians were based in Tapay's largest hamlet, where they stayed with villagers who made a living offering them accommodation and food. During the day a flow of trucks and four-wheeled SUVs rushed in and out of the hamlet which had only become familiar with the presence of cars in Tapay after the road from Cabanaconde was completed in 2015. The presence of the many engineers and workers sent a tangible sign to the villagers that the mining company was keeping its promise, which had a notable effect on their perception of it. Comparing the talks that I was having with some of the villagers about the mine with the conversations we had both before and after it opened in December 2016, I found that several had changed views. Some who previously had been critical of the mine were now positive, while others who initially embraced it had turned skeptical in 2019.

Still, it was my sense that the channel's prospect of enhancing Tapay's water supply had opened a new window of imaginations and expectations, which are critical drivers of the effort to adapt to climate change. Unlike conventional development that changes for the better, climate adaptation implies changes that people didn't ask for but that nevertheless are inevitable. To engage marginal populations in adaptation initiatives it is therefore of paramount importance that they feel there is hope. In Tapay the readiness to adapt to climate change is inextricably linked to the desire for water and the new opportunities and livelihoods it can generate—a need that has existed for a long time. Since I did my restudy in 2011, Peru's Ministry of Agriculture has organized campaigns in Tapay to instruct the villagers in how to protect fruits against pests, and although climate change hasn't been part of the campaigners' glossary, they have contributed to the creation of a new climate awareness and, more importantly, a hope in the future. On numerous occasions villagers have told me they plan to benefit from the warmer climate to change fruit growing by replacing their old apple, pear, peach, and fig trees with mandarin, apricot, and orange trees. Of particular interest, they have pointed to the prospect of planting avocado trees on the dry slopes above Tapay's settlements which traditionally have been used as grazing grounds for animals during the rainy season and more lately for cochineal production. If irrigated regularly, the villagers have explained to me, these otherwise barren areas could be suitable for producing avocados on a large scale, which would lead to a shift from a self-subsistence economy toward the commercial production of cash crops. All they need, the villagers have told me over and over, is water.

Global Climing

Climate change is impelling Mauricio to reconsider his climing practices. When climbing Tapay's mountains to inspect the fields, he is no longer sure how to read its landscape, weather, and water supply. Landslides, falling stones, and erosion due to heavy rainfall have always posed challenges to the villagers; equally, many senior villagers recall how people previously suffered during years of drought. Still, precipitation has become more irregular just as the temperatures are rising which are compelling the villagers to change crops and fruit trees and find new grassland for the animals. And even though the villagers have complained about water scarcity since I first came to Tapay, most agree that it has never been as bad as now, which is making them rethink their irrigation practices. In other words, the environment has become more vulnerable, the weather more unpredictable, and the water flow more unreliable, and even though some villagers try to make sense of what is happening today by recalling their parents' and grandparent's accounts of similar occurrences in the past, there is a general sense that climate change exceeds the scope of human memory and that the wisdom of deep climing no longer serves as guide to sustain their livelihoods.

A growing number of villagers are also questioning social climing as means of coping with water shortage, assuring a good harvest, and preventing disease and other misfortunes. While some claim climate change is a sign that the offerings fail to meet the deities' demand and that the only way to deal with Tapay's environmental crisis is to bring them more gifts, others ascribe it to the villagers' moral decay and new modern lifestyle. Yet others belonging to Tapay's community of *hermanos* interpret climate change in religious terms. They reject the belief in earth beings altogether and argue that the rapid environmental change Tapay is experiencing is a call from God to convert to Protestantism. In a similar vein, some villagers read climate change as an omen of the mountain deities' future death or even of the end of the world. A common message in all these interpretations is that their transhuman relation of reciprocity with *pachamama*, the *gentile*, and the *apu* is broken, and that the villagers need to rethink the meaning of social climing.

Importantly, climate change is disrupting the convergence of deep climing and social climing. In his daily observations of Tapay's landscape and weather, Mauricio engages not only in a nonhuman universe gauged by deep time and deep space, but also in an anthropomorphic universe of earth beings to whom he ascribes more-than-human powers. In Andean cosmology the two worlds are inseparable and when Mauricio registers the geophysical and atmospheric transformations in Tapay's environment he simultaneously attributes social meaning to them by paying respect and bringing offering gifts to *pachamama*, the *gentile*, and the *apu*. In fact, it is only by configuring soil, plants, animals, rocks, and stones as physical elements that co-exist with humans in the present time, and by understanding rain, wind, and frost as climatic phenomena that occur locally within Tapay's territory, that deep time and deep space become comprehensible to Mauricio. Nonetheless, the earth beings' inability to solve Tapay's current water shortage has made him question the effect of not only social but also deep climing as means of engaging with the environment.

Paradoxically, climate change affects Tapay at a time when the villagers have been connected to the outside world by road and may look to a brighter future with more water and the possibilities of producing cash crops for the national and world market. Thus, even though the villagers' climing practices are losing leverage as means of addressing Tapay's environmental problems, other changes are inducing them to re-clime the landscape and climate and interpret climate change and its impact within a new framework of social and political transformation. The construction of a road to Tapay, the installation of running water, electricity, and sanitation in the households, and the introduction of other modern services such as television and telephones have provided the villagers with a feeling of being part of a wider world and a sense of citizenship in a nation-state which until recently seemed utterly remote. Moreover, the daily visits of foreign tourists and, not the least, the presence of a mining company in Tapay have created an awareness of the national and global forces that drive its environmental change and cause water shortages, rising temperatures, irregular precipitation, and natural disasters. Of critical importance is the construction of the new channel that may become a game-changer in Tapay's effort to adapt to climate change by enabling it to shift from a self-subsistence economy to a market-oriented economy based on cash crops and the salaries the villagers make as workers in the mine.

The new opportunities that the state, the tourists, and the mine are offering have not only confounded the villagers' climing practices and challenged their climate perception. They have also altered their idea of the earth beings with whom they co-inhabit the environment and who they believe control the weather, and are making many ask whether the *apu* still command the power to regulate the flow of water. In response to the multiple changes—climatically, socially, as well as politically—that Tapay is experiencing, the villagers have started to re-clime the environment and the climate. Such a shift from an amalgamation of deep and social climing toward what I have labeled global climing implies moving from a local lifeworld that perceives the physical surroundings as a signature of a prehuman time and the climate as an mirror of an extra-human space that humans and other earth beings share, to a global world that is driven by commercial and political interests and that views all the planet's physical elements and living beings as an economic resource to be exploited in the same way humans exploit other humans. For centuries the offering ceremonies to *pachamama*, the *gentile*, and the *apu* have been emblematic of Tapay's climing practice, but as the climate changes and the water supply dries up a growing number of villagers are turning their backs to the earth beings while seizing upon the hope that the mine and the channel it is building will enable them to create an extractive economy based on commercial agriculture. This raises the question of whether Tapay will continue to make offerings to Mt. Seprigina or if it will stop performing the ritual. Or whether, alternatively, the villagers will start to pay tribute to the channel instead of the mountain and in this way recognize the mining company as the new agent that controls the water flow, as happened in the neighboring community of Cabanaconde after it began to receive water from a state-built channel (Paerregaard 2014; Paerregaard et al. 2020). On the other hand, the recent surge of participants in mountain pilgrimages and offering ceremonies to mountain deities in Peru (Paerregaard 2020b, 2021, 2023) suggests that the idea of the transhuman reciprocity which fuels social climing is gaining momentum among new sectors of the Andean population and paving the ground for a new climate consciousness and planetary awareness. Adopting global climing as a new strategy to deal with rapid environmental change and growing water shortages while embracing the national and global world may therefore not exclude the continuous use of deep and social climing as means of engaging in a more-than-human environment.

Conclusion

Climate change is affecting everybody and there is hardly a place on the planet where people don't face the challenge of adapting to its consequences. The world's mountain regions are among the first to suffer from its impact (Gagné 2018; Haberman 2021a; Halperin 2020; Moulton et al. 2021) and the Andes is no exception (Burman 2017; Paerregaard 2023; Salas Carreño 2021; Stensrud 2021). But even though Andean people are among the major victims of a changing climate, their climate perception does not always concur with the Western scientific explanation of climate change. Thus, many disagree with the view that a changing climate is caused by human activities in other parts of the world (Brügger et al. 2021; Jurt et al. 2015; Paerregaard 2020a). And, as evident in the interviews I conducted with the villagers of Tapay, in some places people even think that climate change is a natural rather than an anthropogenic phenomenon. While recognizing that extreme temperatures, irregular precipitation, and water shortages are becoming frequent, they attribute these occurrences to nature's cyclical changes or interpret them as the punishment that nonhuman powers have inflicted upon them due to their moral decay and adoption of a modern lifestyle (Paerregaard 2018b).

But my data do not only shed light on Andean people's understanding of the forces that cause climate change. They also bring to the fore the worldview that shapes their climate perception and that drives them to adapt to climate change. To explore the ideas underpinning Tapay's environmental engagement, I have employed the term "climing" which has two complementary meanings. While deep climing refers to the villagers' sense of a time and a space beyond human reach, social climing highlights their interaction with a universe of earth beings they perceive as controlling the access to the resources upon which their livelihoods depend. The two forms of climing converge given that it is the villagers' configuration of their environmental engagement as a social relationship that renders deep time and deep space comprehensible to the human mind. My material also reveals that climate change is urging the villagers to re-clime their environmental engagement. To benefit from the development that the state and a mining company have created in Tapay, the villagers are planning to shift from their traditional self-subsistence production to a market-oriented production. My argument is that such shift will be driven by global climing that positions humans as the planet's stewards and prioritizes their needs at the cost of other beings and elements. While climate change is exposing Tapay's vulnerability to rapid environmental change, then, other changes are prompting it to adapt to its impact by replacing a local climing practice with a global climing practice that opens to new ways of exploiting its environment and producing for the world market. Climate change adaptation is rarely something people ask for. And often it is something they hadn't even imagined being possible. But the irony of climate change is that it doesn't give you the choice of saying no if the opportunity to adapt presents—even if it entails altering climing practices.

Yet, as a way of engaging with geophysical elements, atmospheric matter, and biological life forms, climing always involves a transhuman relationship and while the shift from deep and social climing to global climing implies moving from a confined indigenous lifeworld to a modern globalized world propelled by economic interests and influenced by Western scientific ideas, it also means adopting a new planetary awareness of anthropogenic change and of belonging to an all-dominating species on Earth inhabited by multiple modes of existence. By not only recognizing the impact of climate change on their environment and adapting to its consequences but also acknowledging their own contribution to global warming, Andean people are increasingly assuming this new anthropogenic identity and accepting their role as both the planet's destroyer and manager of its future (Paerregaard 2020b). A growing body of literature within environmental humanities that explore how ecology and religion intersect in a time of climate change shows how this predicament unfolds and how a new planetary ethic emerges in a variety of regional and social contexts (Bauman 2014; Haberman 2021b; Mickey et al. 2020; Smyer Yü 2021). The insights of this chapter speak to this scholarship by bringing to the fore the existential, social, and global complexity that shapes and reshapes Andean people' climing practices as they transform them in response to climate change and the globalization of their lives and livelihoods.

References

Bastien, Joseph. 1978. *Mountain of the Condor: Metaphor and Ritual in an Andean Ayllu*. Prospect Heights: Waveland.

Bauman, Whitney. 2014. *Religion and Ecology: Developing A Planetary Ethic*. New York: Columbia University Press.

Brügger, Adrian; Robert Tobias and Fredy Monge-Rodríguez. 2021. "Public Perceptions of Climate Change in the Peruvian Andes." *Sustainability* 13(3): 26–77.

Burman, Anders. 2017. "The Political Ontology of Climate Change: Moral Meteorology, Climate Justice, and the Coloniality of Reality ion the Bolivian Andes." *Journal of Political Ecology* 24(1): 921–938.

Carey, Mark and Philip Garone. 2014. "Forum Introduction." *Environmental History* 19(2): 282–293.

de la Cadena, Marisol. 2015. *Earth Beings: Ecologies of Practice across Andean Worlds*. Durham: Duke University Press.

Gagné, Karine. 2018. *Caring for Glaciers: Land, Animals, and Humanity in the Himalayas*. Seattle: University of Washington Press.

Gelles, Paul. 2000. *Water and Power in Highland Peru: The Cultural Politics of Irrigation and Development*. New Brunswick: Rutgers University Press.

Guillet, David. 1992. *Covering Ground: Communal Water Management and the State in the Peruvian Highlands*. Ann Arbor: University of Michigan Press.

Haberman, David L. (ed.). 2021a. *Understanding Climate Change Through Religious Lifeworlds*. Bloomington: Indiana University Press.

Haberman, David. 2021b. "Angry Gods and Raging Rivers: The Climate Change of the Central Himalayas." In *Understanding Climate Change Through Religious Lifeworlds*, edited by David Haberman, 235–259. Bloomington: Indiana University Press.

Halperin, Ehud. 2020. *The Many Faces of a Himalayan Goddess: Hadimba, Her Devotees and Religion in Rapid Change*. Oxford: Oxford University Press.

Irvine, Richard. 2014. "Deep Time: An Anthropological Problem." *Social Anthropology* 22(2): 157–172.

Irvine, Richard. 2020. *An Anthropology of Deep Time: Geological Temporality and Social Life*. Cambridge: Cambridge University Press.

Jurt, Christine; María Dulce Burga; Luís Vicuna; Christian Hüggel and Ben Orlove. 2015. "Local Perceptions in Climate Change Debates: Insights from Case Studies in the Alps and the Andes." *Climatic Change* 133(3): 511–523.

Masuda, Shozo; Izumi Shimada and Craig Morris (eds.). 1985. *Andean Ecology and Civilization: An Interdisciplinary Perspective on Andean Ecological Complementarity*. Tokyo: University of Tokyo Press.

Mickey, Sam; Mary Evelyn Tucker and John Grim (eds.). 2020. *Living Earth Community: Multiple Ways of Being and Knowing*. Cambridge, UK: OpenBook.

Moulton, Holly; Mark Carey; Christian Huggel and Alina Motschmann. 2021. "Narratives of Ice Loss: New Approaches to Shrinking Glaciers and Climate Change Adaptation." *Geoforum* 125: 47–56.

Paerregaard, Karsten. 1992. "Complementarity and Duality: Ecological, Social and Ritual Oppositions Between Agriculturalists and Herders in an Andean Village." *Ethnology* 31(1): 15–26.

Paerregaard, Karsten. 1994a. "Why Fight Over Water? Power, Conflict, and Irrigation in an Andean Village." In *Irrigation at High Altitudes: The Social Organization of Water Control Systems in the Andes*, edited by David Guillet and William Mitchell, 189–202. Washington: Society for Latin American Anthropology.

Paerregaard, Karsten. 1994b. "Conversion, Migration, and Social Identity: The Spread of Protestantism in the Peruvian Andes." *Ethnos* 59(3, 4): 168–186.

Paerregaard, Karsten. 1997. *Linking Separate Worlds. Urban Migrants and Rural Lives in Peru*. Oxford: Berg.

Paerregaard, Karsten. 2013. "Bare Rocks and Fallen Angels: Environmental Change, Climate Perceptions and Ritual Practice in the Peruvian Andes." *Religions* 4(2): 290–305.

Paerregaard, Karsten. 2014. "Broken Cosmologies: Climate, Water and State in the Peruvian Andes." In *Anthropology and Nature*, edited by Kirsten Hastrup, 196–210. London: Routledge.

Paerregaard, Karsten. 2018a. "Power as/in/of Water: Revisiting the Hydrological Cycle in the Peruvian Andes." *WIRE's Water* 5(2): 1–11.

Paerregaard, Karsten. 2018b. "The Climate-Development Nexus: Using Climate Voices to Prepare Adaptation Initiatives in the Peruvian Andes." *Climate and Development* 10(4): 360–368.

Paerregaard, Karsten. 2019. "Liquid Accountability. Water as a Common, a Public and a Private Good in the Peruvian Andes." *Water Alternatives* 12(3): 488–502.

Paerregaard, Karsten. 2020a. "Communicating the Inevitable Climate Awareness, Climate Discord and Climate Research in Andean Communities." *Environmental Communication* 14(1): 112–125.
Paerregaard, Karsten. 2020b. "Searching for the New Human: Glacier Melt, Anthropogenic Change and Self-Reflection in Andean Pilgrimage." *HAU: Journal of Ethnographic Theory* 10(3): 844–859.
Paerregaard, Karsten. 2021a. "Lubricating Water Metabolism: How Mountain Offerings Contribute to Water Sustainability in the Peruvian Andes." *kritisk etnografi. Swedish Journal of Anthropology* 4(2): 83–98.
Paerregaard, Karsten. 2021b. "Recasting the Sacred: Offering Ceremonies, Glacier Melt, and Climate Change in the Peruvian Andes." In *Understanding Climate Change Through Religious Lifeworlds*, edited by David Haberman, 261–283. Bloomington: Indiana University Press.
Paerregaard, Karsten. 2023. *Andean Meltdown. A Climate Ethnography of Water, Power, and Culture in Peru*. Berkeley: University of California Press.
Paerregaard, Karsten, Susann Ullberg and Malene Brandshaug. 2020. "Smooth Flow? Hydrosocial Communities, Water Governance, and Infrastructural Discord in the Peru's Southern Highlands." *Water International* 45(3): 169–188.
Pandit, Maharaj K. 2017. *Life in the Himalaya: An Ecosystem at Risk*. Cambridge: Harvard University Press.
Rasmussen, Mattias B. 2015. *Andean Waterways: Resource Politics in Highland Peru*. Seattle: University of Washington Press.
Salas Carreño, Guillermo. 2021. "Climate Change, Moral Meteorology, and Local Measures at Qoullurit'i, a High Andean Shrine." In *Understanding Climate Change Through Religious Lifeworlds*, edited by David Haberman, 44–75. Bloomington: Indiana University Press.
Salomon, Frank. 2018. *At the Mountains' Altar: Anthropology of Religion in an Andean Community*. London: Routledge.
Smyer Yü, Dan. 2021. "Situating Environmental Humanities in the New Himalayas: An Introduction." In *Environmental Humanities in the New Himalayas. Symbiotic Indigeneity, Commoning, Sustainability*, edited by Dan Smyer Yü and Erik de Maaker. London: Routledge.
Smyer Yü, Dan and Erik de Maaker (eds.). 2021. *Environmental Humanities in the New Himalayas: Symbiotic Indigeneity, Commoning, Sustainability*. London: Routledge.
Smyer Yü, Dan and Jean Michaud (eds.). 2017. *Trans-Himalayan Borderlands: Livelihoods, Territorialities, Modernities*. Amsterdam: Amsterdam University Press.
Stensrud, Astrid. 2021. *Watershed Politics and Climate Change in Peru*. London: Pluto Press.
Travick, Paul. 2003. *The Struggle for Water in Peru: Comedy and Tragedy in the Andean Commons*. Stanford: Stanford University Press.
Urton, Garry. 1981. *At the Crossroads of the Earth and the Sky: An Andean Cosmology*. Austin. University of Texas Press.
Zen, E-an. 2001. "What is Deep Time and Why Should Anyone Care?" *Journal of Geoscience Education* 49(1): 5–9.

3

PLURIVERSAL TUNDRA

Storying More-than-Human Ecologies across Deep, Accelerated, and Troubled Times[1]

Astrid Oberborbeck Andersen

Introduction

This chapter takes us to Kangerlussuaq in West Greenland and its landscape of folded rocks, silt-rich riverbeds, and extensive polar tundra—all shaped by the repetitious spreading and retreating of glaciers, over millions of years, and the effects of changing temperatures and climates. Periods of warming followed by periods of cooling. These rhythmic actions of glacial periods are present and visible in today's terraced terrain. Past presence and movements of ice are tangible in the east-west patterned lines in the landscape, and in the topology and morphology of the land. Terraces are formed by post-glacial rebound, also known as glacial isostatic adjustment: the land masses here, as across the Arctic, are still in the process of rising after the removal of the huge weight of ice sheets at the end of the last glacial (Knudsen et al. n.d.). Sediment materials from the Greenland ice sheet, mainly silt, are in continuous formation, transported by glaciers and by seasonal meltoff. This is a land in constant movement, topographically and geologically, although winter means temperatures of deep freeze, and although a glance of a human eye across this vast and open landscape will not reveal much movement, apart from those of raven, ptarmigan, hare, fox, caribou, or muskoxen.

The human settlement known today as Kangerlussuaq (Greenlandic for big fjord) sits at the bottom of a 190 km long fjord—to which the place owes its name—and houses close to five hundred human inhabitants. Around the fjord, a polar tundra landscape extends in all directions. In the present-day, this is one of the largest ice-free areas in Greenland. Ice-free above ground, that is. Resting on continuous permafrost, hosting rivers fed by meltwater from the inland ice and glaciers, and dotted by small freshwater lakes, this landscape can be seen—and storied—as one large and dynamic water body. To the east, the

DOI: 10.4324/9781003347026-5

tundra is bounded by the vast Greenlandic ice sheet. Land meets ice sheet, yet the water body continues, although in a different, solid shape. Covering 80% of the Greenlandic territory, this is the world's second largest body of ice, after the Antarctic ice sheet. Known to the Greenlandic population as *sermersuaq* (the big glacier), and in old times perceived as a strange territory inhabited by potentially hostile and dangerous beings (Grønnow 2009), the ice sheet is also relationally formed and animated. In geological and glaciological terms, the ice sheet is a dynamic body of flowing and deforming ice (Elberling 2021), a combined land and water body in continuous transformation. Although the Inuit populations of what is now known as Kalaallit Nunaat (Greenland) traditionally spent most of the year on coastal settlements, the tundra landscapes around Kangerlussuaq have been inhabited and hunted in seasonally during summer for several thousand years (Grønnow et al. 1983; Andersen 2022).

Glaciologists and geologists have been visiting to study land and icescape formations in this part of west Greenland since the early 1900s and due to easy access to the ice, it is one of the most studied glacial areas on Earth. This polar region is characterized by strong seasonal variation with cold temperatures and long nights during wintertime, ice thaw, water flow, and long and sunny days in the summer. In recent years, the seasonal processes of melt and freeze have accelerated due to a general warming and other changes in the climate (Oksman et al. 2022). This is a landscape that spans multiple and contrasting temporalities and rhythms. The time-depth and the coming and going of glacials are tangible in traces in the rocks, and rising temperatures are causing processes to speed up. It is as challenging as it is urgent for scholars in anthropology and the environmental humanities to develop conceptual and methodological tools able to grasp, analyze, and convey (or story) this temporal complexity. With this chapter, I aim to contribute to this work, by seeking ways to story the tundra as a more than human ecology across deep, accelerated, and troubled times. I do so in conversation with the other chapters of this book, under the framework of *multipolar clime studies* (Smyer Yü, this book). As highlighted in the introduction by Dan Smyer Yü, "clime" is a concept that situates climatic changes experientially and affectively in particular places, and captures "the diverse environmental embodiments of global climate change in the mountains, plateaus, and floodplains of the Himalaya and the Andes, and in the Arctic tundra." This experiential, affective, and situated approach to climate change is much needed, also to point to the divergent, sometimes conflicting ways of understanding the changes happening. "Clime" acknowledges that places are made by local as well as global forces, forces that are simultaneously natural and societal, and that tangle human and more-than-human agencies in complex ways. As this chapter further highlights, climes are heterogeneous and constituted by temporal complexities.

Not only in temporal terms is the landscape around Kangerlussuaq complex. Through time the land has been known and appropriated through different and radically diverse human perspectives, each of which has acknowledged and related to very different forces and modes of existence (ontologies) in the landscape.

In biological and climatic terms, Kangerlussuaq is located in a tundra ecosystem. The tundra is one among several types of landscapes or biomes characteristic of polar and high-altitude regions, where the growth of trees is hindered by low temperatures and continuous permafrost (Bliss et al. 1973); roots cannot enter the frozen ground, so only moss, lichen, and low shrubs can grow here. Permafrost—a main characteristic of a tundra biome—is soil or sediments that have maintained below the freezing point for at least two consecutive years (Elberling 2021, 48); it appears in areas with an average temperature below −5 °C. Permafrost thickness varies across the Arctic, and so does the active layer: the upper level that thaws during summer and freezes in the winter. The frozen ground impedes drainage and increases flows of water on the surface in the melting season (Bliss et al. 1973). With temperatures getting warmer during summertime, due to climatic changes, the thickness of the active layer of permafrost is observed to be increasing. Previously frozen solid, when meltwater in the soil drains away, parts of the landscape can collapse, erosion increases, and decomposition processes in the soil accelerate (Elberling 2021).

Seeing and classifying the land as "tundra" renders the Kangerlussuaq area comparable to that of other Arctic and polar regions. Tundra, the scientific word with Sami and Russian origins, allows for comparison of different landscapes and contexts through their ecological and climatic conditions. Among Greenlandic Inuit inhabiting Kangerlussuaq today, the land is experienced and talked about as *nuna* (land). There is no Greenlandic word for tundra. Rather, landscapes are named, storied, known, and described by their biophysical features and relational emplacement—they are specific lakes (*tatsit*), mountains (*qaqqat*), rivers (*kuut*), or fjords (*kangerlut*). Among such features can be *qeriuaanartup masarsua*, which literally means the swamp of eternal frost and translates to permafrost. In biological and geological terms, a tundra landscape is characterized by exactly that: a biome defined by a lack of trees, partly due to extensive permafrost (Bliss et al. 1973). To Inuit inhabiting this area, the landscape—*nuna*—is one of more than human sociality and relationality[2]. Animals and other non-humans are regarded as persons, that is, with sentience, consciousness, and intentionality of their own, and mountains, glaciers, and lakes are often experienced as "animated" (see also Nuttall 1992, 2016; Fienup-Riordan 1999; Brightman et al. 2012 for other Arctic regions).

This landscape is formed and deformed by geological and climatic forces working over millions of years to make Kangerlussuaq what it is today. It is also a landscape that has emerged and continues to come into being through a diversity of human perspectives, and that is experienced, lived, and managed through various, sometimes contrasting, and incompatible ways. Over the following pages, I take the landscape around Kangerlussuaq, and its water bodies, as a case in point to question and examine the kinds of methodological and conceptual tools necessary to grasp the multiple and complex temporalities and ontologies lived in Arctic landscapes like that of Kangerlussuaq.

To further complicate this ecology, the present-day settlement of Kangerlussuaq is built on the infrastructure of a former US Air Base (Bluie West Eight, later Sondrestrom Air Base), which was established in 1941 during World War II as part of an extensive US defense system across Greenland and the western Arctic. In 1992, the airport and base facilities were handed over to the Greenlandic home rule; from the 1960s and onward, the airport had been serving as the most important transportation hub in Greenland, around which a community later emerged. With good logistics and easy access to the Greenland Ice Sheet, Kangerlussuaq also serves as a transit point for a variety of scientific surveys, especially within the natural sciences and related to monitoring and studying climatic conditions and changes. The dormant military remnants make Kangerlussuaq, as Greenland in general, a hot-spot for climate science—a cue to understand not only Arctic but global climate systems and how they change. In societal, cultural, and material terms, Kangerlussuaq is characterized by "throwntogetherness" (Massey 2005); a place where multiple worldviews, meanings, interests, trajectories, ontologies, and ecologies converge: a stratified or layered ecology, resting on grounds—permafrost and World War II—in the process of continuous transformation. While there is no doubt or disagreement among people in Kangerlussuaq that the climate is changing, there seem to be several *climes* at play, which are changing in multiple ways. Or, to follow Karsten Paerregaard (this book) who in his chapter highlights *climing* as a verb and a practice, there are multiple ways in which the changes in this land are actively being *climed*. Clime change in Kangerlussuaq is plural, depending on whose experience, whose history, and whose practices are referred to, or storied.

In this chapter, I ask how we can begin to story climes as more-than-human ecologies across deep, accelerated, and troubled times; and across "throwntogether" processes and perspectives. I take it as a premise and starting point that it is urgent for scholars and practitioners of climate and clime change and ecosystems management to expand the registers through which human-environment dynamics are grasped and engaged. Particularly so in (post)colonial contexts where varieties of knowing- and of living environments are silenced, invisibilized, or even erased when it comes to generating knowledge about and regimes for managing environments and their livelihoods. My addressing these questions draws inspiration from empirical ethnographic engagement with locations in Greenland, and from scholars within the environmental humanities and anthropology who have worked on similar questions in other contexts and geographies around the world (e.g., Fienup-Riordan 1999; Cruickshank 2005; Blaser 2009; Van Dooren and Bird Rose 2016).

This story from the Arctic provides a particular viewpoint in the multipolar studies conversation that this book offers on the diverse meanings and consequences of climate change in the Himalaya, Andes, and the Arctic. "Clime" works as an analytical device that helps us connect climate change with local experiences, and to story human-environment relations across distinct geographies and socio-ecological contexts in the Himalaya, Andes, and the Arctic. And it provides a conceptual gateway for making local and situated human and more-than-human experiences matter in the interdisciplinary totality of research about climate change, recognizing—of course—that local and global processes are continuously shaping one-another (Smyer Yü in this book).

I attend to the landscape around Kangerlussuaq and the bodies of water by ethnographically storying tundra climes with different human actors and their perspectives: geologists who reveal the deep time of the tundra and point to ruptures in water-ice-land dynamics; entrepreneurs and policymakers who dream of converting climate change into a source of innovation and profit as global warming accelerates the melting of the ice and increases the availability of freshwater; Inuit hunters and their families who show us the landscape as one of more than human sentience and sociality and offer us the concept of *sila* as a conceptual and ontological gateway to understand the relational Arctic world and the changes happening.

Resulting from these perspectives and ways of storying, I suggest, is a need to understand and acknowledge the tundra as *pluriversal*—as "a world where many worlds fit" (Escobar 2011, 139). I draw on scholars who challenge the idea of a universe as a one-world world where only one version, worldview, and ontology—Western or "modern"—is acknowledged to be true (Law 2015; Blaser 2009; Escobar 2011; de la Cadena 2010). In Kangerlussuaq, many ways of making sense of the changing environment exist, some of which are conflicting, embedded in contrasting ways of knowing and living climes in diverse temporalities and ontologies. The pluriversal tundra is a world of many worlds, made of and lived by different human existences, but also by other than human agencies. The pluriversal tundra is lively, and in continuous process of becoming, as a more-than-human ecology.

Storying the pluriversal tundra leads to questions of how to practice an anthropology of the tundra in the interdisciplinary field of environmental humanities. It leads to new methodological challenges such as how to grasp deep, accelerated, and troubled times from different human as well as nonhuman perspectives. It also leads to questions of the ethics and politics of storying the tundra from more-than-human perspectives. Whose perspectives get included, and whose voices are silenced? What does it take to story—or clime —with contrasting, even conflicting, perspectives? And whose times are deep, accelerated, and troubled in the first place? We shall return to these questions after a presentation of my analytical take on time, temporality, and storying (Figure 3.1).

Conceptual Openings: Storying Across Timescales

To story the tundra, as anticipated in the opening of this chapter, requires attention to diverse timescales and awareness of the complexity of temporality at play. By highlighting the times of the tundra as *deep*, *accelerated*, and *troubled* respectively, I aim to draw attention to a temporal multiplicity and show that the Anthropocene present and future is shaped by complex temporalities and notions of time. The three do not result in neat, clear-cut, and all-encompassing categorizations. Rather, I suggest that all moments and occurrences hold and are

FIGURE 3.1 Different water bodies meet in this land-ice interface.
Photo: Astrid Oberborbeck Andersen.

simultaneously shaped by deep, accelerated, and troubled time; and the distinction is an analytical more than an empirical one.

To make this argument about time, and to story the tundra with an attention to timescales and complex temporality, I draw on anthropologists Tim Ingold (1993) and Richard Irvine (2020) who story and theorize northern landscapes through temporal lenses, and through these seek to expand and extend the temporal dimensions of presence and human-centered timescale to which ethnography tends to be methodologically and analytically confined. In *An Anthropology of Deep Time: Geological Temporality and Social life*, Irvine (2020) theorizes with early scientific geology and the East Anglian fenlands and through these paves the way for conceptually relating the timeframe of human practices and understandings to the deep time of earth history. I follow Irvine in his claim that "The 'human' cannot be divorced from the planet that supports it, and so our existence needs to be understood in relation to the deep time of earth history" (15). As it shall become clear later, and as Irvine also anticipates when asking "whose time is deep? Is it a universal or a particular notion?" (17), there is a need to diversify and multiply "human" as well as "planetary" in that quote. The notion of "pluriverse" helps us do so by challenging the idea of a "one-world world" (Escobar 2011; Law 2015), and thus also one notion of and dimension to "deep."

When walking around in Kangerlussuaq and on the tundra, the different times and complex temporality immediately spring to view. The open-air dump outside of the settlement is a case in point. Military waste alongside that of today's settlement—old domestic appliances, hides and bones, and other leftovers from hunting. The dump is located by the fjord, on the silt foothills of the tundra. Deep, accelerated, and troubled times are simultaneously working on this confined piece of land and its materialities. Integrating time and temporality to our mode of storying climes helps us see that the tundra landscape is shaped by forces that work across various timescales and temporalities, and these are, in turn, shaped by the elements, materialities, and forces of the tundra, and the relations between these. Time and temporality are constituents of the tundra and settlement ecology; becoming and transformation are continuous.

Deep refers to the history and temporality of the earth on a long-term timescale, like the one produced by glacial periods and geological agency and visible in landscape formations. Deep time is perceived and present in the present and the future, as much as in the past. By "deep" I also want to push attention to deepness and rootedness of different perspectives and modes of existence, and to mythical presences.

Accelerated refers to the speeding up of biophysical processes. On the tundra, and in the Arctic more generally, the changing seasons result in acceleration and deceleration of flows and processes: spring and summer time mean warmer temperatures, and the short summer means that things must happen fast. Seasonal acceleration has always been a factor in the annual cycle of the Arctic, and acceleration is a word often used by glaciologists and ecologists when describing such processes (Elberling 2021; Bliss et al. 1973). In recent decades, there has been an acceleration of acceleration, and thus the term is also used to highlight how global warming and climate change result in the speeding up of the melting of ice, leading other processes to speed up to formerly unseen levels (Krupnik and Jolly 2002; Oksman et al. 2022). These locally observable biophysical accelerations are tied to the "the Great Acceleration"—the dramatic increase in carbon-based energy consumption, economic and population growth, and human-caused emissions of carbon dioxide happening since the end of Second World War, and further speeding up in the 1990s (McNeill and Engelke 2016; Eriksen 2016, 2021).

Troubled draws inspiration from Donna Haraway's (2015) "staying with the trouble", understood as an imperative not to turn our backs to the harsh realities of the times of ecological crises and urgencies in which we live and that need to have their manifold stories told; stories about making-with, and across species (Haraway 2015, 37). Staying with the trouble also orients us toward the future: how are we to make ruined ecologies livable through weaving, sowing, and worlding otherwise (Gan et al. 2017)?

The above gives us analytical and ethical entry-points to approach the tundra as a more-than-human and multi-temporal ecology. How to work ethnographically across timescales and ontologies is also a methodological challenge, however, and that is where "storying" demands a little more attention.

In presenting academic texts as stories, storytelling and story*ing* as a verb have become regular and accepted as knowledge practice within the humanities and anthropology over the last decades. In "The Temporality of the Landscape" Ingold (1993) connects temporality with landscape through stories. Working within the traditions of anthropology and phenomenology, Ingold generates a perspective on landscapes centered on human experience and offers a "dwelling perspective," one that highlights the landscape as "an enduring record of—and testimony to—the lives and works of past generations that have dwelt within it" (Ingold 1993, 152). This perspective gives priority to lived experience and the understandings that human people achieve through their everyday involvement in the world (152). To Ingold "the landscape tells—or rather *is*—a story. It enfolds the lives and times of predecessors who, over the generations have moved around in it and played their part in its formation" (152). Stories, Ingold highlights, should not be evaluated based on their truth-value, but rather on the capacity of the storyteller to draw listeners into the storied landscape. This approach to stories and landscapes, helps us attend to landscapes and beings who are other-than-human through stories, yet although extending to the environment, the storying remains human-centered and based on an authorized scholarly voice.

Thom van Dooren and Deborah Bird Rose (2016) propose a related, but slightly different path. They suggest "lively ethography" as a method for "knowing, engaging, and storytelling that recognizes the meaningful lives of others and that, in so doing, enlivens our capacity to respond to them by singing up their character or ethos" (Van Dooren and Bird Rose 2016, 77). Through a series of ethnographic vignettes, mainly about nearly extinct species living in Hawaii, they pave the way for co-becoming across species and for a renewed awareness of the world as animated and inhabited by other-than-human forms of worldly responsiveness, which may—they argue—enhance relational imagination and response-ability towards other beings (91). Ethnography has the capacity to story across differences, also radical ontological and epistemological differences.

Other scholars and writers have gone further in storying and authoring with voices positioned differently. Julie Cruickshank (2005) draws on native (Tlingit and Athapaskan) oral traditions from the Yukon and Alaska, climate histories, and colonial records to create a compelling and rich account of the glaciers in northwestern North America as acting beings, and how particular encounters resulted in the fragmentation of reciprocal relationships between the natural and the social world in that context (Cruickshank 2005).

Valuable experiences of co-authorship across epistemological and ontological differences have been published, and are highly inspiring (e.g., Fienup-Riordan 2000; Krupnik et al. 2010; Bawaka Country et al. 2017). Different stories exist, as do ways of storying. Voice and representation matter, and many, including indigenous scholars, have highlighted how different ways of "narrating the Arctic" (Bravo and Sörlin 2002) has been productive in the upholding of unequal power relations (see Pfeifer 2018; Todd 2018). Not all stories and narratives travel with the same ease. With inspiration from the above and from the conversations had with other authors of this book, I now return to Kangerlussuaq and to walking on the tundra.

Storying Waterscape Movements with Glacial Time

In late July in 2020, when I arrived in Kangerlussuaq to carry out ethnographic fieldwork, I was lucky to share the place of accommodation with three experienced Danish geologists. For more than two decades, they have traveled to Kangerlussuaq to study the long-term dynamics of proglacial landscapes. They walk the tundra and carefully observe how the margin of the Greenland Ice Sheet and inland glaciers interact with the landscape next to them. These observations from year to year are used to learn about the long-term interaction and evolution of land-ice formations. Through their direct observations over decades and practices of "rubber boots methods" (Bubandt et al. 2023), these geologists seek to reveal the deep time of the tundra, and to point to ruptures of concern in present and future water-ice-land dynamics (Carrivick et al. 2018).

The summer of 2020 was a particularly wet one, and continuous rain for over a week had chased the three geologists away from their campsite near Isunnguata, the glacial area that they studied this summer. Everything was soaked, they complained—their tent, their things, the ground, the vegetation—and they sought shelter in the settlement.

One evening, we took a walk to the bridge over the river Qinnguata Kuussua to look at the level and volume of water flow. The main source of this river is the meltwater outflow from Russell Glacier and the Greenland Ice Sheet. Late July is the peak of the meltoff season, and the water was cascading beneath us with tremendous flow and force. In 2020, the Kangerlussuaq area received double the amount of precipitation (242.6 mm) that is normally registered (121 mm/year). These days, in July, locals flocked to the bridge at all times of the day to look at the river, startled by its unusual force.

Several weather patterns are unpredictable in new ways in Kangerlussuaq, and other water bodies than the glacier and the river are acting in unfamiliar ways. Lakes evaporate more during summertime due to warmer temperatures and the fjord that used to freeze in December or January and is used by hunters for transportation on dog sledge or snowmobile, has not been passable until late February or March. Geological and climatic studies show that although annual variations occur there is a tendency that the melt season is becoming longer and more intense (Van As et al. 2018), the active layer in the permafrost is increasing in depth (Hollesen et al. 2015), and, in August 2021, rain was registered for the first time on the Greenland Ice Sheet, resulting in alarming headlines in news worldwide.

Local non-geologists also note the lack of normality. In the words of Aaron Skifte, a seasoned and professional guide with decades of experience bringing tourists and film crews to the ice sheet and guiding them around[3]: "This situation with the ice and the water is extremely unnormal. In 2000, the bodies of the glaciers around here touched the moraines…." Now, there is an empty space and a clear line marking the absence of ice. The studies of the three geologists resonate with these observations. Since the 1990s, the glaciers in this area, like those in other areas in Greenland, have been retreating fast (Knudsen n.d., 4; Wood et al. 2021).

The glacial temporality through which the geologists story the landscape, makes us see the agential forces of geological entities and the effects on landscape formation. The scales of register and narration are millions of years, yet cycles are being broken in unprecedented ways. This is a storying of a global, geological clime, or climate, which is crafted through meticulous observations, ice cores, models, sea level maximums and minimums, among others.

Through these observations and the geological register, the tundra, with its different water bodies, emerges as a *water*scape, where the seemingly bounded water bodies of the cryosphere—ice sheets, glaciers, rivers, fjords, lakes, and permafrost—become graspable as interconnected. Geographer Erik Swyngedouw (1999) defines "waterscape" as a water landscape and uses it to refer to natural flows as well as human interventions in or uses of water. Characterizing "waterscape" as a hybrid, he argues that nature is produced through changing societal and ecological conditions (Swyngedouw 1999, 444, 445). Anthropologists Ben Orlove and Steven Caton define "waterscape" as "the culturally meaningful, sensorially active places in which humans interact with water and with each other" (Orlove and Caton 2010, 408). Both definitions highlight a quality intrinsic to water flows: they tie together society and nature in different configurations, producing particular human relations as well as specific natures, not as separate entities but as deeply entwined and interdependent. The glaciers, the glacial lakes, the water flows, the permafrost thawing and freezing, and the streams of water made by meltoff and rain, each make interconnected water bodies, and together form a waterscape. By referring to the tundra as a waterscape, I wish to highlight the landscape as one shaped and constituted by waters of different kinds, just as much as it is by "land," and—in addition—by the human perspectives dwelling in and storying it in different ways. The tundra, defined by its perpetual permafrost—or swamp of eternal frost—is as much a waterscape as it is a landscape. The land remains land, but it is water—and ice as it melts—that causes landslides and ruptures in the roads and other infrastructures. Water bodies change with the seasons, and it is the form of water that makes the difference in terms of how/if at all one can travel the tundra: snow, sea ice, and meltoff.

The geological gaze on the tundra stretches our notion of time and view on the land- and waterscape. It teaches us much but, not only about this place on earth but on the earth's climate system. It does, however, also leave out certain forms of consciousness and sentience that also shape and inhabit this landscape (see also Brandshaug, this book).

Turning Changes Into Profit: "Taste the Beginning of Time"

In 2017, the government of Greenland published a new "Strategy for Export of Ice and Water," promoting the conversion of the increasing glacial meltoff from Greenland's large freshwater reserve into profit. The ice sheet alone, the strategy highlights, constitutes about 10% of the world's total freshwater sources. The extraction and export of water and ice is integrated into the ambition and plans of the government of Greenland to become financially more self-sustaining.[4] With the goal of eventually liberating Greenland from all colonial ties and gain economic and political independence from Denmark, Greenlandic politicians and leaders are continuously looking for new resources to tap into (Andersen and Flora 2019). Oil, uranium, and rare earth minerals have all been on the agenda for extraction. Climate change helps push these plans by opening new sea routes. The increase in meltoff from glaciers and the Greenlandic Ice Sheet, in combination with increasing lack of and pressure on freshwater sources in other parts of the world, has stimulated dreams of global export of Greenlandic water and ice: "Greenlandic ice cap water heading for the world market," is one of the slogans highlighted by the government strategy. In 2019, a call was sent out inviting companies and investors to apply for licenses for commercial exploitation of Greenland's water and ice, and several companies are now engaged in water extraction.[5]

One such company is owned by Morten Jacobsen, a Danish water extraction entrepreneur in Kangerlussuaq. In July 2020, during dinner and conversations with the refuged geologists, Morten showed up with three bottles of beer, home-brewed with glacier water. He was curious to hear the geologists' opinion on the exact age of the water he was tapping. As he planned to produce beer, vodka, and gin, Morten was eager to create a good story about his products for marketing purposes. By drinking ice cap beer, the ice sheet would be tied to a global, and consumable, history, and provenance. "Taste the beginning of time" was one of several slogans of his company.

The commercialization of natural resources, although welcomed by many in Greenland, is premised on extractivist ways of valuing and viewing the environment. These perspectives conflict with other ways of knowing and living with the land, which sometimes leads to outspoken controversies and tensions. While entrepreneurs and policymakers dream of converting climate change into a source of innovation, profit, and—ultimately—independence, others relate to the land and its "resources"—or beings—oriented by different modes of existence, or ontologies.

Storying with *Sila*: A World of More-than-Human Sentience

"Are we leaving tomorrow?" Aviana asked her father who was an experienced fisherman and hunter of sea mammals (seals, walrus, whales) as well as terrestrial animals (caribou, hare, and muskox). "*Immaqa* (maybe)" Amâsa replied to his daughter and looked at her with warm contemplation in his eyes. "*Sila kisimi* (only the weather can decide)." Their tent was packed, as were food provisions and all the necessary gear to stay in the hunting grounds for weeks. Aviana was eager to leave for Angujaartorfik, the tent site deep in the fjord that her family had visited summer after summer through countless generations. Aviana was twelve years old and knew the answer to her question even before the phrase came out of her mouth. She had heard those words so many times before, yet her impatience and eagerness had not allowed her to keep silent. "*Immaqa. Sila kisimi.*" "Maybe. Only the weather," would be the direct translation. Or, "the weather itself," meaning that the weather decides. Aviana continued packing the things to bring along, and deeply wished for the weather to stay good for the coming day.

Sila is the West Greenlandic, and pan-Inuit, word for weather, but also for a person's sense or reason, consciousness, experience, state of mind, and that which is outside. *Sila* also means world order or universe; a force that connects persons and entities to the world around them. *Sila* overlaps in interesting ways with the English concept central to this book: "clime."

Sila, like clime, highlights a "powerful link between weather and place" (Carey and Garone 2014, also the introduction to this book). But *sila* also exceeds this powerful link. In Greenlandic, there is no exact word for climate, nor for clime. *Sila*, in contrast, permeates lived and local worlds—the here and now of a person and people; it constitutes their inside and outside, and extends from these to uphold the universe (see Nuttall 1992, 2016). For those who travel the landscape or go hunting, *sila* demands constant assessment of the conditions and changes of weather, wind, ice, and temperature, on a daily, often even hourly basis.

Naduk Petersen, a female teacher in Kangerlussuaq who grew up hunting and fishing, told me that respect for *sila* and nature means many things. It is all-encompassing when one moves through the landscape (*nunami*): "If the weather (*sila*) is bad, and you know you shouldn't be going into the land, and for some reason you do it anyway, and [if] you fall, then you know that somehow, you've been disrespectful. I mean, the mountain doesn't hit you, *sila* doesn't push you to fall as such. But you know that you have not behaved as you should." These recognitions by Naduk reveal *sila*, the mountains, and the landscape in general to be sentient and responsive entities. As most Greenlandic Inuit who have grown up hunting and fishing and spending a lot of time *silami* (outside), Naduk perceives *sila* and all beings as sentient, with a kind of will of their own.

Notions of *sila* vary regionally across the Arctic, and across the descriptions by different scholars. Some early ethnographic accounts highlight *sila* as a spirit that upholds the universe (Oostermann 1952). More recently, anthropologist Mark Nuttall has described *sila* as "the breath of life" (Nuttall 2016, 304). *Sila*—the world, weather, outside and consciousness—refers to a responsive and relational world. This means that a change in climate as we know it is not an external thing happening outside of or separate from humans. People in Greenland are used to *sila* constantly changing, but the unfamiliar changes experienced nowadays equal a world changing in all its dimensions, including cultural and spiritual ones. Climate change in a *sila* perspective, can be seen as the world's ethical response to improper human actions (Leduc 2007, 242) (Figure 3.2).

To make these claims I draw on scholars Timothy Leduc (2007), Lene Kielsen Holm (2010), Mark Nuttall (2016), and Pitsoelak Pfeifer (2018) who share a concern about building a ground on which interdisciplinary research on climate change can integrate non-scientific and indigenous ways of knowing. I share the ambition of these scholars, and more, and this is also a central aspiration of this

FIGURE 3.2 *Inussuk* near Angujaartorfik. An *inussuk* is a human-made structure in the landscape, built to show the way for travelers (maybe on a day with bad weather (*sila*) or visibility).

Photo: Astrid Oberborbeck Andersen.

book: a call for more inclusive ways of knowing, narrating, storying, and managing human-environment relations and how they change.

As mentioned above, Aviana, Amâsa, Naduk, and others, do not primarily experience the landscape as tundra, but as land (*nuna*), with mountains (*qaqqat*) and their rivers (*kuut*), and lakes (*tatsit*). What would it take to understand, categorize and compare landscapes, biomes and their changes on *sila* and *nuna* terms? How would sila story tundra and the different humans and beings inhabiting the land? What is lost when attempting to understand this landscape by the generalized category of tundra?

It may seem that I am getting too caught up in perspectives, words, their meanings, and (im)possible, or difficult translation. There is a point to make, however, through these detailed attentions to meanings and translations. It is well expressed in this lengthy quote by Mark Nuttall (2016):

> We need to listen carefully when people tell us that the weather does not appear to be normal for the season, or that the habits of birds, seals, and caribou are changing. Yet we need to think about what these statements actually mean in terms of their rhetorical and metaphorical senses too.
>
> In one sense, there are taken-for granted assumptions about what is real and true and observable about the world; while in another sense,

> we know we need to remain attendant to understanding the multiplicities of meanings deriving from circumstances that are particular to social, cultural, and political interpretations and contexts.
>
> *(293)*

Nuttall's words help me qualify the necessity for careful attention to and unpacking of words and meanings in context. In addition to Nuttall's emphasis on indigenous and local perspectives on changing environments, I will add that we should pay equal attention to the work of non-native terms and discourses in different contexts, and to the concepts we bring to the landscapes as part of our knowledge ecology: such as for example, climate change, tundra, or clime.

Clime, *Sila*, and the Pluriversal Tundra

"Tundra", the scientific word with Russian and Sami origins, renders the landscape around Kangerlussuaq comparable to that of other Arctic regions; it allows for comparison of different landscapes through their ecological and climatic conditions. This comparison, however, requires *translation*—which is never one to one, but always converts that which is being translated into something slightly different when it travels across contexts. The *transference* of concepts and their *translations,* I suggest, demand that we pay close attention to what meanings and values change with the translations, but also that we carefully note what else travels with the concepts we use to compare different worlds.

How can we name, conceptualize, and describe a landscape and ecological relations that are so distinct to different actors? And how can we qualify and define the changes happening to environments, and their significance to different actors and beings?

I suggest that we think of the landscape around Kangerlussuaq and Angujaartorfik as a *pluriversal tundra.* Following Arturo Escobar (2011):

> The pluriverse can be described as "a world where many worlds fit." At their best, it can be said that the rising concepts and struggles from and in defense of the pluriverse constitute a post-dualist theory and a practice of interbeing.
>
> *(Escobar 2011, 139)*

The pluriverse performs a critique on the notion of universe, the idea that we (human beings as one "we") live in the same, singular, "one-world world" (see also Law 2015). The pluriverse is premised on relational worldviews or ontologies, meaning a recognition that all beings are relationally constituted or becoming, and open to a multiplicity of interpretations, articulations, enactments, and beings.

Acknowledging the tundra as a pluriverse allows us to grasp the distinct registers in which different actors and their perspectives engage with, story, and produce the landscape, and it allows us to see the multiplicity of worlds emerging from it. Not all becoming is co-becoming (Van Dooren and Bird Rose 2016) or making-with (Haraway 2015); making and becoming, or emergence, often happens in diverse, mutually conflicting registers. Like "tundra," pluriverse is a concept that has emerged elsewhere than in the Arctic, and has traveled, and thus it demands an effort of translation. Escobar's pluriverse grows out of South American ontologies and "epistemologies of the south." South American social, political, and climatic contexts are far from Greenland, the tundra waterscape, and its epistemologies and ontologies of the north. How, might you ask, does the pluriverse work in the Arctic tundra? And how does the Arctic version differ from the South American?

First, unlike "tundra," which is a descriptive and classificatory term, "pluriverse" is designed for a multiplicity of ontologies (in the plural) and openness, for inclusion of and room for differences rather than reduction and assimilation. There is of course a sense in which the great number of accounts coming from climate science, and glaciological and geological research in Greenland contribute to the production of the Arctic as a part of the one-world world. Indeed, according to climate science, what happens in the Arctic is significant for the entire earth system, or, as once stated by members of the Inuit Circumpolar Council (ICC), "The Arctic is like a barometer of the globe's environmental health" (quoted in Sejersen 2015, 166). And this is why bringing the pluriverse to the Arctic is necessary: to highlight that there are stories and modes of being and making-otherwise, a relational world that is not seen through the perspectives of climate science, and that this "Anthropos-not-seen" prevents us from co-becoming and making-with, in more complex and inclusive "we's" (de la Cadena 2019).

The "ontological struggles" (Blaser, 2009; Escobar 2016) in South America, and the political expressions they take are different than those in Greenland and the Arctic. This is not the place to expand on political expressions of Arctic epistemologies and ontologies in general; suffice it to say that, while some of the dynamics and trajectories of "discovery," colonization, and continuous narration/representation by Europeans and foreigners are similar in the Arctic to those in South America, the forms that politics and struggles take are different. What I would like to highlight here is the agency of the ice, the mountains, and the beings living here, and the potential of storying with *sila* for cultivating attentiveness to relational forms of being and worldmaking, in "complex we's" (de la Cadena 2019).

In another sense, the storying with glacial time and the partial oneness of the earth brings dimensions to pluriversal thinking (and doing) that can be said to be Arctic, or northern. Glacial time and temporality de-centers, or re-centers, the human in a way that opens our attention to temporal complexities and beyond-human timescales of the earth. Glacial time and temporality is an aspect to the pluriverse, and to climes, that enables us to compare the Arctic with lives lived and changing waterscapes in the Himalaya and the Andes.

Approaching tundra climes, and any environment, as a pluriverse—with the added dimensions of temporal complexity and earthly agency—allows for conceptual (epistemological and ontological) openings rather than stabilization, or closure. Accepting that environments and the lives and relations that unfold in them are in continuous becoming (Nuttall 2016) will demand more inclusion, negotiation, and diplomacy when it comes to their management, but will also allow for diverging voices and perspectives to be heard when decisions about landscapes are being made, and—possibly/hopefully— with less violence to a multiplicity of livelihoods and beings, including those more than human.

"Clime" as analytic and entry point enables the traveling across different regions and the comparison of the experiential and affective impacts of climate changes in the Andes, the Himalaya, and the Arctic with rich nuances in life-worlds, realities, and also with inconsistencies and tensions. Storying climes demands attention to experiences situated in diverse geographical, ecological, and political contexts, and attention to the diversity in perspectives, positions, and voices—human and nonhuman. It is remarkable that even climate scientists are beginning to pay attention to modes of storying, or storylining (Shepherd and Truong, this book). Clime, like pluriverse, is an open concept, inclusive of a multiplicity of ontologies and epistemologies, and, I add, temporal complexities.

The pluriversal tundra is an analytic—or an optic—that enables us as scholars to see and understand human-environment relations in ways open to multiplicity. As proposed by Arturo Escobar (2011, 2017) and Marisol de la Cadena (2019), it is also an imperative, or a suggestion for paving the way for a more just world, a different kind of worldmaking, and thus a conceptual and ontological infrastructure for designing livability in the future. *Sila*, as a particular way of climing (Paerregaard, this book), is both a one-world world and a pluriverse, it is a central epistemology and ontology of the Arctic landscape, also known as tundra, and engaging *sila* teaches us to understand and respond to the world/weather/outside/mind/consciousness as one of more-than-human sentience, embracing complex temporalities and we's.

Finally, storying pluriversal ecologies, and climes, means to weave together a complexity of beings and perspectives—across timescales—and to make space for epistemological and ontological multiplicity, in the Arctic, Andes, Himalaya, and elsewhere.

Notes

1 The research for this chapter is funded by The Independent Research Fund Denmark, grant number 9037–00140B, as part of the research project "Muskox Pathways. Resources and Ecologies in Greenland" (2019–2023).
2 Kîsta Bianco Kjær, a student assistant at Aalborg University, has been helpful in working through these detailed translations of words and worlds.
3 The names of research participants have been pseudonomized.
4 https://www.businessingreenland.gl/~/media/Erhverv/Is%20og%20Vand/Markedsfoering/Strategi%20-%20eksport%20is%20og%20vand%20UK.pdf?la=en
5 See https://www.businessingreenland.gl/en/Udbud/Afsluttede-udbud/2019/3-kvartal/Invitation-til-at-soege-om-tilladelse-til-kommerciel-udnyttelse-af-Groenlands-vand-og-is

References

Andersen, Astrid Oberborbeck. 2022. Muskox multiplications: the becoming of a resource, relations and place in Kangerlussuaq, West Greenland. *Acta Borealia*, 39(1), 75–94. https://doi.org/10.1080/08003831.2022.2060619

Andersen, Astrid Oberborbeck and Janne Flora. 2019. Puzzling Pieces and Situated Urgencies of Climate Change and Globalisation in the High Arctic: Three Stories from Qaanaaq. In A. B. Stensrud, & T. Hylland Eriksen (Eds.), *Climate, Capitalism and Communities: An Anthropology of Environmental Overheating*. London: Pluto Press, 115–132.

Bawaka Country, Laklak Burarrwanga, Ritjilili Ganambarr, Merrkiyawui Ganambarr-Stubbs, Banbapuy Ganambarr, Djawundil Maymuru, Sarah Wright, et al. 2017. Co-Becoming Time/s: Time/s-as-Telling-as-Time/s. In *Methodological Challenges in Nature-Culture and Environmental History Research*, edited by Thorpe J., Rutherford S., and Sandberg L.A. London: Routledge, 1–13. https://doi.org/10.4324/9781315665924

Blaser, Mario. 2009. The threat of the Yrmo: The political ontology of a sustainable hunting program. *American Anthropologist*, 111(1), 10–20. https://doi.org/10.1111/j.1548-1433.2009.01073.x

Bliss, L.C., G.M. Courtin, D.L. Pattie, R.R. Riewe, D.W.A. Whitfield, and P. Widden. 1973. Arctic tundra ecosystems. *Annual Review of Ecology and Systematics*, 4(1), 359–399. https://doi.org/10.1146/annurev.es.04.110173.002043

Bravo, Michael and Sverker Sörlin. 2002. *Narrating the Arctic: A Cultural History of Nordic Scientific Practices*. Canton, MA: Watson Publishing International.

Brightman, Mark, Vanessa Elisa Grotti, and Olga Ulturgasheva. 2012. Introduction. animism and invisible worlds: The place of non-humans in indigenous ontologies. In *Animism in Rainforest and Tundra: Personhood in Contemporary Amazonia and Siberia*, edited by Mark Brightman, Vanessa Elisa Grotti, and Olga Ulturgasheva. Oxford and New York: Berghahn Books, 1–50.

Bubandt, Nils, Astrid Oberborbeck Andersen, and Rachel Cypher. 2023. Introduction: Rubber boots methods outline for a multispecies study of the anthropocene. In *Rubber Boots Methods for the Anthropocene: Doing Fieldwork in Multispecies Worlds*, edited by Nils Bubandt, Astrid Oberborbeck Andersen, and Rachel Cypher. Minnesota: University of Minnesota Press, 1–35.

Carey, Mark and Philip Garone. 2014. Forum introduction. In *Forum: Climate Change and Environmental History*, edited by Mark Carey, Philip Garone, Adrian Howkins, Georgina H. Endfield, Lawrence Culver, Sam White, Sherry Johnson, and James Rodger Fleming. *Environmental History*, 19(2), 282–293, https://doi.org/10.1093/envhis/emu004

Carrivick, Jonathan L., Jacob C. Yde, Niels Tvit Knudsen, and Christian Kronborg. 2018. Ice-dammed lake and ice-margin evolution during the Holocene in the Kangerlussuaq area of west Greenland. *Arctic, Antarctic, and Alpine Research*, 50(1), 1–11. https://doi.org/10.1080/15230430.2017.1420854

Cruickshank, Julie. 2005. *Do Glaciers Listen? Local Knowledge, Colonial Encounters, and Social Imagination*. Vancouver: UBC Press.

De la Cadena, Marisol. 2010. Indigenous cosmopolitics in the andes: Conceptual reflections beyond 'politics.' *Cultural Anthropology* 25(2), 334–370. https://doi.org/10.1111/j.1548-1360.2010.01061.x

De la Cadena, Marisol. 2019. An invitation to live together: Making the 'complex we.' *Environmental Humanities*, 11(2), 477–484. https://doi.org/10.1215/22011919-7754589

Elberling, Bo. 2021. *Det Isfrie Grønland*. København: Gyldendal.

Eriksen, Thomas Hylland. 2016. *Overheating: An Anthropology of Accelerated Change*. London: Pluto.

Eriksen, Thomas Hylland. 2021. Climate change. In *The Cambridge Encyclopedia of Anthropology*, edited by Felix Stein. http://doi.org/10.29164/21climatechange.

Escobar, Arturo. 2011. *Encountering Development: The Making and Unmaking of the Third World*. Princeton: Princeton University Press.

Escobar, Arturo. 2016. "Thinking-feeling with the Earth: Territorial Struggles and the Ontological Dimension of the Epistemologies of the South." *Revista de Antropología Iberoamericana*, 11(1), 11–32.

Escobar, Arturo. 2017. *Designs for the Pluriverse: Radical Interdependence, Autonomy, and the Making of Worlds*. Durham: Duke University Press.

Fienup-Riordan, Anne. 1999. Yaqulget Qaillun Pilartat (What the Birds Do): Yup'ik Eskimo understanding of geese and those who study them. *Arctic*, 52(1), 1–22.

Fienup-Riordan, Anne, W. Tyson, P. John, M. Meade, and J. Active. 2000. *Hunting Tradition in a Changing World: Yup'ik Lives in Alaska Today*. New Brunswick, NJ: Rutgers University Press.

Gan, Elaine, Anna Tsing, Heather Swanson, and Nils Bubandt. 2017. Introduction: Haunted landscapes of the anthropocene. In *Arts of Living on a Damaged Planet: Ghosts and Monsters of the Anthropocene*, edited by Anna Lowenhaupt Tsing, Heather Anne Swanson, Elaine Gan, and Nils Bubandt. Minneapolis: University of Minnesota Press, G1–14.

Grønnow, Bjarne. 2009. Blessings and horrors of the interior: Ethno-historical studies of inuit perceptions concerning the inland region of West Greenland. *Arctic Anthropology*, 46(1/2), 191–201.

Grønnow, Bjarne, Morten Meldgaard, and Jørn Berglund Nielsen. 1983. Aasivissuit: The Great summer camp. archaeological, ethnographical and zooarchaeological studies of a caribou-hunting site in West Greenland. Meddelelser om Grønland. *Man and Society*, 5. Copenhagen: Commission for Scientific Research in Greenland.

Haraway, Donna. 2015. *Staying with the Trouble. Making Kin in the Chthulucene*. Durham and London: Duke University Press.

Hollesen, Jørgen, Henning Matthiesen, Anders Bjørn Møller, and Bo Elberling. 2015. Permafrost thawing in organic arctic soils accelerated by ground heat production. *Nature Climate Change*, 5(6), 574–578. https://doi.org/10.1038/NCLIMATE2590

Holm, Lene Kielsen. 2010. Sila-Inuk: Study of the impacts of climate change in Greenland. In *SIKU: Knowing Our Ice*, edited by Krupnik I., Aporta C., Gearheard S., Laidler G., and Kielsen Holm L. Dordrecht: Springer. https://doi-org.zorac.aub.aau.dk/10.1007/978-90-481-8587-0_6

Ingold, Tim. 1993. The temporality of the landscape. *World Archaeology*, 25(2), 152–174.

Irvine, Richard. 2020. *Anthropology of Deep Time: Geological Temporality and Social Life*. Cambridge: Cambridge University Press.

Knudsen, Niels Tvis, Jesper Hannibalsen, and Christian Kronborg. n.d. *The Geology in the Area between Sisimiut and Kangerlussuaq*. Unpublished manuscript.

Knudsen, Niels Tvis. n.d. Land Forming Processes in the Proglacial Landscape at Kangerlussuaq, West Greenland. Unpublished manuscript.

Krupnik, Igor and Dyanna Jolly. 2002. *The Earth Is Faster Now: Indigenous Observations of Arctic Environment Change*. Fairbanks, AK: Arctic Research Consortium of the United States.

Krupnik, Igor, Claudio Aporta, Shari Gearheard, Gita Laidler, and Lene Kielsen Holm. 2010. *SIKU: Knowing Our Ice*. Dordrecht: Springer. https://doi-org.zorac.aub.aau.dk/10.1007/978-90-481-8587-0_6

Law, John. 2015. What's wrong with a one-world world? *Distinktion: Scandinavian Journal of Social Theory*, 16(1), 126–139, DOI: 10.1080/1600910X.2015.1020066

Leduc, Timothy B. 2007. Sila dialogues on climate change: Inuit wisdom for a cross-cultural interdisciplinarity. *Climatic Change*, 85, 237. https://doi.org/10.1007/s10584-006-9187-2

Massey, Doreen. 2005. *For Space*. London: SAGE.

McNeill, J.R. and Peter Engelke. 2016. *The Great Acceleration: An Environmental History of the Anthropocene since 1945*. Cambridge, MA: Harvard University Press.

Nuttall, Mark. 1992. *Arctic Homeland: Kinship, Community and Development in Northwest Greenland*. London: Belhaven Press.

Nuttall, Mark. 2016. Living in a world of movement: human resilience to environmental instability in Greenland. In *Anthropology and Climate Change: From Encounters to Action*, edited by Crate S.A. and Nuttall M. London/New York: Routledge, 292–310.

Oksman, Mimmi, Anna Bang Kvorning, Signe Hillerup Larsen, Kristian Kjellerup Kjeldsen, Kenneth David Mankoff, William Colgan, Thorbjørn Joest Andersen, et al. 2022. Impact of freshwater runoff from the Southwest Greenland ice sheet on Fjord productivity since the late 19th century. *The Cryosphere Discuss*, 1–10. https://doi.org/10.5194/tc-2021-373.

Orlove, Ben and Steven C. Caton. 2010. Water sustainability: Anthropological approaches and prospects. *Annual Review of Anthropology*, 39, 401–415.

Ostermann, Hother. 1952. The Alaskan Eskimos, as described in the posthumous notes of Dr. Knud Rasmussen. *Rep 5th Thule Expedition 1921–1924*, 10(3). Copenhagen: Nordisk Forlag.

Pfeifer, Pitsoelak. 2018. From the credibility gap to capacity building: An inuit critique of Canadian Arctic research. *Northern Public Affairs*, 6(1), 29–34.

Sejersen, Frank. 2015. *Rethinking Greenland and the Arctic in the Era of Climate Change: New Northern Horizons*. London: Routledge.

Swyngedouw, Erik. 1999. Modernity and hybridity: Nature, regeneracionismo, and the production of the Spanish Waterscape, 1890–1930. *Annals of the Association of American Geographers*, 89(3), 443–465.

Todd, Zoe. 2018. Métis storytelling across time and space: Situating the personal and academic self between Homelands. In *Activating the Heart: Storytelling, Knowledge Sharing and Relationship*, edited by Christensen J., Szabo-Jones L., Cox C., and Boisselle A. Waterloo: Wilfrid Laurier Press, 172–189.

Van As, Dirk, Bent Hasholt, Andreas P. Ahlstrøm, Jason E. Box, John Cappelen, William Colgan, Robert S. Fausto, et al. 2018. Reconstructing Greenland ice sheet meltwater discharge through the Watson river (1949–2017). *Arctic, Antarctic, and Alpine Research*, 50(1), 1–10. https://doi.org/10.1080/15230430.2018.1433799

Van Dooren, Tom and Deborah Bird Rose. 2016. Lively ethography. Storying animist worlds. *Environmental Humanities*, 8(1), 77–94. https://doi.org/10.1215/22011919-3527731

Wood, Michael, Eric Rignot, Ian Fenty, Lu An, Anders Bjørk, Michiel van den Broeke, Cilan Cai, et al. 2021. Ocean forcing drives glacier retreat in Greenland. *Science Advances*, 7(1), 1–10. https://doi.org/10.1126/sciadv.aba72

PART II
Water Climes

4

ENO

Eco-Spiritual Water Climes of Dibang Valley, Arunachal Pradesh[1]

Razzeko Delley and Ambika Aiyadurai

Introduction

The Himalaya plays a critical role in the hydrological cycle of the earth and in sustaining innumerable biological species and their habitats, in addition to many socio-cultural and economic systems. It is experiencing new risks due to climate change. The rising temperature of the earth and its impact on the Himalayan ecosystems show the fragility of mountain ecosystems (Pant et al., 2018). Similarly, the impact of climate change on the Himalaya heavily disturbs human cultural environments, as demonstrated by Pandit (2017). The indigenous history of water, local memories of climate change, and indigenous climate knowledge can tell us a lot about the conservation values related to indigenous animistic landforms and water bodies. These histories are potentially powerful agents of environmental peacebuilding and the emerging new environmental ethics. Moving away from the positivist understanding of climate science, in this chapter, we bring narratives of people highlighting how they experience climate variability.

This chapter is concerned with the cultural meanings of water, as embodied in indigenous histories of the Idu Mishmi people in Arunachal Pradesh, India. It is a climate study mediated through the emerging notion of *climes*, as brought forth by Carey and Garone (2014, 284). The authors underscore the powerful link between weather and place, emphasizing their significance for both cultural and scientific constructions. We prefer to use the word "clime" over "landscape" or "place" because the former, synonymous to place, embodies climate, while the latter has no direct etymological association with climate at all, as Smyer Yü points out in the introduction to this book. Intended to mediate and communicate climate and climate change for an interdisciplinary audience, the

DOI: 10.4324/9781003347026-7

application of clime is found in our alternative use of "riverine climes" and "water climes" to refer to human and nonhuman habitats shaped by rivers, streams, ponds, and lakes, which embody local and regional climate histories, climate knowledge, climate making, and climate change. We use the Idu Mishmi term *eno*, a plural noun for spiritual spaces, which encompasses physical environments, along with their association with social, cultural, and psychological worlds. At the same time, we also translate the indigenous word *eno* as eco-spiritual climes for the purpose of emphasizing the Idu Mishmi water climes as nature's provision as well as embodiments of spiritual, ritualistic, and symbolic meanings of the local life-cycle. Thus, the coinage of the term eco-spiritual climes provides us with an opportunity to identify and articulate unique meanings of clime and clime change among the Idu Mishmi.

By recounting commonly practiced rituals of the Idu Mishmi, we will show how *eno*, as eco-spiritual water climes, connect the earth, humans, animals, plants, and spirits. Our storytelling is intended to demonstrate how the Idu Mishmi revere, invoke, and offer sacrifices to *eno*, physically identified as the sacred mountains, rivers, streams, and lakes during the healing rituals known as *machiwu*. In addition, we will also recount the shapeshifting nature of water as a "disruptor" and "barrier" during floods and landslides, and discuss how it is understood and narrated by local people. Finally, by situating the people's understanding of their eco-spiritual water climes in the regional economic development discourses that often marginalize their experiences and responses, we argue that modern infrastructure such as hydropower and highway projects are severely altering the lifeworlds of the Idu Mishmi and eroding their cultural and natural heritage.

Our collaborative research was undertaken in two districts of Arunachal Pradesh: Dibang Valley and Lower Dibang Valley, which are the primary dwelling places of the Idu Mishmi. Razzeko Delley is an Idu Mishmi, fluent in the Idu Mishmi language that is spoken daily, and familiar with the shamanic chants that are understood by only a handful of Idu Mishmi members. He is trained in English literature and teaches in a college in Roing (Lower Dibang Valley district). Considered an "outside" researcher, Ambika Aiyadurai has been working in Dibang Valley on the social dimensions of wildlife conservation, especially on human-wildlife relations (Aiyadurai, 2021). She has interdisciplinary training in wildlife sciences and anthropology. Given her limited knowledge of the Idu Mishmi language, she conducted interviews in Hindi, the lingua franca of Arunachal. We carried out independent studies in various locations at different points in time. Delley has recorded the rituals mentioned in the chapter and provided translations of the shamans' rituals (Delley 2018, 2021). This chapter is a combination of the first-hand indigenous knowledge of water in folkloric, animist, ecological, and political terms. By co-authoring, we bring together our respective indigenous roots, long years of working in the region, and a shared interest in recognizing the importance of indigenous environmental knowledge and values.

Animist Riverine Climes of the Idu Mishmi

The Dibang Valley district of Arunachal Pradesh, the easternmost state of India (see Figure 4.1), has numerous rivers descending from the mountains—tributaries of the Brahmaputra—that make their way to the plains of Assam. The two districts where the Idu Mishmi predominantly live are named after the Dibang River, originally known as Talon to the Idu Mishmi, which is fed by several perennial rivers like the Ithu, Emra, Dri, Mathu, Eze, Iphi, and Aso, to name a few. Out of the twenty-five districts in the state, nineteen are named after the rivers that flow through their respective districts. Dibang Valley is a riverine world, too, animated by both humans and nonhumans, including the local spirits and deities.

Recent works on the Idu Mishmi highlight the integral sociality and spirituality of animals, plants, spirits, and physical environments (Aiyadurai, 2018; Delley, 2018; Nijhawan and Mihu, 2020). The Idu Mishmi follow an animistic belief system that incorporates various shamanic rituals for the well-being of both individuals and the community as a whole (Chaudhuri, 2008). Shamans (*igu*) are considered as spiritual mediators between humans and nonhumans and between humans and the spiritual world (see Figure 4.2). This human-nonhuman interaction can be understood as what Anna Tsing calls "more-than-human sociality" (Tsing, 2013, 38). We commence our description and discussion of the Idu Mishmi's eco-spiritual water climes as animist climes in Dibang Valley, in which human networks go beyond people to encompass nonhuman species, spirits, and landforms.

Animism as a social-spiritual system is critically important to understand if we are to also understand the Idu Mishmi riverine climes as unions of the material and the spiritual worlds. An animist perspective affords us to recognize that the

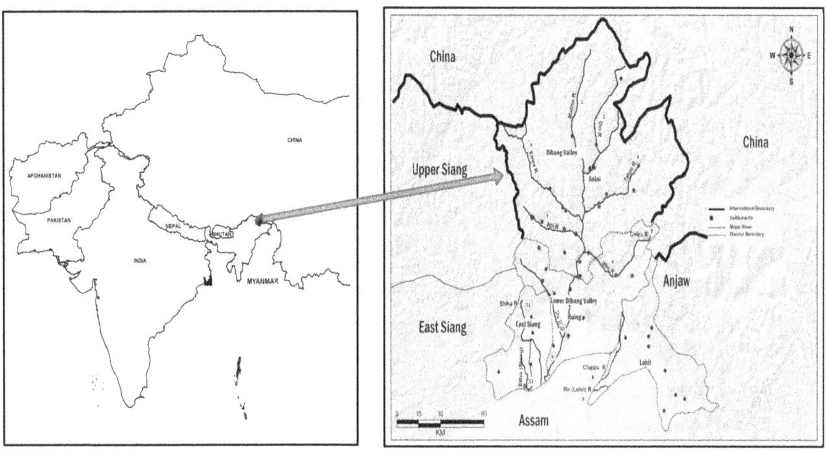

FIGURE 4.1 Rivers of Dibang Valley.
Map prepared by Hiranmoy Chetia and K. K. Rashid.

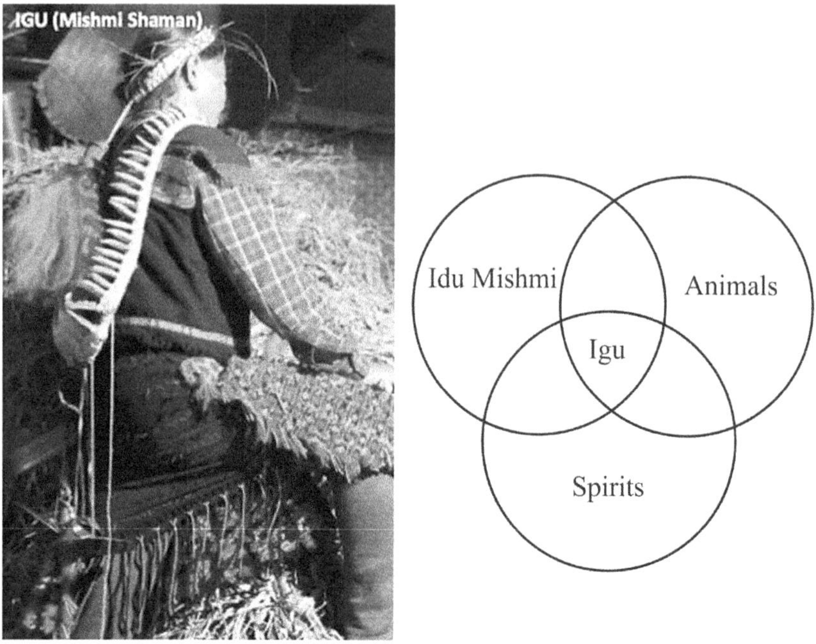

FIGURE 4.2 Idu Mishmi social worlds.
Photograph by Ambika Aiyadurai.

same life-essence takes different bodily forms or transmigrates from one form to another. This is apparent in the lifeworlds of the Idu Mishmi, their companion spirits, animals, plants, and animate landforms. This more-than-human animist sociality encompasses a network of human-nonhuman associations (Figure 4.2). In this respect, animism in Dibang Valley is multidimensionally understood in terms of human-animal, human-plant, plant-animal, and human-earth relations. Approaching the Idu Mishmi, particularly from the new-animist perspectives of Descola (2013), Viveiros de Castro (1998), and Smyer Yü (2020), we see that the life-essence moves between the bodies of humans, animals, plants, and landforms. To emphasize, the earth is not merely a physical, nonspiritual world but is lively with animate landforms and water bodies synonymous with spirits and deities. When *eno*, as eco-spiritual water climes, are seen in this light, they are recognized as a set of animist climes because of the spiritual contents in their ecological forms. It is thus crucial for us to understand *eno* with the approaches of animist ontologies and relational epistemologies (Naveh and Bird-David 2013; Smyer Yü 2020). This is especially important because epistemologies of indigenous peoples have been long rendered invisible, irrelevant, and have been marginalized in the face of dominant Western colonial ways of knowing, especially in the modern Indian context. As highlighted by Burman (in this book), in the discussions on climate change mitigation, there is little or no acknowledgment of relational indigenous life worlds. He suggests that our understanding of climate change must recognize other-than-human subjectivities. Here, we take the opportunity to reclaim and re-include the "other knowledge and other worlds" of the Idu Mishmi into what Escobar (2019) and de la Cadena and Blaser (2018) call a "pluriverse" or "a world of many worlds."

Our animist approach is centered on water bodies in their liquid, moving states, such as rivers, streams, and lakes. Water (*machi*) plays a pivotal role in the cultural canvas of the Idu Mishmi. *Machi* in the Idu language means both water and river. The rivers that share the Dibang Valley with the Idu Mishmi have their own identity. The shamanic chants manifest different aspects of water and its sources that relate to the identity of the Idu Mishmi. One such concept is *chithu huluni*, which literally translates to the twelve river valleys of the Dibang Valley (*chithu*: water source; *huluni*: twelve; see Table 4.1). Different Idu Mishmi clans identify themselves based on the twelve river valleys where they have traditionally resided. Their ethnolinguistic identity is centered on these rivers, gives them not only a sense of emplacement but also helps in maintaining social relations. For example, those living in the Ithu River Valley are called Mithu, while those residing by the side of the Dri River are the Mindri; likewise, the Mitaloh are those Idu clans who live in the Talon River Valley, and so on (see Table 4.1).

In the Idu Mishmi cosmological and ecological understanding, these river valleys are not merely geographical locations but also embodiments of the Idu Mishmi people's cultural heritage and spiritual traditions. These sites embody various attributes, such as being places that nurture plants, animals, birds, and fishes, or being host to various spiritual entities that reside in these water bodies. People connect with them through shamans who invoke spirits that help protect individual members, communities, and villages. The Idu Mishmi shows us a world comprising "geographically and ecologically specific places, in which human and nonhumans spin together their shared but negotiated and contended lifeworlds" (Smyer Yü, introduction).

TABLE 4.1 Names of the river valleys by which Idu Mishmis identify themselves

S.No.	*River Valleys*	*Idu Mishmi Linguistic Clans*
1.	Mathu	Mimathu
2.	Dri	Mindri
3.	Talon	Mitaloh
4.	Emra	Mimrah
5.	Ahi	Mihi
6.	Ithu	Mithu
7.	Chiyu	Chiyuchi
8.	Chipu	Chipuchi
9.	Ilu (Lohit)	Ilu-ah
10.	Ihi	Ihi-ah
11.	Shiku	Shiku-ah
12.	Etho (Siang)	Ethochi

Eno, the Eco-Spiritual Climes of the Idu Mishmi

For the Idu Mishmi, lakes, mountains, and rivers carry deeper cultural and spiritual meanings. They go beyond just being water bodies and landforms, as they are part of the larger, integrated social, ecological, and cosmological networks. Thus, the term *eno*, which the Idu Mishmi value so highly, can be understood as eco-spiritual climes; *eno* makes environmental flows and weather, and nourishes life communities; it is physically incarnate in the high mountains as spiritual guardians hosting human villages. Similarly, the rivers nearest to human dwelling places have connections with resident river spirits, the mountain ranges that mark human boundaries have mountain spirits, and the sources of drinking water in the form of streams and rivulets are also places that host spiritual guardians.

Eno thus hosts spiritual guardians to safeguard the people and bestow well-being and prosperity. Through rituals, people make offerings and observe taboos as regulatory mechanisms to minimize human offenses against the spirits. When annoyed, the spirit guardians punish people by inflicting diseases upon them and creating problems in their lives. This is taken very seriously among the Idu Mishmi, as angry spirits are known to withdraw the favors granted to people. When this relationship goes sour, a shaman is called, and rituals are conducted to address the conflict and restore the people's health and prosperity. This kind of contractual relationship between spirits and humans is maintained to ensure that humans and nonhumans live together peacefully (Figure 4.2). During these rituals, the names of important rivers and lakes are chanted frequently, as each water source is associated with certain supernatural entities and is revered by the tribe. Table 4.2 explains the water bodies that the Idu Mishmi consider sacred and whose names are invoked in different rituals. From time to time, the Idu Mishmi also offer sacrifices to these places through shamanic rituals. Apart from being a source of well-being and prosperity, these sites are *mraba,* the stations or stopovers for the soul while on its quest for the next world.[2]

From birth to death, water and water bodies shape, drive, and maintain diverse social and cultural associations. Therefore, *eno* could be seen as the integral union of the physical and the spiritual worlds, as Smyer Yü points out in the introduction. *Eno* provides a complementary understanding to climate science's numerical models, projections, and abstract discourses of climate change by combining the human and nonhuman realms and entities.

Eno represents the intertwinement of the ecological and spiritual liveliness of the earth, joining nature and culture and humans and nonhumans, and transforming the lifeless sites that feed the numerical models of climate change into living spaces with social connections, stories, and histories of humans and nonhumans. As *sila* is the sentient essence of the Inuit homeland (Andersen, this book), *eno*, as we ascertain, spiritualizes the dwelling places of the Idu Mishmi, too. The Idu Mishmi sense of climate and climate change is found in weather, seasons, rainfall, snowfall, etc., as well as in the negotiated human-spirit relations.

Similarly, Brandshaug (in this book) discusses the relations of the Yanqueños of Southern Peruvian Andes with their surroundings. For them, materials and phenomena are sentient beings, and as such are named as persons with feelings and emotions. *Achachilas,* the mountain beings, play a significant role in the lives of Yanqueños, who are mindful of their agentive presence. People, therefore, share more-than-human habitats for their living and sustenance. We show one such inter-relationship of place, water, humans, and spirits in the form of a ritual, called *machiwu*.

Machiwu: Water—The Healer and the Clime-Maker

Delley suffered from severe back pain in 2015. This was not the first time that he had approached Mola Mili, a shaman who is frequently requested to perform various rituals, including *machiwu*, for health-related issues. *Machiwu* spiritually cleanses diseases and evil entities from the body and uses water to cure pain.

TABLE 4.2 Eco-spiritual water sources, *eno* revered by the Idu Mishmis

Eco-Spiritual Water Bodies	*Forms*	*Status/Role and Characteristics*
Aduya	River	Place of wealth and prosperity, invoked while making offerings for a prosperous life in many shamanic rituals.
Doko	Lake	Considered a spiritual site that bestows wealth, fertility and prosperity. It is associated with fishes and fertility.
Ilu	River	Also known as the Lohit River. This is where many *mraba*, paths to the next world, lie. According to the Idu Mishmi myth, by the bank of this river, humans and monkeys parted due to a quarrel.
Ilu Iphini	River	Ilu Iphini is the place of Andizu, the almighty guardian shaman of the Reh ritual. This is the venue for the spiritual dance *Chita Nna* performed during the Reh ritual. In this *mraba,* the river Ayume Pawu wu also flows. The dance is to safeguard the host family and people attending the ritual from enemies and spirits.
Kandi/ Andomo	River	Known as the river of death, Kandi separates the human world and the ancestral world. The soul crosses in a boat to the next world through the river Kandi.
Mitungu	Lake	Spiritual place of *mithun*.[1] The place is invoked, and offerings are made to it for the bestowing of *mithun*.
Dondowu	Lake	Spiritual place of pigs.[2] The place is invoked, and offerings are made to it for bestowing of pigs.

1 *Mithun* (*Bos frontalis*) is a semi-domesticated cattle that is economically important and culturally valuable for the Idu Mishmi.

2 Pigs (*Sus domesticus*) have a special place in the lives of Idu Mishmi. They are reared for meat, or for sale in the markets, but more importantly for ceremonial slaughter during important occasions.

Mola Mili carried water in a container (a bamboo-culm jug) and bestowed spiritual power on the water by invoking the water's spiritual guardian, Asosoaru (see Figure 4.3). An example of how people, water, place, and spirits are interconnected will be clear in the following story, which Mola Mili narrates during the ritual:

> There was a time when everything between the rising sun and the setting sun was burnt down. All the living entities and the threads of life from east to west perished. Only rocks and stones remained. All the trees and plants perished in the great inferno. Water was much needed to stop the fire, but the water was nowhere to be found. Taking clues from birds and animals, shamans explored the possibility of finding water sources. Crabs and birds helped the shamans to channel water to revive and sustain the human world.
>
> The ancestral shamans asked, "Who shall channel back the water to the world?"
>
> First, Shili Arudi, a bird with clouds, vapour, and mist in her beak and claws, helped bring water, but with tiny claws and beak, her services were not sufficient.[3] Another bird, Bruwu-ethathi offered her help, but again, the water seemed insufficient.[4] Innyi-toroli, the crab, was ready, and the shaman asked, "Can you burrow to the water and bring it to humanity and save us?" The ancestors realized that the water found by a burrowing crab wouldn't reach the human world. At this point, Lomo-matome, the keeper-spirit of plantains, gave birth to Asosoaru, a son embodied in cloud, mist, and vapour. He was declared by the ancestors to be the one who should channel water back to the world by carving out rivers and streams, and by digging ponds, lakes, and puddles. Thereafter Asosoaru began digging out water with his powerful staff called Mecho-briyi. Thus, the rivers and streams flowed. Lakes and ponds got filled. Life came to the puddles. Rivers flowed into the human world. Asosoaru counted the river sources, named them, and became the guardian of the water sources and water entities.

After invoking the origin of water, the shaman calls down the spiritually potent herbs from the sky and earth. After these invocations, the water in the container has acquired spiritual attributes. Thereafter the shaman employs this spiritual water to cure people of diseases, cleanse houses of evil entities and vibrations and bestow prosperity to people.

Water, according to the Idu Mishmi, bears medicine from the sky and the earth, and possess healing power to cleanse even the deepest chambers of the earth, along with other beings. Indigenous histories take us back to the time when there were no humans on earth. This pre-human world gives us a sense of Deep Time and demonstrates the ancient power of water. Idu Mishmis' "Deep Time Earth," as Smyer Yü calls it, speaks through the words of indigenous humans, their histories, and rituals that embody places, water, animals, birds, plants, and people. These are indigenous climes where water and its forms play a key role

FIGURE 4.3 Mola Mili, a shaman cleansing a house using consecrated water.
Photograph by Razzeko Delley.

as clime-makers. In addition to water that is essential during ritualistic practices that are carried out to protect the lives of the Idu Mishmi, the ambiguous nature of water as a disruptive agent, in the form of heavy downpours, floods, and landslides, is also discussed.

Water Stories

In Idu Mishmi cosmology, water (*machi*), in the form of rain (*ara*), is an ambiguous entity that functions both positively and negatively, the latter when it disrupts the normal way of life. Dibang Valley is marred by incessant rain that causes havoc in monsoon months. The lives of the people come to a standstill as heavy rains result in landslides. This disrupts agriculture, hunting, and fishing. There is no equivalent or exact term for weather in the Idu Mishmi language; instead, *ara*, meaning rain, is used interchangeably. For instance, an Idu will say, "Tomorrow I will go farming but don't know how it will rain," or "We are planning to go on a hunting expedition, but it will depend on the rain." Aiyadurai, during her fieldwork, often heard people say in Hindi, "*baaa… mausum kharab hogaya*" (the weather has gotten worse). *Mausam* is a Hindi word for "weather," but often the reference is made to rain.

Rain can disrupt the daily activities of the people, including the all-important rituals. Among all the natural phenomena, rain is the one that brings the most misery, as the Idu people say. In situations of crisis, the Idu Mishmi approach shamans for respite. These powerful *igu* attempted to control or halt rains through the ritual called *ara-hri*. There is also a ritual to bring rain and storm, but it is the rain-stopping ritual that is sought more by the tribe. *Ara-hri* is particularly carried out during the cultivation season in which felled trees and slashed plants are to be burned. If there is a delay in burning the fields, it may result in a too brief cultivation season, with hunger or starvation a possible outcome. The rain-stopping ritual is also done before conducting big ceremonies to prevent rain from disrupting them, and also to erect barriers for outsiders who could be equally disruptive.

Sometimes, the shamans, in anger, can also cause heavy rains. Heavy rains also occur when someone meets with an unnatural death that is believed to be caused by spirits. Incessant rain, called *tsi,* occurs when people break taboos and violate customs. Using different terms for different kinds and intensities of rainfall indicates the importance and anxieties people have about rain. For example, there is a complex lexicon referring to just rain: a*yo/ara* (rain), *indu* (incessant rains), *arachi* (drizzling in fine weather), *ara-mbu* (cloudy with slight drizzle), *tsi* (incessant rain because of breach of taboo).

In all these rituals, the shamans' role in narrating stories of life, sustenance, and final departure to the next world is crucial. In all the life stages, water stories appear to provide healing, protection, and care, and also to provide sustenance for both the living and the dead. By telling these stories, the Idu Mishmi connect, materially or otherwise, with the river valleys where people derive their identities from the landscapes and the powerful symbols of myth that connect newborns to the present generation and to the past and the future. These cosmo-ecological narratives provide alternate worldviews of knowing and understanding in multi-time, multi-scale, multi-space, and multispecies perspectives. Revisiting the past through oral narratives, the Idu Mishmi's folklore and stories provide locally-rooted knowledge that cautions them of the dangers of understanding their surroundings in a unidimensional, disconnected, and simplistic way. The unidimensional and simplistic way of understanding water as merely a resource, separable from its climes, is causing the environmental and social challenges in the form of development projects, causing these climes and their waters to be simultaneous recipients and agents of climate change (see Smyer Yü, introduction).

Environmental Changes and Community Memories

Idu Mishmi community memories document environmental changes. For example, the Mayudia Mountain Range in Lower Dibang Valley is a significant doorway to the plains from the high mountains and is considered a difficult and dangerous pass. Historically, each year, dozens of people died while attempting to cross the snowy mountain path in the winter. Even now, with roads and four-wheelers, this route is still dangerous as accidents continue to take place.

In December 1900, when the British undertook punitive expeditions against the Idu Mishmi residing in Ithu Valley, the treacherous snowy route took the lives of two dozen porters (Reid 1942). Similarly, in the living memories of the Idu Mishmi, there were several instances where many lives were lost due to heavy snowfall. Pikho Mega, a seventy-nine-year-old woman from Elope, a village beyond Mayudia, recounted in an interview the memories of the dangerous path: "Earlier, every year, the Idu Mishmi country would echo with news of several people dying in the snow while passing through the snow-laden mountains. When someone got stuck, it was impossible to pull them out. The more you struggled, the more you got pulled into the snow. Therefore, on many occasions, people would just bid goodbye to their stranded companion and continue on their journey. The stuck people also accepted their fate and handed their ornaments to their relatives. Ibizu, the Himalayan Pika would eat the stuck people after they died." More casualties were witnessed when roads were built in the late 1970s and early 1980s. Many laborers from the plains were not used to the extremely cold climate and perished in the snow.

Being the native of a village in middle Dibang Valley beyond the Mayodia Range, Delley has witnessed changes in the snow levels of the area. Every year, when he was younger, heavy snowfall would block the road despite its constant removal with bulldozers. Returning from school at Roing for vacation one March, Delley remembers the snow extending along the road for miles and miles. The first vivid memory Delley has of a dangerous snowfall experience was in 1987 during one such March journey, accompanied by his uncle and cousin. Their four-wheeler jeep could not maneuver through the treacherous snowy terrain. They spent the night inside the vehicle itself. At night, stepping out of the vehicle, Delley saw the mountains glittering in the bright light of the full moon, and it looked like a sunny day. Each year presented a similar ordeal for Idu Mishmi, who lived beyond the Mayodia Range, as Delley relates. The Border Road Task Force (BRTF), the organization responsible for maintaining the route, would erect twelve to fifteen feet long bamboo poles on each side of the whole stretch of road. In many locations, the snow would rise above the poles. When traveling along the route, elders would trace the pug marks of *achangu*, the leopard cat. They said that the *achangu* never forgets the way and could thus be relied upon when the landscape was so changed by the snow. The snow would remain up until the first week of May. As a student returning to school in May, Delley and his classmates would find that there was still snow in many pockets of Mayudia. In contrast, these days, snowfall has declined considerably. There are no reports of road blockages due to rapid snowmelt in the mountain pass, since it snows for just seven to twelve days a year.

Similar scenarios were experienced by the people of Anini, another important town in the region. After nearly two decades without it, the town finally experienced snowfall in 2022. Delley's and other Idu Mishimi' testimony of decreasing snowfall over the last four decades corroborates the scientific assessment of climate change in the eastern Himalayas (Wester et al., 2019).

Changing Perspectives and Shifting Perceptions

The modern way of life has changed the Idu Mishmi animist belief system and related practices, affecting the status of revered and feared water bodies. With their beliefs shaken and tempered, many lakes that were taboo in the past have become tourist destinations. Emudu Trekkers, a group of young Idu Mishmi men, have been organizing treks in the area since 2020. Tourists are largely from outside Dibang Valley and Arunachal Pradesh. The organizers make sure that they do not make any loud noise near the lakes, which may disturb the spirits; however, there are concerns expressed in the villages that the increasing number of tourists may be harmful to the cultural, ecological, and spiritual significance of the place. Similarly, Mehao Lake and Sally Lake were, in earlier times, feared and revered as the home of the serpent spirit Beka. Accidents that happen in these sites have been attributed to Beka, and both lakes were considered taboo for the Idu Mishmi and other residents in the district. Even talking loudly was said to bring the wrath of the spirits and invite a heavy downpour. Yet, today these places have become popular tourist spots.

Rivers that were previously feared have been bridged and, at times, even diverted. The turbulent rivers, deep mystic lakes, and high spiritual mountains continue to be considered colossal entities that are now witnessing unchecked infrastructural development, raising concerns and anxieties among the people. One such project was the 3,000 MW Dibang Multipurpose Project, which was approved by the Indian Ministry of Environment and Forests in 2014. The latest is a major controversy over the Etalin Hydropower Project (3,097 MW), in Dibang Valley, which saw a heated contest between various groups in 2020–2021. Similarly, an environmental activist in Dibang Valley pressed charges against the National Highways and Infrastructure Development Corporation Limited (NHIDCL) stating that the construction and the demolition waste was not being disposed of properly and was polluting the river, in turn harming aquatic species and their habitat (Riba, 2021).

> Eighty-year-old Ethadu Dele from Elope Village, recounted:
>
> When road projects started from Roing to Hunli, we used to bet among ourselves that the company wouldn't be able to cross Mayodia Pass. But the *gonlos* (powerful mountain spirits) seem to be considerate to the *mecha*s (outsiders). Their dynamite blasting and machines seemed not to bother the *gonlos*. Similarly, we thought that building bridges by digging the spirit-possessed banks of Ithun and Talon wouldn't be possible. But today, the modern machines have overcome the wrath of these mighty river spirits.

The Idu Mishmi have witnessed a steady infringement upon their revered spiritual natural entities by modern machinery in recent times. In days to come, such entities risk being wiped out by development projects like dams and highways. While some of the Idu Mishmi welcome these megaprojects, seeing roads and electricity as essential in remote areas, others oppose these projects, fearing the loss of their lands, livelihoods, and indigenous heritage, both natural and cultural. Meanwhile, the new tourist economy has also raised concerns among the local residents about disturbances to their natural and spiritual homeland.

Conclusion

Centered on *eno*, the eco-spiritual worlds of the Idu Mishmi, we have narrated stories from the Dibang Valley and initiated discussions on indigenous experiences, eco-spiritual knowledge, and communal memories of their intimate relations with the animate, living earth as well as the precarious living conditions faced by humans and nonhumans in the eastern Himalaya. We find similarities between *eno* and the indigenous Greenlandic concept of *sila* as both weather and the spiritual essence upholding the physical world (Andersen in this book). Like *sila*, *eno* also unites the spiritual and the physical worlds by linking weather and place, water and human well-being, humans, and animals, and humans and plants. Both *sila* and *eno* embody indigenous ecological knowledge, climate histories, and climate change. We see *eno* as a series of eco-spiritual water climes comprising riverine, multispecies habitats. As spiritual spaces, *eno* are home to supernatural beings and local deities incarnated as mountains, rivers, streams, and lakes. Spiritual life is found not merely within the community's own cultural consciousness but is also externalized in the surrounding environment; thus, the physical world and human psychological world are deeply entwined by environmental flows and the earth-based pantheon as an integral part of human life. The documented ritual performances and folk stories aforementioned attest to the fact that the Idu Mishmi riverine/water climes have been both materially and spiritually nourishing for the well-being of the Idu Mishmi since ancient times. At the same time, they are undergoing modern anthropogenic changes. This chapter shows the people's spiritual, material, and affective relations mediated through place-embodied local deities and through ecological relations of humans with nonhuman beings. As their lives are transformed by development and climate change, the Idu Mishmi's cosmological meanings, and relations to water climes could contribute to the growing literature on environmental humanities intersecting ecology, religion and climate change.

Notes

1 We would like to acknowledge Firebird Foundation of Anthropological Research for granting the fellowship to document Idu Mishmi Shamanic rituals and Social Science Research Council Transregional Research Junior Scholar Fellowship under Andrew W. Mellon Foundation grant for fieldwork in Dibang Valley and Lower Dibang Valley.
2 The journey of the soul is an arduous one covering two hundred and sixty-seven such spiritual stations. These *mraba* are places of ordeals and challenges and also the places of rejuvenation and rewards. The ordeals, challenges, and rewards are determined by the deceased's past conduct in the human world.
3 A species of crane.
4 The blue whistling thrush.

References

Aiyadurai, Ambika. 2018. "Human-Animal Relations: A View from the Mishmi Hills." *Seminar,* 702 (2): 1–8.

Aiyadurai, Ambika. 2021. *Tigers are Our Brothers: Anthropology of Wildlife Conservation in Northeast India.* London: Oxford University Press.

Carey, Mark and Philip Garone. 2014. "Forum: Climate Change and Environmental History." *Environmental History,* 19 (2): 282–293.

Chaudhuri, Sarit. 2008. "Plight of the Igus: Notes on Shamanism among the Idu Mishmis of Arunachal Pradesh, India." *European Bulletin of Himalayan Research,* 32: 84–108.

de la Cadena, Marisol and Mario Blaser. 2018. *A World of Many Worlds.* Durham: Duke University Press.

Delley, Razzeko. 2018. *Idu Mishmi Shamanic Funeral Ritual Ya.* New Delhi: Bookwell.

Delley, Razzeko. 2021. *Igu: Study on Four Idu Mishmi Shamanic Rituals. Idu Mishmi Cultural and Literary Society. Arunachal Pradesh.* New Delhi: Bookwell.

Descola, Philippe. 2013. *Beyond Nature and Culture.* Chicago: University of Chicago Press. https://doi.org/10.7208/chicago/9780226145006.001.0001.

Escobar, Arturo. 2019. "Thinking-Feeling with the Earth: Territorial Struggles and the Ontological Dimensions of the Epistemologies of the South." In *Knowledges Born in the Struggle: Constructing the Epistemologies of the Global South,* edited by Boaventura de S Santos and Maria P Meneses (pp: 41–57). New York: Routledge.

Naveh, David and Nurit Bird-David. 2013. "Animism, Conservation and Immediacy." In *The Handbook of Contemporary Animism.* Chapter 2. Slough: Acumen Publishing.

Nijhawan, Sahil and Achili Mihu. 2020. "Relations of Blood: Hunting Taboos and Wildlife Conservation in the Idu Mishmi of Northeast India." *Journal Ethnobiology,* 40 (2): 149–166.

Pandit, K. Maharaj. 2017. *Life in the Himalaya: An Ecosystem at Risk.* London and Cambridge, MA: Harvard University Press.

Pant, G.B, Pradeep, P Kumar, Jayashree, V. Revadekar, and Narendra Singh. 2018. "Climate and Climate Change: An Overview." In *Climate Change in the Himalayas* (pp: 1–38). Cham: Springer Link.

Reid, Robert. 1942. *History of Frontier Areas Bordering on Assam from 1883–1941.* New Delhi: Eastern Publishing House.

Riba, Karyir. 2021. "NHIDCL Accused of Gross Negligence and Violation of Environment Laws." *The Arunachal Times.* 09 September, 2021.

Smyer Yü, Dan. 2020. "The Critical Zone as a Planetary Animist Sphere Etho-graphing an Affective Consciousness of the Earth." *Journal for the Study of Religion, Nature and Culture,* 14 (2): 271–290.

Tsing, Anna. 2013. "More-than-Human Sociality: A Call for Critical Description." In *Anthropology and Nature,* edited by Kirsten Hastrup (pp: 27–42). New York and London: Routledge.

Viveiros de Castro, Eduardo. 1998. "Cosmological Deixis and Amerindian Perspectivism." *The Journal of the Royal Anthropological Institute,* 4 (3): 469–88. https://doi.org/10.2307/3034157.

Wester, Philippus, Arabinda Mishra, Aditi Mukherji, and Arun Bhakta Shrestha. 2019. *The Hindu Kush Himalaya Assessment: Mountains, Climate Change, Sustainability and People.* Cham: Springer.

5

WATER CLIMING

A Cosmopolitical Ecology of Water in the Southern Peruvian Andes[1]

Malene K. Brandshaug

The landscape of Yanque, a farming district in the Colca Valley of the southern Peruvian Andes, consists of dwelling places, farmlands, high mountains, small lakes, and glacier-fed rivers. Yanqueños, the people from Yanque, are known for their care and respect for natural landforms and environmental flows that they recognize as sentient beings. Water is one such sentient being who takes the form of springs, lakes, and streams that are named persons with feelings, thoughts, and will. Important for animals, crops, and life-sustenance for small-scale farmers, these waters take part in creating a shared more-than-human habitat.

Building on the idea of clime as the embodiment of place and weather (Fleming 2014; Carey and Garone 2014), and on the conceptualizations of *climing* in the Peruvian Andes and the Himalaya by Paerregaard, Gamble, Bhutia, and Amrith and Smyer Yü, in this book, this chapter focuses on "water climing." The use of clime in verb-form in the case studies of my peer authors effectively captures the terrestrial understandings of climate change moving through the physical environment and compelling humans and nonhumans to take adaptive responses for self-preservation. Likewise, I seek to put forth an understanding of climing as more-than-human practices and give special attention to water as a lively agent of climing. In Yanque, water climing entails being attuned to subtle cues in the landscape of water's whereabouts and movement, it includes being attentive to weather patterns, and it involves communication and reciprocal relations between Yanqueños and particular sentient beings, namely "earth-beings" (de la Cadena 2015). Marisol de la Cadena (2015, xxiii) translates the Quechua word *tirakuna* (composed of the word for earth, *tierra*, in Spanish, and the Quechua word *kuna*, which indicates plurality) to "earth-beings." *Tirakuna*, in many places across the Andes, is used as a collective designation for the sentient beings who are manifested in the materiality of the landscape. Various kinds of landscape formations and water bodies are thus not merely physical forms but also individual,

DOI: 10.4324/9781003347026-8

living beings. Similarly, a range of supernatural and spiritual beings in the Himalaya and Tibetan Plateau embody themselves in physical landforms and water bodies, as described by Bhutia, and Delly and Aiyadurai, in this book. In this chapter, I use the term "earth-beings" to refer to the collective of other-than-human sentient beings in Yanque, such as Pachamama (earth-mother), various *apus* (mountain-beings), Tata Mallku (water-father), Yakumama (water-mother), and other smaller water-beings in the near landscape.

Hence, what I seek to add to the scientific descriptions of *climas*, the Spanish term for climate that dominates in Peru's water management regime, is a focus on emplaced engagements with weather and landscape—in particular, interactions between humans, earth-beings, and water-beings. This interplay is not merely ecological, material, or climatological, but is also of a social, affective, and political character. Thus, by describing water climing in Yanque, I suggest that climes can be sentient or animate, consisting of biotic and abiotic entities with a kind of vitality that resembles that of humans (Descola 2013; Ingold 2011; Kohn 2020; Rose 1996; Viveiros de Castro 1998).

Moreover, water climing in Yanque refers to a more-than-human negotiation to ensure the presence of water at the right time and place, thus enabling life in Yanque. Hence, I argue, water climing involves a "cosmopolitics" (de la Cadena 2010; Stengers 2013)—a more-than-human politics involving local power relations of humans and earth-beings, in which humans are expected to give offerings to and treat earth-beings with respect, and in which humans and waters engage in two-way relations where humans try to teach and guide the waters that have their own powers to flow as desired. Challenges with lack of water—what Yanqueños refers to as *No hay agua* (there is no water)—must therefore be understood as not only a result of less rain and less glacier melt-off due to climate change, but also a result of local environmental changes that are influenced by cosmopolitics of humans and earth-beings through and through—involving both state politics of water extraction and unjust distribution of water and less conventional politics overlooked by the state that are tied to local more-than-human negotiations in Yanque.

The chapter concludes with thoughts on what it could mean to live well, relationally and sustainably, in the Peruvian Andes in the Anthropocene. By coining the term "water climing," I call for understandings of climate and climes that include social and sentient aspects. In the case in question, this is done by making visible lively earth-beings and more-than-human affective relations in Andean climes. Moreover, inspired by thoughts on new environmental ethics for sustainable futures grounded in sentient relations to the earth (Bauman 2014; Kohn 2020; Smyer Yü 2021a, 2021b; Stengers 2013), I reflect on whether the recognition of the sentience of water and more-than-human water climing within Peruvian governmental water management could offer an alternative that paves the way for negotiated, fair, and sustainable co-living and co-becoming among human and nonhuman life communities that are confronted with extractive activities, climate change, and the subsequent lack of water.

FIGURE 5.1 Water stream and water-being, and mountain and *apu* (mountain-being).

Andean Climes

While the term climate often refers to weather patterns over a stretch of time, clime is used as its place-based counterpart. In contrast to the often abstracted notion of climate, a clime is embedded in the geography and ecology of a specific landscape. As emphasized by Fleming, the notion of clime points to the interrelation between person, landscape, place, and weather (Fleming 2014, 339). Hence, the concept not only refers to emplaced weather or "weather as regions," but is also linked to human experiences, imaginations, perceptions, and constructions of climate (Carey and Garone 2014, 294, 295).

Building on the abovementioned conceptualizations of clime, Paerregaard (this book) introduces the idea of *climing* to describe the ways in which Andean people imagine climate change, socialize with spiritual elements of the landscape, and engage with "climatically and ecologically diverse climes." Paerregaard's *climing*, then, invites us to explore the climatic agency of people in the environment and their perceptions and reactions to changing climes. My coinage of water climing is intended to pay attention to clime-making practices, as it underscores that climes, as weathered places, are not static or given but are in constant processes of "becoming" (Ingold 2011). Further, water climing opens up fresh perspectives for noticing the particularities of the Yanque water clime where water is a lively agent of climing. Based on my ethnographic work, this clime is also a sentient one where earth-beings in the forms of waters and mountains communicate and have affective relations with Yanqueños. The actions of both earth-beings and humans influence the movement, volume, form, and location of water. Such sentient, more-than-human water climing is not visible in the Peruvian state's understanding of *climas*. To elucidate my point, it is interesting to look at the differences and possible overlaps between the ways in which climate, climes, or *climas* are used within the governmental schemes and among Yanqueños in Peru.

In Spanish, the word "*clima*" is equivalent to the English words "climate" and "clime." When the Peruvian government through webpages, documents and public meetings mentions "*el clima*," "*cambio climático*" (climate change) and "*calentamiento global*" (global warming), these notions commonly refer to global weather patterns as conceptualized within climate science. The more regional "*climas*" or "*microclimas*," however, are more closely tied to ideas of "climes" understood as "weather as regions" (Carey and Garone 2014, 295) since they emphasize how climatic and geographic aspects work together in specific places (SENAMHI 2020). Nevertheless, missing from the governmental categorization of different "*climas*" in Peru are human persons and their experiences and imaginations, as fore-fronted in both Fleming's (2014) and Carey and Garone's (2014) descriptions of the term "clime." Also missing from the governmental categorization are other-than-human beings who are present in the life of Yanqueños and many other Peruvians, and who actively participate in climing.

For my interlocutors in Yanque, for instance, *clima* not only encompasses physical features of the landscape in which they live or weather as climatological phenomena, but also encompasses sentient beings who embody themselves in physical places such as mountains, springs, or lakes, or more generally in the earth (Pachamama) and water (Yakumama) of the cosmos. Hence, while the Spanish word *clima* invites us to think "climate" (global) and "clime" (local) together as "glocal" (Pálsson and Swanson 2016, 167), narrow understandings of *climas* still dominate the governmental spheres in Peru. Therefore, there is a need within these spheres to recognize the more-than-human water climing in Yanque and elsewhere in Peru. Paying attention to such vernacular, affective climing together with clime and climate as understood in climate science is generative of interdisciplinary understandings of climatic implications from practices, relations, and nonhuman beings that are normally overshadowed by dominant notions. Before I go on to describe water climing in Yanque, it is useful to look at the dominant narratives of Andean *climas*.

Within the governmental schemes in Peru, climate-science inspired notions of climate zones and micro-climates are normally used to describe variations in weather—particularly with reference to average temperature and precipitation—placing water at its center-stage. SENAMHI (2020)—Peru's National Service of Meteorology and Hydrology—classifies thirty-eight distinct *climas* in Peru, which are defined by the interplay between geographic and climatic factors in the coastal, Amazonian, and Andean regions. These three climatic zones are differentiated by altitude, temperature, humidity, and precipitation. *La costa* (the coast) is classified as arid and temperate with almost no rain, *la selva* (the rainforest) as moist and temperate with high rainfall, and *la sierra* (the mountains) as semiarid and chilly with a three-month rainy season and a nine-month dry season. Moreover, the Department of Agriculture and Irrigation Development even divides the Andean *sierra* into the categories *clima frio o boreal* (cold or boreal weather) of the mountain valleys, *clima frigido* (frigid weather) of the high tundra, and *clima de nieve* (snowy weather) of the highest altitudes toward the mountain tops (MIDAGRI 2015). Additionally, within these climatic zones, several even more distinctive *climas* are identified. Hence, the Andean vertically oriented landscape encompasses many *microclimas* in one region, which, combined with skillful irrigation, allow for cultivation of many types of plants—from fruits and grains, corn and fodder, to potatoes and root vegetables. In other mountainous landscapes, such as in the Himalayas, similar vertically defined microclimates exist, coupled with elevation specific biodiversity (Pandit 2017, 84, 100). In the Himalayas and the Andes alike, local knowledge of the climatic variations in the different altitudes are used to cultivate a wider range of produce.

The altitude of the Yanque territory, for instance, stretches from about 3,000 m at the bottom of the Colca Valley to snowy mountain peaks at almost 6,000 m. In the rainy season, snow accumulates on glaciated mountains and rain gathers in lakes, which together make up natural water reserves that provide water to the valley town and fields in the form of run-off and melt-off in the dry season. These tropical glaciers play an important hydrological role in the region, especially for irrigation. Moreover, Andean glaciers are, according to climate science studies, embedded in climatological and atmospheric processes regionally and globally. The shrinking of Andean glaciers is well documented, and studies show that "[t]his region experienced the largest cumulative specific mass losses since 1961" (Dussaillant et al. 2019, 802). Moreover, in Peru "air temperatures [and] the freezing level height (FLH) has risen over the last decades" (Schauwecker et al. 2017, abstract). This reduces the accumulation of snow and decreases the mass of the glaciers (Schauwecker et al. 2017). Moreover, there is a close relation between shrinking glaciers and decreasing snow surface albedo (Dussaillant et al. 2019; Schauwecker et al. 2017)—albedo being the ability of the earth's surface to reflect solar energy. Since snow is a particularly reflective surface, when there is less of it, more solar energy gets absorbed as the earth is directly exposed to sunlight. Hence, the glacier reduction not only affects the fresh water supply of millions of people living by the Andean cordillera, but the decrease in snow surface albedo contributes to global warming (Dussaillant et al. 2019). The same kind of albedo processes are found in the glaciated Himalayas, and "the net albedo of the Earth is showing a decreasing trend" (Pant et al. 2018, 43). Climate studies show, then, that glacier mass is not only sensitive to rising temperatures and rising freezing level height, but that mass reduction contributes to global warming as the glaciers' climatic cooling function decreases. Hence, a feed-back loop is created where warmer temperatures cause glaciers to melt, which in turn decreases the albedo and, thus, causes more warming.

According to studies of the Chila Mountain Range near Yanque, 98% of the glaciers have disappeared in the past fifty years (INAIGEM 2016). The mountain Mismi, for instance, used to be covered with a permanent glacier, but now only has small spots of ice left. This has severely diminished the water sources of the Colca Valley that are used for irrigation by small-scale farmers. When delving into the hydrological and ecological aspects of climate and climes, questions arise of how these *climas* are experienced, lived, and engaged with, and how they come into being in local places. Immediately pertinent to my case study, my question is how the Yanque "*climas*" are *climed* in the sense of more-than-human water climing.

Water climing depends on global climatological, hydrological, and ecological processes and local ecological, social, and political practices. Here, I focus on that which is locally embedded—water climing as life-enabling, more-than-human negotiations in Yanque—while recognizing that climes are "glocal" (Pálsson and Swanson 2016, 167), meaning that they are scalarly complex. In order to draw out the more-than-human aspects of water climing in Yanque, and the ways in which humans and nonhuman sentient beings are deeply entangled, I give special attention to the "aguasociality" of water (Brandshaug 2021). Aguasociality refers to the more-than-human relations that water makes as a result of its material properties as well as by being a living being with sentience and sociality. Hence, the concept puts forth a thicker notion of water that enables my aim of adding important social, affective, and political aspects to understandings of Peruvian "*climas*."

Ethnographic Descriptions of Water Climing in Yanque

On February 8, 2016, *lunes de carnaval* (Carnival Monday), I arrived at the small adobe house where Cintia and Hector (the *yaku alcalde* of Huarancante[2]) had prepared for the ceremonial start of the carnivals and the ritual *Mallku T'inkay* (water offering). Cintia served me a plate of chicken, rice, and potatoes, and asked me to sit down on a bench along the walls of the small room where several other friends, relatives, and *comuneros* (members of the community) were gathered in a circle. The room was decorated with flowers, and a table to my right was filled with quinoa, broad beans, and maize plants. When all had finished their plates of food, two men lit some incense in a small hanging ceramic bowl, blew three times on three coca leaves, and put them in the smoking mass. One of them started the round of ritual thanksgiving to the earth-beings to save and secure the agricultural production. First, he made three circular movements with the smoking incense in front of each of the bundles of plants on the table. Then he stepped outside and did the same in the directions of different mountains. The woman sitting next to me whispered to me that he "está agradiciendo al cerro, a los apus" (he is thanking the mountain, the mountain-beings). After him, all the men and then the women did the same. At the end, everyone was served large glasses of chicha (maize and barley brew), which was shared with earth-beings – such as *Tata Mallku* (water-father), *los apus* (the mountain-beings), and *Pachamama* (earth-mother) by sprinkling it on the ground and in the air in the direction where the earth-beings were located. The woman next to me explained that these beings needed the chicha in return for the water they gave to humans.

Then, most people turned to decorate a group of youths with the quinoa, broad beans, and maize plants from the table by tying them to the upper body in a *lliklla* (colorful woven cloth). As two flautists showed up and started playing, the procession toward the *chakra* (field) where the water ritual was, set off. A large Peruvian flag was waving in the front. Cintia took my hand and that of another woman, and we dance-walked with small steps to the rhythmic flute music. When we arrived at the right spot in-between many small plots of land, a table was set down for the plants and the Peruvian flag was planted in the earth on the other side of the narrow dirt road. The *yaku alcalde*, the president of the water-users commission and a couple of other male authorities sat down behind the table, facing Yanque Anansaya's primary *apu* and the flag. They were served *chicha* in big glasses and they offered *t'inkas* (libations) to the plants, *Pachamama*, and the *apus*, before they drank up. The rest of us stood or sat along a stone fence encircling one of the nearby *chakras*.

Meanwhile the ritual expert (*yana*) and his two helpers performed the *Mallku T'inkay* a bit further away. *Mallku* is the ritual denomination of water in Quechua, and *T'inkay* means to make an offering. They kindly asked for abundant water— from Tata Mallku (the great water-being) and Tata Huarancante (Anansaya's highest elevated mountain, a central water source and *apu*)—while offering gifts. The names of all the other important *apus* around Yanque were mentioned, as was Pachamama, who all were central to securing enough water for the crops and for life. Everyone present was drinking and offering alcoholic beverages to the earth-beings—asking for plentiful water.

When the call for water was over, the sun had gone down, the cold was sneaking in, and it started to rain. Still, the music continued, and everyone danced and drank long into the night.

(Author's fieldnotes, 8 February 2016)

FIGURE 5.2 Water ritual on Carnival Monday.

This ethnographic vignette from my fieldwork in 2016 touches on an important part of "water climing" in Yanque. The *Mallku T'inkay* is primarily directed toward the great water-being Tata Mallku and the more locally specific Tata Huarancante. It is one of many seasonal rituals performed to pay respect to and engage in relation with powerful earth-beings. That year the farmers had been struck by drought and waited for the rains that did not significantly set in until mid-February. Normally, *el tiempo de lluvias* (the rainy season) lasts at least for the three months from January through March. Irrigation—another important climing practice that is normally only necessary in the dry season—had to be organized in January and parts of February and March, normally the rainy months of the year. The altered irrigation pattern due to late rains could be read as changing seasonal water climing practices in relation to the actual experiences of *clima*. Such situations of changing *climas* and unpredictable water availability remind Yanqueños of the vitality of good communication and reciprocal relations with earth-beings who, to a certain degree, have powers to influence the Yanque *clima*.

Huarancante, mentioned above, and Mismi are two of the most important water sources for Yanque's two communities Anansaya and Urinsaya, respectively. In the semiarid Peruvian Andes, meltwater from the snow and ice on the two highest mountain tops are important water reserves utilized to make crops grow and to sustain people's lives as small-scale farmers. These are often referred to as Tata Huarancante and Tata Mismi; they are the two persons who watch over

Yanqueños and give them water in return for offering gifts, care, and respectful behavior, for instance, during carnivals as described above. It follows that Huarancante and Mismi are not only mountains, but also *apus* or *apukuna*—powerful "earth-beings" with their physical manifestation in the form of mountains (de la Cadena 2015, xxiii). Moreover, the water from Huarancante or Mismi is not only an element given to Yanque farmers by an *apu*, but also a living being itself. Hence, Yanqueños engage in personal relationships with related mountain-beings and water-beings who are manifested in hills, knolls, peaks, springs, streams and lakes, all of whom can be understood as being emplaced beings of the earth-mother, Pachamama (Allen 2002, 32, 33) and the greater water-beings Tata Mallku (the ritual name for water) or Yakumama (the common name for water). Through irrigation and canal maintenance and through ritual offerings, *pagos al agua* (payments to water), and ritual drinking with the earth-beings, Yanqueños engage in reciprocal relationships of respect and care with each other and with nonhuman sentient beings to make water flow. Consequently, Huarancante, Mismi, and human Yanqueños, together with other sentient beings, are part of the more-than-human collective of what I call water climing.

The water from Huarancante and Mismi—with the sources located on each side of the valley—flows through several long canals, each up to 25 km long, that previous generations of Yanqueños dug. The canals are supported by rocks, plants, and root-formations. Along the way, several smaller springs and streams, many of which are also named earth-beings, join the canals whose water makes up a substantial part of the water supply to Yanque. Part of water climing in Yanque also entails maintaining these canals through repair and harnessing as much water as possible from the higher altitudes to the valley. This requires humans not only to socialize with many earth-beings, but also to walk the high-Andean landscape closer to the mountain peaks, looking for springs, streams, or *bofedales* (wetlands). The location of water is always unpredictable, given the shifting courses of water streams and the occasional seismic movements of the earth that can close up existing springs and open new ones in unexpected places. Additionally, the presence, amount, and form of water are affected by the sun, rainfall, and temperature. Hence, this water clime is characterized by the emergence and disappearance of water—shifting from sudden riches to increased scarcity in the long run—which leads to persistent efforts to search for and to capture water. Sometimes reservoirs (*qochas*) are built in flat areas between surrounding peaks and knolls where rainwater and mountain streams gather. Then, water, through the making of small paths in the ground, is guided and led toward the larger canals that move it down the hills with the pull of the gravity.

However, water can still "escape" (*escapar*) from human containment. Although the canals are custom-made and adapted to the topography of the landscape, plants, sticks, dirt, and stones can hinder the water from flowing inside the canals. Water is then diverted and spreads out in other directions away from the canals. At other times, the earth is so dry that it soaks up all the water; the heat and strength of the sun makes it evaporate; or cold temperatures at night make it freeze at high altitudes. For farmers, water climing means being attuned to subtle cues from the earth in terms of water's whereabouts, speed, sound, decline or rise in a canal or an open field. It also requires attentiveness to the movement and shape of the clouds and the humidity of the air. Water moves with the terrain, soil, plants, and stones it encounters. Hence, water's travel route and the ways in which it runs depend on its relation to these material and ecological members of the earth. Yanqueños, on their side, gain an advantage from having relational knowledge of the moving water. Although water can act unexpectedly, the human practices of water climing are based on the accumulated knowledge of weather patterns and water ways. Besides, as water is not merely a physical entity but is also a sentient being with will and intention, humans respect and care for water as well as fear its powers and possible anger.

For instance, when farmers in Yanque irrigate their fields, they are aware of water as an active agent that responds to human actions and that can possibly behave according to its own motivation. In Yanqueños' relationship to water lies an incentive to use water with respect and to be attentive to its responsiveness. Yanqueños are compelled to maintain smooth relations with water by offering gifts to it, hoping that they will continue to be granted water for irrigation. Hence, water climing depends on the behaviors of water and on wider more-than-human relations. I thus argue that water climing in Yanque relies on negotiations between humans and the earth-beings embodied as mountains, glaciers, rivers, streams, and springs, who co-create the *clima* of Yanque, a shared more-than-human habitat. Such negotiations are ongoing as this collective water climing becomes more challenging due to the changes in weather, precipitation, and temperature that Yanqueños are experiencing. These negotiations thus remain important in efforts to ease the suffering of humans and nonhumans from lack of water.

A Cosmopolitical Ecology of "*No hay agua*"

A much-heard expression in Yanque and the Colca Valley is "No hay agua" (there is no water). One could say that lack of water is how the Anthropocene makes itself present to Yanqueños, and that *No hay agua* is characteristic of the *clima* in Yanque. *No hay agua* does not mean that there is no water at all, but that there is just not enough. In practice, it means that there is not enough water to sufficiently meet the needs of the crops, not enough for everyone who desires water, or not enough to grow particularly water-demanding plants. It can also mean that water is badly timed, that it appears in wrong places or in wrong forms—frozen because of low temperatures or rapidly evaporating because of high temperatures. Or in general, that there is too little water to make a proper living out of small-scale farming. This situation leads to suffering and quarrels, but also to more-than-human collaborations to ensure as much water as possible, as described above. Stories and traces in the landscape indicate a many centuries-long history of water scarcity in the region, shown in the millennia old intricate systems of irrigation ditches, stone canals, and reservoirs in an arid landscape. *No hay agua* is existential, something that has always been. As expressed by many Yanqueños: "Water has always made us suffer."

Yanqueños' experience of too little water in the area over the past decades is coupled with more unpredictable and extreme weather conditions such as hotter sun, intense heat, and irregular rains, rapid ice-melt, and severe droughts. To cope with these changes, they continue to engage in water climing, cultivate a general sense of attentiveness to elements in the landscape, and negotiate with earth-beings. In a sense, *clima* in Yanque is negotiable and open to alterations put forth by powerful earth-beings, whose actions Yanqueños can influence through strengthening reciprocal relations, communication, and interactions with them. As such, the cosmopolitics of water climing is synonymous with more-than-human politics. Thus, it is not as simple as saying that there is less water due to global warming, as the powerful explanatory models of the Peruvian state would have it, because the environmental processes are deeply embedded in politics. Therefore, it compels all participants to recognize that *clima* is concurrently related to a local more-than-human politics including earth-beings and influenced by conventional human politics—especially by prevailing state politics.

For instance, the Peruvian government initiates mega-infrastructure projects that move water from the Andes to the dry coast in the name of economic growth and development. Such modernization efforts, more recently called neoliberal development, disfavor Andean populations (Oré et al. 2009). Canal Majes, a large-scale concrete canal passing through the Colca Valley, is connected to the mega-irrigation projects Majes I and Majes-Siguas II. Both of these projects collect water from the highlands in large dams above the valley and transport great volumes to irrigate the coastal farmlands (Stensrud 2019; Ullberg 2019), "to make the desert green" as the slogan goes. Yanque and a few other communities receive a small percentage of water from the canal. They have to work hard for this water by smoothing relations with engineers and state representatives, fulfilling all kinds of bureaucratic obligations, paying water tax, and engaging in negotiations over water where they do not have much leverage (Brandshaug 2019). The people of the Colca Valley are marginalized in relation to coastal populations and, consequently, the volume of water they receive is never certain. Moreover, state water management follows the logic of progress and efficiency that builds on modern ideas that separate humans and nature and see humans as masters of nature. According to this idea, water is valued as a resource for human use, and not as a being engaged in affective relations with humans. It follows, then, that Peru's water governance regime does not give room to the skills, knowledge, practices, and values of large parts of the rural Andean population. Affective relationships between humans and sentient beings are not recognized as real. Thus, building on modern thought intrinsically linked to coloniality, Peruvian state water governance and its extractive projects exclude certain actors, and put pressure on water, the environment, human life, other-than-human life, and more-than-human collectives (Stensrud 2019). The state also contributes to creating the situation of *No hay agua*.

No hay agua is, to an extent, echoed in governmental water politics, but here it appears as "water scarcity," understood as a mainly natural phenomenon, although linked to anthropogenic climate change. Although climate is considered to be something that can change, it is still viewed as given and abstract, something that is linked to processes and actions far away and is therefore out of reach for Yanqueños and even the state. Hence, the dominating discourse of climate change and water scarcity points to these perceived natural phenomena as the cause behind an increasing lack of water in many parts of Peru, and not to the politics of extraction and unjust distribution of water within the governmental water management regime. What happens then is a naturalization and depoliticization of water scarcity (Brandshaug 2020, Chapter 2; Boelens and Seemann 2014). So, although knowledge of global climatic processes and their local ecological consequences are invaluable, there is a need to understand how they are intrinsically interwoven with political processes and practices—an important point within the field of political ecology (e.g., Bryant and Bailey 1997). My point then is that water climing is an entanglement of many processes and practices such as climate change, state politics, local practices, and more-than-human affective relations. The *clima* in Yanque must therefore be understood as a cosmopolitical process characterized by *No hay agua*. Decreasing water is not only a result of less rain and less melt-off from glaciers due to climate change, but also of local environmental changes that are influenced by cosmopolitics through and through—involving both state politics and less conventional cosmopolitics tied to more-than-human negotiations in Yanque. I argue that such relations must also be accounted for by scholars and governments alike.

Yanqueños find themselves in a precarious situation when confronted with the cosmopolitical ecology of the Anthropocene. The intimate more-than-human relations between humans, water bodies, water-beings, the earth, *apus* (mountain-beings), soil, plants, canals, and stones, are fragile and vulnerable to environmental change, extractive politics, (post)colonial relations, and marginalization in water governance. Just as nonhuman species are under the threat of extinction (Rose et al. 2017) and other-than-human beings are disappearing in the wake of extractive activities across Latin America (Blaser and de la Cadena 2018, 2), Mismi, Huarancante, and water-beings are endangered in Yanque. However, characterizing this situation as one qualified only by increasing precarity and vulnerability is not an exhaustive description of this *clima*. As my eighty-year-old friend Marcela points out, in Yanque it is more important than ever to strengthen the sentient relations with earth-beings and rely on their support for abundant water. Could we understand the sentient more-than-human relations as a form of wealth in terms of relationships and care? As the climate is changing, Yanqueños are becoming more explicitly aware of the significance of water climing. They collectively reflect more on their forms of water climing, and by doing so they become amplified and maybe also strengthened. The fact that Yanqueños never lost sight of affective more-than-human relationships is, perhaps, what gives them a sense of strength when meeting current and future challenges.

Likewise, Yanqueños' ability to understand that there are many ways of relating to water, affords vigor. They see how state engineers, scientists, or the government value water. Making use of existing opportunities to receive water from a state canal, they sometimes relate to water as a resource. At other times, however, they rely on Huarancante and Mismi, and relate to water as a sentient being. They stay open toward what the world will become and how one can relate to the changes in diverse ways for the sake of sustainable living. Similar to their Himalayan counterparts, Yanqueños' ecological knowledge and sentient relations with their land and water can be an intrinsic part of the "planetary common ground for building new environmental ethics for a sustainable future" (Smyer Yü 2021a, 3). A broader understanding of waters as sentient beings and the caring attentiveness to *apus* and water-beings as the basis of Yanqueños water climing can create room for more-than-human diversity in a way that the building of a cement canal does not offer. Thus, the recognition of more-than-human sentience and local ways of "water climing" by the water management regime of the Peruvian government can potentially offer alternatives to extractive economic activities that have contributed to the degradation of native landscapes and to increasing inequalities.

Future Co-Living

As the world is confronted with the climate crisis and the crisis of modernity (Blaser 2013) and the Anthropocene reminds those who had forgotten about the close entanglement of humans and nature (Latour 2013), the need for sustainable futures is more pressing than ever. Confronted with a continued destruction of landscapes, indigenous activists and environmental protection activists join forces in regional and global movements. In the academic world, scholars are looking to indigenous life projects (Blaser 2004, 26) and their environmental ethics as more sustainable alternatives to modernist development (Blaser and de la Cadena 2018; Kohn 2020 Rivera Andía and Ødegaard 2019; Smyer Yü 2021a), as they are encouraged to "begin to think about what types of futures we want to cocreate" (Bauman 2014, 3).

Dan Smyer Yü calls for a new animism that entails an interdisciplinary and holistic effort that aims toward "a thicker notion of the earth" (Smyer Yü 2020, 285) for building new environmental ethics in the Anthropocene. Similarly, Isabelle Stengers encourages us to think with the presence of Gaia—an earth that is alive, which could be a way toward "a new ecopolitics of life" (Green 2020, 72). Across the planet, there are many examples of ecologically-appreciating relationships of this kind, in which the liveliness of the earth itself and biotic and abiotic entities such as water, stones, plants, and animals, are sentient beings with a vitality akin to humans (Descola 2013; Kohn 2020; Rose 1996; Viveiros de Castro 1998). Especially in the Anthropocene, there is a need for a new "kind of ethics that is appropriate to a time in which separating humans from nonhumans is no longer practically or metaphysically conceivable" (Kohn 2020, 56). Thus, an important task of anthropologists in the Anthropocene is to make room for nonhuman persons who also live in the places we study, but who, in many instances, are not accounted for (Kohn 2020, 56).

Following this, the story of water climing in Yanque based on more-than-human affective bonds has the potential to make a valuable contribution to geological, hydrological, and climatological understandings of Andean *climas*. The recognition of such affective climing could have implications for how climate change is understood and how water is managed within the Peruvian water governance regime. Given the fact that peoples' affective relations to the earth and nonhuman beings are widespread across the Andes and Amazon, the Peruvian state is compelled to include such life ethics and ways of being in the world into its economic activities and environmental policies.

Although, recognizing the liveliness of water and the manifold more-than-human relations may not be "the one solution" to achieve more sustainable, future co-living across indigenous–nonindigenous and human–nonhuman divides, it could be a step toward an alternative to the current environmental destructive politics—one that recognizes the reality of many Peruvians and prevents the disappearance of other-than-human persons and more-than-human collectives. The fact that the logics of extractivism and the hegemonic narratives of climate change are deeply rooted in Peruvian state politics, underscores the need to narrate stories of water climing and other vernacular, sentient, more-than-human practices in the search for new livable futures.

Notes

1 The ethnographic data material used in this book chapter was gathered during fieldwork for my PHD in 2016. Some of the ethnographic details may therefore overlap with that of my PDH dissertation *Liquid landscapes: Human-water interactions and water scarcity in Yanque, Peru* (2020).

2 *Yaku alcalde* (water mayor), or *regidor*, is one of the community's water allocators who is chosen for a year at a time. Hector is that year's water mayor of the water source and mountain-being Huarancante, which means that he is responsible for allocating water from Huarancante between the community farmers every morning in the irrigation season. He is also in charge of arranging water rituals to the earth-beings who reside in the landscape surrounding the community Yanque Anansaya—one of which is performed the first Monday of the carnival celebration in February.

References

Allen, Catherine J. 2002. *The Hold Life Has: Coca and Cultural Identity in an Andean Community*. Washington, DC: Smithsonian Institution Press.

Bauman, Whitney. A. 2014. *Religion and Ecology: Developing a Planetary Ethic*. New York: Columbia University Press.

Blaser, Mario. 2004. "Life Projects: Indigenous Peoples' Agency and Development." In *In the Way of Development: Indigenous Peoples, Life Projects and Globalization*, edited by M. Blaser, H. A. Feit, and G. McRae, 26–44. London and New York: Zed Books.

Blaser, Mario. 2013. "Ontological Conflicts and the Stories of Peoples in Spite of Europe: Toward a Conversation on Political Ontology." *Current Anthropology* 54 (5): 547–568. https://doi.org/10.1086/672270.

Blaser, Mario, and Marisol de la Cadena. 2018. "Introduction: Pluriverse. Proposal for a World of Many Worlds." In *A World of Many Worlds*, edited by M. de la Cadena and M. Blaser, 1–22. Durham: Duke University Press.

Boelens, Rutgerd, and Miriam Seemann. 2014. "Forced Engagements: Water Security and Local Rights Formalization in Yanque, Colca Valley, Peru." *Human Organization* 73 (1): 1–12. https://doi.org/10.17730/humo.73.1.d44776822845k515.

Brandshaug, Malene. 2019. "Water as More than Commons or Commodity: Understanding Water Management Practices in Yanque, Peru." *Water Alternatives* 12 (2): 538–553. https://www.water-alternatives.org/index.php/alldoc/articles/vol12/v12issue3/541-a12-2-24/file.

Brandshaug, Malene. 2020. *Liquid Landscapes: Water Scarcity and Human–Water Interactions in Yanque, Peru*. PhD Dissertation, School of Global Studies, University of Gothenburg.

Brandshaug, Malene. 2021. "Water, Life, and Loss: Aguasociality and Environmental Change in the Peruvian Andes." *Kritisk etnografi – Swedish Journal of Anthropology* 4 (2): 51–66. https://uu.diva-portal.org/smash/get/diva2:1626741/FULLTEXT01.pdf.

Bryant, Raymond L., and Sinead Bailey. 1997. *Third World Political Ecology*. London: Routledge.

Carey, Mark, and Philip Garone 2014. "'Forum Introduction.' Climate Forum: Climate Change and Environmental History." *Environmental History* 19: 282–293.

de la Cadena, Marisol. 2010. "Indigenous Cosmopolitics in the Andes: Conceptual Reflections beyond 'Politics.'" *Cultural Anthropology* 25 (2): 334–370. https:/doi.org/10.1111/j.1548-1360.2010.01061.x.

de la Cadena, Marisol. 2015. *Earth Beings: Ecologies of Practice across Andean Worlds*. Durham: Duke University Press.

Descola, Philippe. 2013. *Beyond Nature and Culture*. Chicago: University of Chicago Press.

Dussaillant, Ines, Etienne Berthier, Fanny Brun, Mariano Masiokas, Romain Hugonnet, Vincent Favier, Antoine Rabatel, et al. 2019. "Two Decades of Glacier Mass Loss along the Andes." *Nature Geoscience* 12: 802–808. https:/doi.org/10.1038/s41561-019-0432–5.

Fleming, James R. 2014. "'Climate Physicians and Surgeons.' Climate Forum: Climate Change and Environmental History." *Environmental History* 19: 338–345.

Green, Leslie. 2020. *Rock, Water, Life. Ecology and Humanities for a Decolonial South Africa*. Durham: Duke University Press.

INAIGEM (Instituto Nacional de Investigación en Glaciares y Ecosistemas de Montaña). 2016. *Impacto del cambio climático en las cordilleras glaciares del Perú*. Peru: Ministerio del Ambiente.

Ingold, Tim. 2011. *Perception of the Environment: Essays on Livelihood, Dwelling and Skill*. London: Routledge.

Kohn, Eduardo. 2020. "Anthropology as Cosmic Diplomacy: Toward an Ecological Ethics for Times of Environmental Fragmentation." In *Living Earth Community: Multiple Ways of Being and Knowing*, edited by S. Mickey, M. E. Tucker, and J. Grim. Cambridge: Open Book Publishers. https://books.openbookpublishers.com/10.11647/obp.0186.pdf.

Latour, Bruno. 2013. *An Inquiry into Modes of Existence: An Anthropology of the Moderns*. Translated by Catherine Porter. Cambridge, MA: Harvard University Press.

Ministereo de Desarollo Agrario y Riego (MIDAGRI). 2015. "Clasificación de tipos de climas en el Perú." MIDAGRI website. Accessed January 6, 2022. https://www.midagri.gob.pe/portal/53-sector-agrario/el-clima/367-clasificacion-de-climas#:~:-text=Este%20tipo%20clim%C3%A1tico%20de%20la,anuales%20promedio%20de%20700%20mm.&text=anuales%20y%20temperaturas%20tambi%C3%A9n%20promedio%20anuales%20de%206%C2%B0C.

Oré, María Teresa, Laureano Castillo, Saskia Van Orsel, and Jeroen Vos. 2009. *El agua ante nuevos desafíos. Actores e iniciativas en Ecuador, Perú y Bolivia*. Lima: Oxfam Internacional.

Pálsson, Gísli, and Heather Swanson. 2016. "Down to Earth: Geosocialities and Geopolitics." *Environmental Humanities* 8 (2): 149–171. https://doi.org/10.1215/22011919-3664202.

Pandit, Maharaj K. 2017. *Life in the Himalaya: An Ecosystem at Risk*. Cambridge and London: Harvard University Press.

Pant, G. B, P. Pradeep Kumar, Jayashree V. Revadekar, and Narendra Singh. 2018. *Climate Change in the Himalayas*. Cham: Springer.

Rivera Andía, J. J., and Cecilia V. Ødegaard. 2019. "Introduction: Indigenous Peoples, Extractivism, and Turbulences in South America." In *Indigenous Life Projects and Extractivism: Ethnographies from South America*, edited by C. V. Ødegaard, and J. J. Rivera Andía, 1–50. Cham: Springer International Publishing.

Rose, Deborah Bird. 1996. *Nourishing Terrains: Australian Aboriginal Views of Landscape and Wilderness*. Australian Heritage Commission.

Rose, Deborah Bird, Thom van Dooren, and Matthew Chrulew. 2017. *Extinction Studies. Stories of Time, Death and Generations*. New York: Colombia University Press.

Servicio Nacional de Meteorología e Hidrología del Perú (SENAMHI). 2020. "Mapa Climático del Perú. Ministerio del Ambiente." Senamhi website. Accessed January 6, 2022. https://www.senamhi.gob.pe/?p=mapa-climatico-del-peru.

Schauwecker, Simone, Mario Rohrer, Christian Huggel, Jason Endries, Nilton Montoya, Raphael Neukom, Perry Baker, et al. 2017. "The Freezing Level in the Tropical Andes, Peru: An Indicator for Present and Future Glacier Extents." *JGR: Atmospheres* 122 (10). https://doi.org/10.1002/2016JD025943.

Smyer Yü, Dan. 2020. "The Critical Zone as a Planetary Animist Sphere: Ethnographing an Affective Consciousness of the Earth." *Journal for the Study of Religion, Nature and Culture* 14 (2): 271–290.

Smyer Yü, Dan. 2021a. "Situating Environmental Humanities in the New Himalayas: An Introduction." In *Environmental Humanities in the New Himalayas: Symbiotic Indigeneity, Commoning, Sustainability*, edited by D. Smyer Yü, and E. de Maaker. London: Routledge.

Smyer Yü, Dan. 2021b. "Symbiotic Indigeneity and Commoning in the Anthropogenic Himalayas." In *Environmental Humanities in the New Himalayas: Symbiotic Indigeneity, Commoning, Sustainability*, edited by D. Smyer Yü, and E. de Maaker. London: Routledge.

Stengers, Isabelle. 2013. "The Challenge of Animism." Unpublished Manuscript. John E. Sawyer Seminar: Indigenous Cosmopolitics: Dialogues about the Reconstruction of Worlds. University of California, Davis. http://sawyerseminar.ucdavis.edu/resources/.

Stensrud, Astrid B. 2019. "Water as Resource and Being: Water Extractivism and Life Projects in Peru." In *Indigenous Life Projects and Extractivism: Ethnographies from South America*, edited by C. V. Ødegaard, and J. J. Rivera Andía, 143–164. Cham: Springer International Publishing.

Ullberg, Susan B. 2019. "Making the Megaproject: Water Infrastructure and Hydrocracy at the Public-Private Interface in Peru." *Water Alternatives* 12 (2): 503–520.

Viveiros de Castro, Eduardo. 1998. "Cosmological Deixis and Amerindian Perspectivism." *The Journal of Royal Anthropological Institute* 4 (3): 469–488.

6

OFFERINGS FROM THE RIVERS TO THE MOUNTAINS

Mist and Fog as Connecting Life Force in the Sikkimese Himalayas[1]

Kalzang Dorjee Bhutia

In western Sikkim, trips to "the plains," and especially the city of Siliguri in North Bengal, are a regular part of life. Trips to the hospitals for medical treatment and the markets for purchasing food and building supplies are frequent and necessary for many Sikkimese families. On one such trip several years ago with members of my family, we ventured out from Gezing, the district headquarters, early on a winter morning as the sun rose, hoping to avoid traffic jams and to get to a doctor's appointment by 10am. As we drove down the winding road to Legship, the red sunlight of the early morning was obscured by increasing amounts of mist, rising up in tendrils from the fields around us. By the time we reached Legship, the mist had become fog, and the road was barely visible. We continued on slowly, surrounded by white fog, as my brother, who was driving, frequently sounded his horn to signal to cars coming in the opposite direction on the narrow road ahead that we were on the way. The fog slowed us down, but my father, who was already in his late eighties at that time, was not concerned. He told us all, "Don't hurry! This fog is a good sign for our journey. It is a sign that Kanchendzonga has accepted the offering of the water spirits and other beings who live in our sacred habitat." Kanchendzonga is the third highest mountain in the world, and to Sikkimese communities he is the mountain protector deity who presides over Sikkim and all the many beings—seen and unseen—that reside in the mountains and valleys. The propitiation of Kanchendzonga predates the entrance of Buddhism in Sikkim, and by the time Tibetan yogis were venturing into the region to propagate their Buddhist lineages and Tibetan migrants were arriving to settle in the region, he was well established as a powerful being in need of propitiation and supplication by the diverse cultural communities who lived in Sikkim. Later, Buddhist rituals worked with Indigenous Lepcha cosmologies and traditions to continue to pay respect to the mountain and his retinue (Balikci-Denjongpa 2001).

DOI: 10.4324/9781003347026-9

Part of the power attributed to Kanchendzonga by different cultural communities is his role as a life-giver for the region through his part in the creation of the climate and, especially, for his role in generating waterways that extend far beyond the cultural and political boundaries of Sikkim. This role predates human settlement and goes back through time to the ancient geological formations of the mountain and surrounding region. These formations gave rise to the glaciers of Kanchendzonga which feed the rivers that flow through the river valleys of Sikkim down to the plains to join with the ocean. As historian Sunil Amrith has written, the Himalayan mountain range shapes the Earth's climate, and "Asia's great rivers are a product of this geological history. They flow south and southeast, and they have shaped the landscape that is visible here: the force of rivers descending from the mountains eroded rock, creating the gorges and valleys. Over centuries rivers have carried silt and sediment from the mountains; they have deposited them along Asia's valleys and floodplains to sustain large human populations" (Amrith 2018, 2).

Human dependence on river ways can be seen vividly in Sikkim. Across different religious and ethnic communities, the significance of the rivers is expressed in interconnected forms of ritual propitiation and daily care. In this chapter, I will discuss the relationships between the animated interdimensional world of Sikkim's landscapes and how climate is understood by drawing on examples from my upbringing and later ethnographic fieldwork in west Sikkim. As Mark Carey and Philip Garone have pointed out, there is no one way to define climate in scholarship that engages with interactions between the weather and landscapes (Carey and Garone 2014, 284–5). In this chapter, I will engage with connections between the weather and specificities of the ecologies that make up the place of west Sikkim as inspired by anthropologist Karsten Paarregaard's exploration of the relationship between climbing, and climing, into different altitudes in the Andes, and the understanding of clime as connection with location (Paarregaard, this book). In the Sikkimese context, the climate is experienced differently between different altitudes, and these altitudes are acknowledged in local schemas for thinking about the interdimensional inhabitants of places. In ritual texts and oral recitations, the animated landscape is often divided into upper, middle, and lower sections, and this corresponds with the different climates found in Sikkim's lowlands and river valleys, foothills, and mountains, respectively. These divisions are not necessarily hierarchical, but instead are related to types of beings and their respective responsibilities. They also map onto two of the three different climes of the monsoonal Himalayas identified by Sunil Amrith and Dan Smyer Yü elsewhere in this book, where they discuss the "mountain clime" and "riverine clime" that are interlinked with the "oceanic clime" and the way water moves "between the atmosphere, the hydrosphere, and the lithosphere in different seasons" between the Himalayan mountains, the Bengal plains and the Indian Ocean (Amrith and Smyer Yü, this book).

I will lead the reader to clime with the water as it courses from the upper to the middle, and lower parts of the land that connect to the plains, and that forms into the vaporous states of fog and mist as a way to engage with Sikkimese Buddhist conceptions of interspecies health at different levels, or climes. Mist and fog are manifestations of beings known as *lu* (Tibetan: *klu*).[2] In Sikkimese Bhutia language, mist and fog are both known as *hompo*. Unlike in other Northeast Indian areas where there are many terms connected to rain, mist, fog, and other water-related weather conditions (Rahman 2020), *hompo* refers to a wide range of weather conditions. Despite this lack of specificity, the appearance of mist and fog is seen as an important indicator of the health of the environment. There are important ecological reasons for this significance. The foremost is the connection between fog, mist, and the climatic conditions that afford Sikkim's abundant forests and agricultural lands that historically led Sikkimese communities to consider their land as one of abundance and flourishing. For this reason, *hompo* is understood as an indicator of the strength of the life force of the interdimensional residents of the land and as an agent itself. The role of *hompo* as an other-than-human subjectivity for understanding interdimensional coexistence in Sikkim resonates with the other-than-human subjectivities discussed by Anders Burman in his research on Aymara daily practices in Bolivia, which include offerings to the *achachilas*, beings who govern the weather. These offerings help to maintain good relations between dimensions (Burman, this book).

In this chapter, I am focusing on west Sikkimese Buddhist traditions, but Sikkim is a diverse, multiethnic, and multireligious state, so this is only one perspective and it cannot be easily unentangled from other Indigenous traditions. Buddhism is over-represented in academic and popular representations of Sikkimese society, and especially environmental issues (Chettri 2017). Centering on west Sikkimese traditions in this chapter is not a way to valorize Buddhism over other religions or cosmologies in Sikkim, because many of these traditions are interconnected. This chapter has been inspired by my research within my own community, particularly around Pemayangtse in western Sikkim, where the demographically dominant ethnic groups are the Lepcha (or in their own language, Rong), Bhutia (or Lhopo), and Subba (or Yakthung or Limboo or Tsong). Not all of these communities are Buddhist, but many people who do not identify as Buddhist also participate in the ritual traditions outlined here. My intention in this chapter is to outline how Buddhists in this region incorporate their cosmologies and rituals into their understanding of how human flourishing depends on the climate, and how other beings—including the deities and spirits of the animated landscape—are also part of this interconnected system. The climate is understood as an expression of interspecies communication. As a starting point, I will outline who these beings are before exploring how Buddhist ritual life is one of the ways that Sikkimese human communities maintain their relationships with other interdimensional residents of the land. Then, I will turn to an examination of changes in these relationships and the ritual life that aims to re-establish good relations and interspecies flourishing in this era of rapid change and challenge in the eastern Himalayas.

Sikkim's Climate as a Form of Interdimensional Communication

Communicating with Kanchendzonga was part of daily life for my father, a lifelong practitioner of Sikkimese Buddhism from the Bhutia ethnic community, who grew up alongside Lepcha and Subba communities. He saw the rivers, lakes, and other water bodies of Sikkim as indicators of Kanchendzonga's health, and thereby, the well-being of the different interdimensional beings who lived in the state. The idea of the landscape as a sacred habitat (*gnas*) of many beings is also found elsewhere around the world. In their chapter in this book, Razekko Delley and Ambika Aiyadurai discuss the different water bodies in the Dibang Valley of Arunachal Pradesh as representations of the Idu Mishmi concept "*eno,* a plural noun for spiritual spaces, which encompasses both physical environments, along with its association with social, cultural, and psychological worlds." *Eno* acknowledges the enlivened nature of place in the Dibang Valley, thereby understanding the agency and impact of place on the well-being of interdimensional communities. In this chapter, the idea of sacred habitat and its different climes also works against a flattening of the agency of the different interdimensional residents of the landscape.

This is especially important when considering the agency of water. Anthropologist Malene Brandshaug has discussed how, in the district of Yanque in the southern Peruvian Andes, water is "often a medium for interactions between humans and mountain beings (Yanque: *apus*)" (Brandshaug 2019, 542, 543). In the Sikkimese context, water is also a medium of communication between the different deities and spirits of the place. Kanchendzonga is at the apex of a complex and rich set of interdimensional co-residents in Sikkim's landscape. Many of these co-residents are forces and spirits that predate Buddhism's entrance into the region. They became Buddhist protectors of the Dharma, known locally as the *yullha zhidak* (*chos skyong yul lha gzhi bdag*), when Guru Rinpoche visited Sikkim in the eighth century CE, thereby setting in motion Sikkim's Buddhist history. At that time, he proclaimed that the region was a hidden land (*sbas yul*) that would provide refuge to Buddhist communities in times of need, and his activities and interactions with local beings led the entire region to be considered a sacred habitat, where many powers and beings, seen and unseen, were resident (Lha Tsering 2002). Other non-Buddhist communities in Sikkim already understood themselves as part of an interdimensional animated landscape. Among Lepcha communities, Sikkim was Mayel Lyang, a paradise on earth (Lepcha 2013, 1, 2), with water also central to Lepcha life ways and community health (Lepcha and Lepcha 2021). Sikkim is thereby engaged with on multiple levels in entangled and overlapping forms of veneration.

Since mist and fog form from water vapor, my father understood them as interconnected with Sikkim's other waterways, and indeed as ways for the different beings resident in the landscape to communicate. As the mist and fog formed in the river valleys in the morning and settled into the valleys of west Sikkim, my father explained that they were offerings settling into the lap of the mountain. He saw them as equivalent to the water offering bowls on Buddhist household shrines

FIGURE 6.1 Clouds that appear as offering scarves to Kanchendzonga. Pelling, Sikkim, India, July 2022.
Photograph by the author.

offered by local Buddhists every morning. According to him, "this fog is the way that the river spirits and other beings make their offering to the mountain."

Fog and mist are part of daily life in every season in Sikkim. During the monsoon months, it is almost constant, but even in the clearer winter months, it appears in the mornings. It can be frustrating for its obscuration of light and for the general dampness that accompanies it, permeating clothing and concrete buildings walls. Anthropologist Saskia Abrahms-Kavunenko has examined how fog, mist, and pollution are understood as intimately interconnected with spiritual well-being—and also with its decline—in Mongolia (Abrahms-Kavunenko 2019). But its appearance is also important, because in Sikkimese Buddhist and connected Indigenous conceptions in Lepcha, Bhutia, and other cultural communities, the regular appearance of fog and mist is an indicator of water health, and thereby environmental health for humans, nonhumans, and other residents of the animated landscape. My father's understanding of its appearance as an offering from the land to Kanchendzonga is based on his perspective of the interconnectedness between different forms of health and his conception of water as the province of *lu*, a specific type of water deity. Because mist and fog are made from water vapor, Sikkimese Buddhists see them as a form of communication between the *lu* and other inhabitants of the land and as indicative of the life force of the land and as an other-than-human subjectivity.

Ritual life is the main way that Sikkimese residents communicate with other non-humans, and within local perspectives, nonhumans also use ritual to communicate with each other (as the water spirits use offering bowls or scarves that manifest as mist and fog for offerings to Kanchendzonga).

Lu as Protectors of Well-Being in Sikkim's Waterways

Although water is a pervasive force that connects the different elements and landscapes of Sikkim, according to the Buddhist cosmology, the different classes of beings that reside in, or are associated with, water are most often grouped into one category: *lu* (Sanskrit: *nāga*). These *lu* are resident in the sacred rivers that flow from Kanchendzonga (Dokhampa 2001), but also in all other watery landscapes and associated elements and weather forms, including fog and mist. As a prominent historical guidebook to Sikkim's animated landscape reported, "if it does not rain on time, this is a sign that the *lu* are agitated" (Author's translation, Tshe ring 2008, 343). *Lu* are a significant class of being far beyond Buddhism. Most often associated with reptiles (including dragons), snakes, and water, *lu* have been propitiated by many different religious and cultural groups, across Indigenous, Hindu, and Buddhist communities, throughout South, Southeast and East Asia for millennia.

In Buddhist cultures, the power of *lu* varies considerably between different locations. What unites them is that they are considered very powerful and need to be propitiated by humans, lest they bring sickness, poverty, or other forms of disaster to human communities. The association between snakes and protection in Buddhist traditions connects back to the life story of Shakyamuni Buddha, when Mucalinda, a large snake sheltered him during his forty-nine day practice that culminated in his awakening. That the snake gave him shelter instead of attacking him was interpreted as a manifestation of the snake's position as a *lu*, and recognition of the Buddha's realization. Since then, *lu* have been considered to be protectors of wealth and Buddhist teachings, but are also recognized as volatile creatures (Vargas-O'Bryan 2013).

Lu in the Himalayas are found across different religious communities. In a comprehensive study of *lu*-related religiosity in the Western Himalayas, Om Chanda Handa divided his study into sections that reveal different elements of the power of *lu*, including as guardians of caves and the weather (2004). In Buddhist traditions of the Himalayas, *lu* are attributed with powers across many areas. In Tibetan Buddhist and Bon texts, they are "believed to dwell in the underworld and embody water sources, stones and trees" and to be "active during the winter season" (Vargas-O'Bryan 2013, 113). Religious studies scholar Ivette Vargas-O'Bryan researched *lu* across several Tibetan and Himalayan Buddhist cultures and pointed out how they represent the coming together of different Indigenous and Buddhist elements and are concerned with maintaining order. *Lu* have long been held to be ambiguous entities, capable of acting as both protectors and supporters of the Dharma and as adversaries of Buddhist humans, as well as bringers of renewal, wealth, and fertility (Vargas-O'Bryan 2013, 115).

The ambiguous nature of *lu* and their intimate connections with well-being can be seen from their significance in Tibetan and Himalayan medical systems. They are particularly associated with skin and mental diseases and are often connected to specific sites as important figures in the foundation of significant sacred places, as well as connected to more local spirits. If humans behave without thinking at these sites, illness or misfortune may result, which demonstrates the association between *lu* and the appropriate moral order (Vargas-O'Bryan 2013, 115, 116). The care that Tibetan medicine-trained doctors exhibit in seeking protection from *lu*, ritual traditions that propitiate *lu*, and the role of *lu* in foundational texts of Tibetan medicine demonstrate the centrality of influence they have in Tibetan medicine (Vargas-O'Bryan 2013).

In Sikkim, *lu* is the most common class of powerful being associated with water, and local spirits can also be included in this category. Lu can appear in trees and other landscapes, but are most commonly associated with water bodies or damp places. Their widespread distribution is evidenced by the common sight of *lubum*, stupas dedicated to *lu* that are conspicuous due to the imagery of snakes on them. *Lubum* are frequently found beside rivers and streams and in damp forests. They are installed to balance energy and ward off snakes, who attack humans for trespassing in their territory and for not showing appropriate respect. People who live near *lubum* leave milk out as an offering and burn incense every morning to propitiate the local snake population.

The Misty Mountains as the Storehouse of Flourishing: The Upper Regions of an Animated Landscape

In her important research on Ladakh, anthropologist Karine Gagné has demonstrated how interspecies care is an essential element of daily life for Ladakhi Buddhists that is expressed through ritual, but also through folklore, agricultural traditions, and other elements of culture (Gagné 2018). Sikkimese Buddhist communities express their concern for *lu*, and for water more generally, by looking after waterways that take the forms of streams, rivers, lakes, and glaciers, by keeping them clean and free from different types of pollution. These waterways are home to multiple species, but at their center are the *lu*, who help balance these environments and must be acknowledged carefully.

Part of this acknowledgement is expressed in a regular and important ritual, *Nesol* (Tib: *Gnas gsol, The Propitiation to the Sacred Habitat of the Valley of Rice*), that venerates Sikkim's sacred habitat, in which the significance of water is invoked repeatedly. The ritual is directed toward Kanchendzonga, the primary guardian deity who presides over Sikkim, and whose refers to the five storehouses of wealth. Kanchendzonga's wealth should not be understood as literally commensurate with conceptions of the mountains as resource frontiers. Instead, this flourishing is understood to come about through living together and honoring the landscape. After Kanchendzonga is invoked in the *Nesol* ritual, all of the other interdimensional residents are then listed along with their abodes which are cared for as sacred places. The *Nesol* outlines all of the waterways, and the propitiants' request for the deities to,

> ...let the rain fall on time; let there be a good harvest; may all war, bad diseases, and famine for humans and animals be warded off; then may there be long life and peace and prosperity all over the world! (Author's translation, from *The Propitiation Rite for the Sacred Habitat of the Valley of Rice ('Bras ljongs gnas gsol)* [Lhatsun n.d., 55–56]).

Securing the regular arrival of rain is one of a list of requests that show the conditions required for humans and other beings to flourish in Sikkim. The Nesol is undertaken frequently throughout the year, on the fifteenth and the twenty-fifth of the month in Sikkim's monasteries, temples, and in private residences.

The Nesol plays a central part in one of the most important rituals of the year, the Pang Lhabsol (*Dpang lha gsol*) festival. Pang Lhabsol takes place during the twelfth and fifteenth days of the seventh month of the lunar calendar (Bhutia 2021a, 280, 281). As part of this festival, lamas, ritual propitiants, and sponsors who have been dispatched from monasteries give offerings of burned incense and prayer in the region below the upper reaches of Kanchendzonga, and especially at Rathong Glacier, the source of the Rathong River in western Sikkim. This expression of care takes place at Putsokarpo (*Pu tsho dkar po*), the White Lake, that is home to some of the uppermost guardian deities in Kanchendzonga's retinue. This annual enactment has in recent years been coopted by the state as an example of nature worship (Vandenhelsken 2011). However, for Buddhist communities, it is instead understood as a reinforcement of their interdependent relationship with the mountains. The opening of the mist around the mountains during Pang Lhabsol is interpreted as a favorable sign for community well-being in the coming year.

Human and Nonhuman Habitations in the Hills: The Middle Regions of an Animated Landscape

In recent years, Rathong Glacier, one of the major glaciers that feed the mighty Rangit River, has receded. Local communities are seeing directly the deleterious impacts of climate change (especially changes in when the monsoon arrives and departs), deforestation, and other forms of unplanned development in their daily lives. For my father and other community members, ecological change was vividly obvious, and not tied to nostalgia or romanticism for the past; instead it was made obvious by their own awareness of how things had changed so dramatically—from changing monsoon timing, to new forms of crop blight, to new health problems for humans and nonhuman animals.

Water access and sanitation have been consequential issues in the expansion of Sikkim's urban centers since the opening of the state to tourism after absorption into India in 1975, and especially since the 1980s. In west Sikkim, Pelling is a tourist center that acts as a salient conduit for understanding the ramifications of unplanned development and growth in the hilly terrain in the lap of Kanchendzonga. Until the late 1980s, Pelling was a small town in the forest on the ridge above Hungri River, close to Pemayangtse monastery and Sangha Choeling retreat center. Within thirty years, hundreds of heavy concrete buildings have appeared along the ridgeline and have began to expand into nearby villages.

Pelling was historically significant due to its association with the nearby Buddhist practice sites perched above it. By the late 2010s, the most famous of these sites had developed into tourist sites. Sangha Choeling was transformed from a remote site of solitude to a colorful tourist destination adjacent to a giant, 135-foot statue of Chenrezig, one of Sikkim's megastatues where pilgrims can pray and tourists can take impressive selfies and mountain photos (and frequently, these groups overlap). Even more popular than the statue is the Skywalk, an impressive glass bridge beside the statue. Pemayangtse is known for its grand shrine rooms and masked dances. While these institutions have long been important, other smaller sites have faded, and the sacred landscape of Pelling is not acknowledged in tourist literature. In the guidebooks of the Tertons, the hill that Pelling is perched on is a *lu*, represented with the body of an elephant (*glang po rin po che*). On the very top of the head of this elephant, down the ridge from Sangha Choeling, there is a throne that locals believe Guru Rinpoche sat in during his visit to the area, and where the great Terton Pema Lingpa (*Pad ma gling pa,* 1450–1521) visited from Bhutan—hence the name Pelling, a shortened version of Pe(ma)Ling(pa).

During the British colonial period, the British installed dak bungalows with a day's walk between them to provide forest officers and other official representatives with places to stay when they ventured into the region for administration and recreation, especially outdoor activities such as mountaineering, hunting, and fishing.[3] After 1975, tourism in west Sikkim was slow to develop, since the mountainous terrain of the region posed challenges to the installation of the mass infrastructure necessary for the large numbers of Indian domestic tourists who began to arrive, seeking mountain vistas and fresh air. To cater to this new market, many Bhutia, Subba, Gurung, Pradhan, and Karki families resident in Pelling began to build large concrete, multi-storied hotels using bank loans supplemented by investments from Bengali hoteliers who used their connections in Kolkata to market their hotels and ensure they were regularly filled to capacity during Sikkim's drier seasons. The hotels were an obvious cosmetic change in Pelling; but along with them, septic tanks, roads, water lines, and electricity lines all had to be installed. This meant more intervention, and the rerouting of waterways that flowed from the head of the elephant down to Hungri Stream. During the building of Pelling, there were a number of significant landslides that hampered construction efforts. These have been attributed to the dissatisfaction of the elephant *lu*.

A vivid example of this took place in the early 1990s, when heavy rain led to a huge landslide that scarred the earth all the way down to Hungri Stream. Local

Buddhist residents did not understand this event to be a natural disaster. Instead, they approached Yangthang Rinpoche (*G.yang thang rin po che*, 1930–2016), a respected Buddhist teacher, for information on what may have caused this event. Yangthang Rinpoche warned them they had had agitated the *lu* and its retinue of protectors through the blasting, noise, pollution, and rubbish that large-scale tourism brought to Pelling. In particular, the streams of Pelling had been sullied with fast food packets, and septic tanks installed without any care as to where the ritual pollution associated with toilets would go. The landslide, and the destruction of property that took place because of it, was a signal that they needed to take care to acknowledge their unseen neighbors.

Yangthang Rinpoche prescribed offering *bumpas*—vases filled with precious consecrated substances—to the *lu* and other *yullha zhidak* (Dharma protectors). These vases were consecrated and then buried along the ridgeline, below the hotels and above Hungri Stream, to stabilize the earth and the energies of Pelling that were disrupted by unplanned "development."

The *lu* of Hungri Stream has continued to sporadically express its dissatisfaction with human activities in the area. On the hill above Pelling there is sportsground and helipad that is surrounded by hotels. The helipad was only installed at the turn of the twenty-first century. Before, the area had been densely forested and had hosted many meditation huts where a number of respected Buddhist teachers resided. These teachers had worked with the lamas and practitioners at Pemayangtse and Sangha Choeling to carry out Nesol and other rituals that acknowledged the elephant and other powerful *yullha zhidak* of the area. Practitioners had built stupa and *mani* walls throughout the forest, and villagers in Pelling would gather wild garlic that was held to have powerful medicinal properties there.

When the forest was cleared and the ground flattened to make way for the helipad and hotels, many *yullha zhidak* were displaced and religious objects, such as *mani* walls, were broken. Although the uppermost part of the hill was not damaged, the tusk of the elephant was, and as a result the mighty elephant *lu* became agitated. Although local residents and lamas were careful to propitiate these beings with consecrations and *Nesol* rituals, the helipad represented the culmination of a slowly building process of intervention in Pelling's landscape that had begun with the building of the first concrete hotels and had slowly ramped up through the 1990s. Hotels were reliant on the abundance of available water; but in 2006, Hungri Stream dried up. This was interpreted by Yangthang Rinpoche as a sign that the elephant *lu* was angry because of the installation of the helipad, and also because of the rubbish that had been disposed of in areas around the stream below Pelling. Yangthang Rinpoche once again prescribed *bumpa* in order to bring balance to the *yullha zhidak*, and especially to the elephant.

In the 2010s, the affordability of airline travel led tourist numbers in Sikkim to rise far beyond previous decades. During the tourist season in Pelling in the years leading up to the 2020 pandemic, there were regularly vivid demonstrations of the need for further infrastructure: enormous traffic jams became normal in the morning and evenings as tourists were picked up and dropped off at their hotels down the sides of Pelling's winding road; and water shortages also became more frequent. Residents frequently had to call in deliveries of water trucks. The drying up of water sources and the occurrence of landslides, floods, and other water-related issues continue to be recognized as the providence of *lu*. These events are all the more strange given the abundance of rain during monsoon months, but since the monsoon is no longer regular, rain water cannot be relied on for regular use throughout the year. Lamas and other powerful religious figures act as intermediaries, prescribing a variety of religious activity to appease the *lu*. It is noteworthy that the ongoing challenges of building heavy concrete structures on erosion-prone soil that house tourists who use abundant amounts of water during the tourist season has not led to broader discussions about sustainable tourist infrastructure. Instead, concrete structures continue to be built and are even appearing in villages near Pelling; and as McDuie-Ra and Chettri have noted, concrete continues to hold positive associations with modernity, development and progress in the hills, even though Sikkim's heavy monsoon leads to mold and seepage in concrete buildings (McDuie-Ra and Chettri 2020). The long-term impact of the increase in concrete-heavy, water-thirsty tourist infrastructure is yet to be seen. The *lu* will surely be watching, particularly since, as guardians of wealth and flourishing, travel providers, hotel managers, drivers, cooks, and other people in the tourist industry rely on their blessings and support.

Foggy Forests as the Source of Abundance: The Lower Regions of an Animated Landscape

An issue connected to the tourist industry that has emerged in recent years has been heavy fog over the newly constructed Pakyong Airport near Gangtok. Local experts have pointed out that this fog may be due to the damming of the Teesta River (Dhungel 2021), which acts as a vivid example of how fog manifests as a sign of the health of Sikkim's waterways and interdimensional prosperity.

Changing fog patterns due to changes along the river are found throughout Sikkim and have other impacts on prosperity beyond affecting the tourist industry. For decades, my family has planted cardamom. The best land for cardamom plantations in west Sikkim is close to the river, due to the mist and fog that naturally form on winter mornings that lead to ideal climate conditions for the growth of cardamom. Cardamom was historically always planted during the rainy monsoon season between May and July, and it took two to three years to be ready for harvesting. It was planted in the forest, a climatic ecosystem in Sikkim that is deeply connected to inter-dimensional flourishing.

This flourishing was due to the abundance of food available in the vast forests of the state. In the guidebooks left by Guru Rinpoche and revealed by Tibetan treasure-revealers (*gter ston*) connected to the Guru between the fourteenth and twentieth centuries, there are frequent discussions of Sikkim's deep forests, including trees, streams, waterfalls, and plants, and their abundance of wild plants that could be used to sustain human communities. Bhutanese treasure-revealer Pema Lingpa's guidebook provides vivid illustrations of the food that is magically available in the jungles of Sikkim that can sustain human life without being cultivated. In it, Guru Rinpoche outlines how

> There are about 155 varieties of fruits with different tastes and nutritional values. [These include] a walnut that tastes like butter; a fruit known as *wallay* … and a grape with the taste of wine. There are fruits called *tingding* with the taste of meat, and *sedey*, which can be eaten as the equivalent of an entire meal; turnips, and thirty-seven other types of root vegetables are available.[4] There are twenty different varieties of garlic. Altogether, among the edible plants, there are 360 varieties available. There are wild radishes, along with *tsolay*, *nyolay*, grapes in the valley. In the trees, among the rocks and hanging from the cliffs there are beehives.
>
> (Author's translation, *Tshe ring 2008, 250*)

In 2019, the "State of the Forest Report" compiled by the national Forest Survey of India in Dehradun reported that forest covers 76.02% of the state of Sikkim.[5] Due to its mountainous topography, Sikkim has several types of ecosystems that are affected by differing amounts of rainfall, temperature, humidity and altitude. These include "sub-Himalayan wet mixed forests, oak-fir forests, moist alpine shrubs and dry alpine shrubs," and an estimated 717 different species of trees (Dash and Singh 2011, 91). According to oral tradition, before deforestation started in the service of infrastructure development in the nineteenth century, the forests were even more diverse and abundant. As abundant as forests were, in Indigenous and Buddhist traditions they were not simply there for resource extraction. They were believed to be home to many *yullha zhidak* who served in Kanchendzonga's retinue to preside over the Hidden Land. Elsewhere in Pema Lingpa's guidebook to Sikkim, there are stern instructions on what appropriate decorum humans must follow when entering into the misty forest. In a conversation between Guru Rinpoche and the royal court, the Guru instructed the king and queens on how to behave if they went to visit Sikkim. He said,

> When you are going from the northern door to the Hidden Land [Sikkim], on the way, continuously maintain your good intention and pray to the Three Jewels. Offer feasts to the Dakinis and oblation to the *yullha zhidak*. Go slowly and quietly, follow the advice of your lamas and recite prayers to Ushishvaijaya [*Gtsug gtor rnam 'joms*]. Plant arrows of long life on the dangerous paths. When you come towards streams, look carefully [at the path] and be careful to move slowly and quietly. When you reach rocky cliffs and [are] in the deep jungle, take care to examine the [safest] path to move forward. Whenever you meet yetis [*Mi rgod*] or other powerful beings, do not panic or be afraid; do not lose your concentration. Avoid burning strong smelling things like meat and disposing of waste [in

> incorrect places]. Always purify yourself and take care to clean up places you have passed through...Make a pledge to [do this] all the gods of the place with oblation.
>
> (Author's translation, *Tshe ring 2008, 261*)

Guru Rinpoche's recommendation stemmed from a central function of rituals for Buddhists in the Himalayas: rituals functioned as technologies for interdimensional communication. Preparing and offering food and arrows and making prayers on the path were ways that pilgrims arriving in the Hidden Land could acknowledge the residents, such as yeti, already present there, and begin to establish a relationship with them.

Similar traditions were present for farmers who resided in Sikkim. From a very early period, local communities engaged in settled agricultural practices. When Buddhism became adopted by local communities after being imported by Tibetan migrants after the thirteenth century CE, ritual traditions also appeared to acknowledge the local *yullha zhidak* in areas that humans settled and cultivated the land. Many of these rituals are part of daily life, and include burnt food offerings in the morning and evening, burial rituals for domestic animals and the support of annual rituals in people's homes, including Nesol performances. These ritual traditions allow for agriculture to flourish, including crops of both local and imported vegetables.

The abundance of local food sources include items that are now part of global commodity chains. In the nineteenth century, the British annexed Darjeeling, part of Sikkim, in order to have good land for cultivating tea. Part of what made it so appropriate for tea cultivation was the climactic conditions, including the mist, fog, and rain. As with Assam, this led the Darjeeling region to be configured in imperial imagination as part of what historian of the Himalayas Jayeeta Sharma has called the "empire's garden" (Sharma 2011). This period also saw the introduction of new crops that could be grown successfully, and cardamom, a native plant, began to be grown on a mass scale. Cardamom has become Sikkim's predominant cash crop, as it grows well due to climatic conditions. Local farmers attribute the success of the crop to the abundant mist and fog.

However, since the early twenty-first century, cardamom has been struck by a blight caused by the fungus *Colletotrichum gloeosporioides* (Gurung et al. 2020). Many attribute this blight to changes in the soil's temperature and climate change, also evidenced by the lack of snow on Kanchendzonga, as the glaciers have visibly and rapidly receded in the last five decades. Additionally, the fog that was once so much part of the forest is now less abundant due to the control of the river by dams. In the last two decades, the sacred status of Sikkim's waterways has made global news, as local communities, especially in western and northern Sikkim, have struggled with the state and global companies over hydroelectric dams that have been installed throughout the region. For these communities—which span ethnic and linguistic groups in the diverse state—hydroelectricity projects pose an existential threat, as, in local Indigenous Lepcha, Buddhist, and Hindu cosmologies, the health of the water is intertwined with the health of the human and nonhuman communities who depend on that water for their well-being; the importance of healthy waterways is shown by the strength of the movements that have developed to protests the dams. While these movements are sometimes positioned as ethnic and/or political movements, their common goals—of calling for sustainable futures in the region without rapacious development and environmental destruction—and their acknowledgement of the important of water for interspecies health are common (Gergan 2017, 2020; Lepcha 2020; Arora 2009; Balikci 2008; Chettri 2017).

These activist movements are enriched by local ways of knowing. Ritual life continues to facilitate interspecies communication. In a recent publication, anthropologist Charisma Lepcha noted how the fluctuations in the timing and intensity of the monsoon have affected Buddhist ritual life, specifically in relation to the Bumkor rituals, where books from the Buddhist canon are taken out and carried around local communities to bring positive impacts to the land (Lepcha 2021). As religious-studies scholar Amy Holmes-Tagchungdarpa has pointed out, the Bumkor takes place in the spring precisely because of the favorable climate that allows for the books to be aired; and it is thereby recognized as a beneficial ritual for many different human and nonhuman participants (Holmes-Tagchungdarpa 2021). The changes to the monsoon have also influenced the fog, mist, and other elements of the climate. These changes are interpreted as fluctuations in the life force of the sacred habitat. However, as long as the mist and fog continue to rise from the river valleys to sit in the lap of Kanchendzonga, all is not lost: instead, these different ways of thinking about flourishing together across dimensions can continue to provide meaning and guidance to communities, and inspiration for activism.

Conclusion

In contemporary Sikkim, the continuity of older ritual forms that allow for communication between humans and the beings residing in the world around them has heightened significance due to local beliefs that hold that part of the reason for the decline of cardamom, the melting of glaciers, and the increase in landslides, earthquakes, and other forms of ecological disturbance is due to human immorality (Gagné 2018; Gergan 2017; Bhutia 2021b). The *yullha zhidak* are concerned, angry, disturbed, and even made ill by human activity, and according to some, they have left Sikkim entirely as a result of being displaced. Changing patterns of fog and mist, which are understood as forms of communication between the deities and spirits, and also between humans and the deities, are also seen as another indicator of systemic health issues. All of these ecological events are understood as a sign that the world is out of balance.

This imbalance is material and spiritual, and the loss of protector deities, as well as their agitation and illness, is of great concern in contemporary Sikkim since without them, flourishing in the Hidden Land is not possible. Ritual traditions are adapting to continue to provide different species with meaning during these uncertain times; but it remains to be seen how rituals can adapt in these climes that are experiencing so many different strains and stresses from human- and beyond-human, forces. The appearance of mist and fog, such an important element of everyday life in Sikkim, is an important marker of the balance of health and the stability of the life force present for the ecosystems in the region. Instead of being understood as an obscuration of light, their appearance is more important than ever as a form of communication between dimensions. Here, climes occurring between different altitudes, and the climatic conditions of these altitudes, can be understood as indicators of well-being. Attentiveness to Sikkimese understandings of fog and mist can be extended to other locales, as changes in climatic conditions that are so common in daily life must be taken seriously. Even if Sikkim's specific vaporous language cannot be exactly understood in other inter-dimensional global settings such as the Andes and the Arctic, listening for similar messages, and being attentive to similar forms of connections, across other watery landscapes is essential for all sentient beings to flourish.

Notes

1 I offer my respect to the many human and nonhuman communities who have enriched this research, especially in the Pemayangtse region of west Sikkim, and especially thank my late father His Eminence Pemayangtse Dorje Lopen Chewang Rinzin, my mother Sonam Ongmu, and my brother Palchen Dorjee for their insights into *hompo*. This research was funded by an ACLS/Robert Ho Research Fellowship.
2 All transliterations of non-English words are from Classical Tibetan language unless otherwise indicated.
3 Dak bungalows were installed by British administrative officers as post offices and resting places for British representatives, including those employed by the postal service, in different areas under British rule in South Asia.
4 This could be local avocado, which is considered to be a meat substitute in Bhutia cuisine.
5 https://fsi.nic.in/isfr19/vol2/isfr-2019-vol-ii-sikkim.pdf

References

Abrahms-Kavunenko, Saskia. 2019. *Enlightenment and the Gasping City*. Ithaca: Cornell University Press.
Amrith, Sunil. 2018. *Unruly Waters*. New York: Hachette.
Arora, Vibha. 2009. "They are all Set to Dam(n) our Future: Contested Development Through Hydel Power in Democratic Sikkim." *Sociological Bulletin* 58 (1): 84–114.
Balikci, Anna. 2008. *Lamas, Shamans and Ancestors: Village Religion in Sikkim*. Leiden: Brill.
Balikci-Denjongpa, Anna. 2001. "Kangchendzönga: Secular and Buddhist Perceptions of the Mountain Deity of Sikkim among the Lhopos." *Bulletin of Tibetology* 38 (2): 5–37.
Bhutia, Kalzang Dorjee. 2021a. "Living with the Mountain: Mountain Propitation Rituals in the Making of Human-Environmental Ethics in Sikkim." *Journal of Buddhist Ethics* 28: 261–294.
Bhutia, Kalzang Dorjee. 2021b. "Purifying Multispecies Relations in the Valley of Abundance: The *Riwo Sangchö* Ritual as Environmental History and Ethics in

Sikkim." *MAVCOR* 5 (2). https://mavcor.yale.edu/mavcor-journal/essays/purifying-multispecies-relations-valley-abundance-riwo-sangch-ritual
Brandshaug, Malene K. 2019. "Water as More than Commons or Commodity: Understanding Water Management Practices in Yanque, Peru." *Water Alternatives* 12 (2): 538–553.
Carey, Mark and Philip Garone. 2014. "Forum Introduction: Climate Change and Environmental History." *Environmental History* 19: 282–293.
Chettri, Mona. 2017. "Ethnic Environmentalism in the Eastern Himalaya." *Economic and Political Weekly* 52 (46): 34–40.
Dash, S.S. and P. Singh. 2011. "Trees of Sikkim." In *Biodiversity of Sikkim – Exploring and Conserving a Global Hotspot,* edited by M.L. Arrarwatia and Sandeep Tambe, 89–124. Gangtok: Information and Public Relations Department, Government of Sikkim.
Dhungel, Pankaj. 2021. "Sikkim Flight Cancellation: Is Fog from Dam-locked Teesta the Reasons?" *EastMojo* October 12. Accessed January 2, 2022. https://www.eastmojo.com/news/2021/10/12/sikkim-flight-cancellation-is-fog-from-dam-locked-teesta-the-reason/
Dokhampa, Rigzin Ngodup. 2001. "Origins of the Bumchu of Dakkar Tashiding (*Brag dkar bkra shis sdings*." *Bulletin of Tibetology* 38 (1): 25–30.
Gagné, Karine. 2018. *Caring for Glaciers*. Seattle: University of Washington Press.
Gergan, Mabel. 2017. "Living with Earthquakes and Angry Deities at the Himalayan Borderlands." *Annals of the American Association of Geographers* 107 (2): 490–498.
Gergan, Mabel. 2020. "Disastrous Hydropower, Uneven Regional Development, and Decolonization in India's Eastern Himalayan Borderlands." *Political Geography* 80: 102175.
Gurung, Kabita, Khashti Dasila, Anita Pandey, and Niladri Bag. 2020. "*Curvularia eragrostidis*, a new threat to large cardamom (*Amomum subulatum* Roxb.) causing leaf blight in Sikkim." *Journal of Biosciences* 45 (113): 1–8.
Handa, Om Chanda. 2004. *Naga Cults and Traditions in the Western Himalaya*. Delhi: Indus Publishing.
Holmes-Tagchungdarpa, Amy. 2021. "The Book as a Generative Agent: The Buddhist Canon as Community Member in Book Procession Rituals of the Himalayas." *MAVCOR* 5 (2). https://mavcor.yale.edu/mavcor-journal/book-generative-agent-buddhist-canon-community-member-book-procession-rituals
Lepcha, Charisma. 2013. "Religion, Culture, and Identity: A Comparative Study of the Lepchas of Dzongu, Kalimpong, and Ilam." PhD dissertation, North-Eastern Hill University.
Lepcha, Charisma K. 2021. "Lepcha Water View and Climate Change in Sikkim Himalaya." In *Environmental Humanities in the New Himalayas*, edited by Dan Smyer Yü and Erik de Maaker. New York: Routledge.
Lepcha, Kachyo. 2020. "The Teesta Hydro Power Projects: A Historical Analysis of the Protect Movement in North Sikkim." Ph.D. dissertation, Sikkim University.
Lepcha, Rongnyoo and Mongfing Lepcha. 2021. "Painting the Genesis of the Lepcha: A World Arising from Water Spirits." In *Environmental Humanities in the New Himalayas*, edited by Dan Smyer Yü and Erik de Maaker. New York: Routledge.
Lha Tsering, Khenpo. 2002. *Mkha' spyod 'Bras mo ljongs kyi gtsug nor sprul pa'i rnal 'byor mched bzhi brgyud 'dzin dang bcas pa'i byung ba brjod pa blo gsar gzhon nu'i dga ston – A Saga of Sikkim's supremely revered four pioneer Nyingmapa reincarnates and their torchbearers.* Gangtok: Khenpo Lha Tsering.
Lha btsun nam mkha' 'jigs med. 2016. *Rig 'dzin sgrog sgrub*. Delhi: Chos spyod dpar skrun khang.

Lha btsun nam mkha' 'jigs med. n.d. Date unknown. *Bras ljongs gnas kyi bsang dang gsol ki'i rigs bzhugs pa'i dbu phyogs lags so.* Privately published, in possession of author.

McDuie-Ra, Duncan and Mona Chettri. 2020. "Concreting the Frontier: Modernity and its Entanglements in Sikkim, India." *Political Geography* 76, 1–12.

Rahman, Mirza Zulfiqur. 2020. "The Languages and Imagination of Monsoon Rains below the Eastern Himalaya." *Australian Himalaya Research Network* Blog, Oct. 29. Accessed January 2, 2022. https://aushimalaya.net/2020/10/29/the-languages-and-imaginations-of-monsoon-rains-below-the-eastern-himalaya/

Sharma, Jayeeta. 2011. *Empire's Garden.* Durham: Duke University Press.

Tshe ring, Bkra shis, ed. 2008. *Mkha' spyod' bras ljongs kyi gnas yig phyogs bsdebs bzhugs.* Gangtok: Namgyal Institute of Tibetology and Dharamsala: Amnye Machen Institute.

Vandenhelsken, Melanie. 2011. "The Enactment of Tribal Unity at the Periphery of India: The political role of a new form of the Panglhabsol Buddhist ritual in Sikkim." *European Bulletin of Himalayan Research* 38: 81–118.

Vargas-O'Bryan, Ivette. 2013. "Falling Rain, Reigning Power in Reptilian Affairs: The Balancing of Religion and the Environment." *Religions of South Asia* 7 (1–3): 110–125.

7

LIFE AND LOSS OF A FELT HABITAT

Exploring the World of *Haor* in Bangladesh

Iftekhar Iqbal

Introduction

> ...as many as the birds
> that flock to land [terram] from ocean deeps, when the cold of the year
> drives them abroad and despatches them to sunnier countries [*terris*].
> They stood there, pleading to be first to make the crossing,
> stretching out their hands in longing for the far shore.
>
> –Virgil, *The Aeneid*

Since their emergence about fifty million years ago, the Himalayas have given birth to different kinds of climes, located in as varied places as its cryospheric zone and the delta of rivers it gave rise to. One such remarkable clime evolved in connection to the world's highest–rainfall area that is triggered by the encounter between the monsoon and the mountain ranges of Meghalaya state in India, the easternmost fringes of the Himalaya. The foremost impact of this heavy rainfall is observed in the wetlands, known as *haor* (হাওর), located in northeastern Bangladesh at the foothills of the Meghalayan ranges (Figure 7.1).

Although the Meghalaya ranges (comprising the Garo, Khasi, and Jaintia Hills), with record rainfall in Sohra (Cherrapunji) and Mawsynram, are often considered the "rainiest" place in the world, the "wettest" place appears to be the *haor*. Because of topographical factors, most rainwater that falls on the Meghalaya and surrounding regions quickly drains into the *haor*, where the drained waters are retained for the entire year in different forms. This provides the *haor* with a relatively rich ecosystem where humans and nonhumans have flourished. A contrasting picture of vitality of human reproduction and sustenance is seen in the fact that in the early twentieth century the *haor*-dominated Sylhet region had

DOI: 10.4324/9781003347026-10

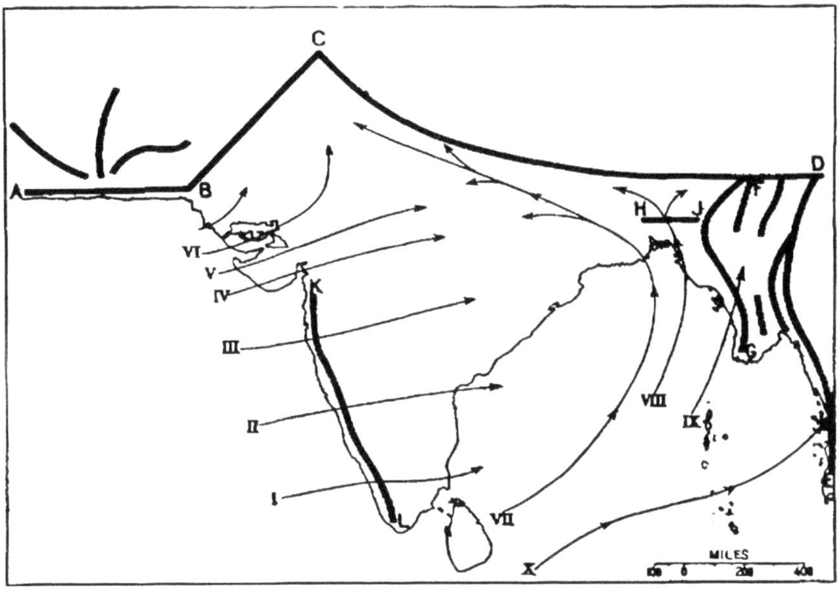

FIGURE 7.1 Monsoon flows and rainfall patterns over the Bay of Bengal-Bengal Delta-*Haor*- Meghalayan ranges.

Source: Alexander McAdie, "Monsoon and Trade Winds as Rain Makers and Desert Makers," *Geographical Review*, 12, 3 (1922), 415.

a population density of 412 persons per square mile, whereas in the Khasi Hills, where Sohra and Mawsynram are located, the population density was 34 persons per square mile (Government of Eastern Bengal and Assam 1907).

Being an important source of evapotranspiration, the *haor* contributes to the rain-making process, which has been important to the long-term sustainability of the region. Most of the moisture that forms the South Asian monsoon comes from terrestrial water spaces followed by the moisture generated in the Bay of Bengal and the Arabian Sea (Mei and Ashfaq 2014). Evaporation from the *haor* regions was noted by officials in the late nineteenth century although no data was systematically generated in the colonial period. Data from the period 1960–1998 shows that average monthly evapotranspiration from the *haor* regions ranged from 2.00 mm/day to 4.80 mm/day, being highest in April and lowest in December (Government of Bangladesh 2012, 74). The *haor* thus complemented the Ganga-Brahmaputra-Meghna's evapotranspiration regime. In effect, the Meghalaya raindom, the *haor* wetlands, and the Bay of Bengal all remain connected and mutually constitutive of each other, aiding the growth of the *haor* as a terrestrial clime that eminently caters to the need and nourishment of a broader biotic community.

Within this larger backdrop, this chapter looks at the *haor* from two connected perspectives. One relates to how they are conceived as a "felt habitat" where both human and more-than-human entities flourish, providing a window to a universal commons in a local setting where a pluriverse of humanity takes center stage (Smyer Yü 2021, 10–19). A second perspective directs to the ideas, practices, and politics that contributed to the ongoing dislocation of the *haor* as a place of affective bonding between human and nonhuman, both animate and inanimate, that characterizes a felt habitat.

In keeping with the conceptual thrust of this book, this chapter benefits from the use of the term "clime" and, by extension, "felt habitat." Developing on the ancient Greek idea of weather conditions occurring according to latitudinal exposure to the sun, some medieval Muslim scholars divided the earth into seven climes and conceptualized a "humoral pathology" corresponding to each of these climes. In this context, people in temperate zones that are less exposed to the sun were considered temperamentally and intellectually inferior whereas people in the Middle East were considered superior (Olsson 2014). Montesquieu advanced a similar climatic argument and in the era of European imperialism such climatic determinism became associated with racism and was a tool for colonial expansion (Lamb 1996, 9). In both these versions, climes were assumed to be part of the latitudinal "mathematical climate" that influenced human behavior—a concept that was appropriated by scholars of both "Occidentalist" and "Orientalist" persuasions.

As opposed to this, since classical Greek times, clime has also connoted local or terrestrial environments, such as swamps, with implications for diseases and well-being (Carey and Garone 2014, 285). It is only recently that environmental humanists have begun to explore how climate knowledge and climate awareness are being perceived, reworked, and communicated in a posthumanist context in which it is possible to study human engagements with nonhuman entities in their shared habitat where nature and culture shape each other—a process that Lorimer (2012) and Paerregaard (2020) call "multinaturalism."

It is in this context of multinatural communicability and relationality, which bridges "mathematical climate" and "terrestrial climate" studies, that this paper uses the term "felt habitat." Depicting the seasonal movement of migratory birds in the *Aeneid*, Virgil uses both "terram" and "terris" to denote land. While "terram" refers to the place where the birds assembled while in transit before crossing the Mediterranean, "terris" represents the warmer region (for example, *terris apricis* or sunny lands) where the flocks of the same birds finally settled for the season away from extreme cold (Frieze 1862, 513; Macdonald 1958, 185). With regard to the *haor*, this chapter uses the term "felt habitat" in the sense of "terris," denoting a local lifeworld of biotic communities that inhabit a place in close conversations with the rotating seasons of the year, as well as the place-based manifestations of the animated earth that speak to both human and nonhuman sensibilities, making way for what Smyer Yü (2021, 239–260) calls "symbiotic indigeneity".

From "Titanic Sunk Face" to "Saucer"

The town of Sohra made headlines in 1861 when it received 915 inches of rainfall that year. But this was not steady over the years: in 1873, the recorded amount was 283 inches and in 1911, it was 460 inches (Government of Eastern Bengal and Assam 1907, 7; Government of India 1911, 2). The unusually high 1861 figure was probably erroneous (McAdie 1922, 415n). More likely, the historical rainfall amount in Meghalaya ranged from about 100 to 150 inches annually from the mid-nineteenth to the mid-twentieth century, with Sohra and Mawsynram locally receiving around 400 inches. Still, this amount of precipitation was enough for the Meghalaya ranges to earn the accolade of being one of the rainiest places on earth, where rainfall was often too much to bear and, in places like Sohra, downpour was tantamount to "oppressions" (Nag 2015).

The chapter by Amrith and Smyer Yü in this book shows how remarkable this mountain-plain relationship bonded by the bounty of monsoon rains is. In its regional context, the Meghalayan rain is like a fleeting tourist, as the rainwater hardly settles in the region's ranges. The low porosity of sandstone doesn't allow the soaking in of the water and the steep slopes drain the rain in no time, diverting it straight to the *haor* in the floodplain that begins right at the bottom of the mountains. The geomorphological arrangement is such that it is impossible to stop the rainwater from reaching the plains from the highlands, as seen through the northern cliffs of the Meghalaya ranges which appear as a "Titanic

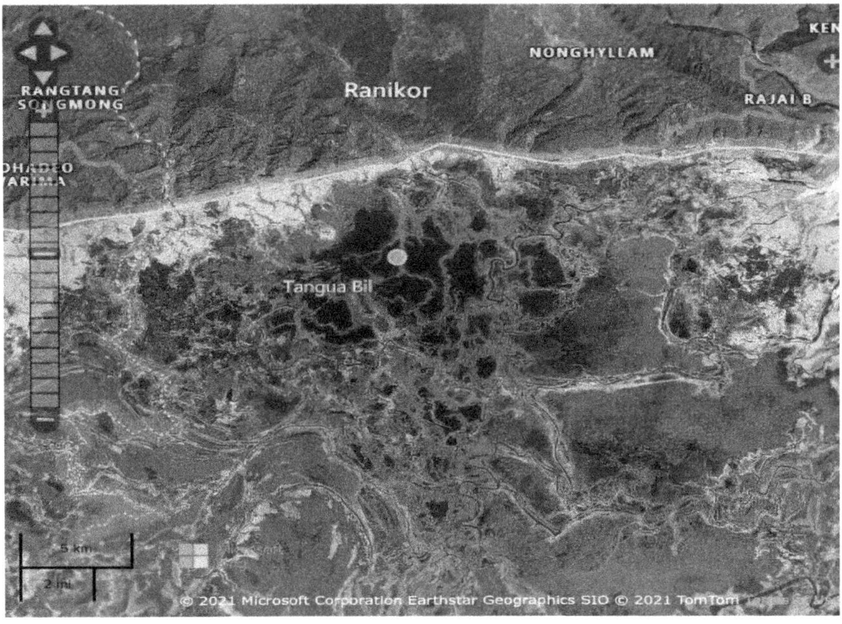

FIGURE 7.2 Tanguar *haor* in Bangladesh at the foothills of the Meghalayan ranges. *Source*: ArcGis.

sunk face," dropping 4,000 feet toward the plain. The water inevitably settles in the *haor* which are "huge saucer-like depressions" (Allen 1905, 1, 2) (Figure 7.2). The rainfall and the river waters that flow from the hill ranges in the east, including the Tippera Hills, which amounted to about 25,000 cubic feet per second of water flow in the early nineteenth century, poured into the *haor* (Fisher 1840, 811). No wonder, then, that during the rainy season and the months that follow, the *haor* look like a vast inland sea, as is reflected in its name, *haor* (a colloquial or a corrupted term of *sagor* or *saiyor*, which means sea in Bangla).

In geological terms, the region emerged from the ocean through the sedimentation process following the merging of the Indian plate with the Eurasian plate (Hunter 1875, 263; Scott 1828, 55). A Bengali author noted in a memoir: "When I had the opportunity to see a sea, I thought that the experience of being in a sea with a boat was just like the similar experience of being in a *haor*. There is no doubt that those *haor* were a large piece of the sea that was left behind. Although imprisoned in the confines of land, these *haor* have fully maintained the mood and modes of the sea" (Gupta 1982, 180).

An 1879 government definition of *haor* ran like this: "Low tracts of land covered more or less with grass and reed *jangal*, almost entirely under water during the rains, but dry, or only partly so, during the rest of the year" (Government of Assam 1879). A more recent version of the definition is from Hugh Brammer, quoted in Chakrabarti (1992, 6):

> ...an inland sea in the rainy season, with substantial waves created by the monsoon winds and miniature beaches formed on any high ground—including settlement mounds—against which the waves break. Water quickly drains from the flood plain ridges after the rainy season, but large areas in the basins stay wet or submerged for most of the dry season.

By definition, in the *haor* the water itself is amphibian, moving in and out of its tracks throughout the seasons of the year. By the second half of the annual monsoon period, roughly through August and September, the run-off from Meghalaya and the rains in the region itself fill the *haor* to their brims, which then indeed look like an inland sea. In the late nineteenth century the Hakaluki *haor*, the largest of all *haor*, extending twenty miles by nine, was passable by foot in the winter, but a "dangerous stormy sea" in the summer. Since it was shaped like a saucer, water stayed beyond the rainy season subject to only slow and partial evaporation. At the end of the process, before the arrival of the next cycle of the monsoon, the water retreats further into the lowest part, leaving a trail of silt all around it. The perennial water bodies that remain in the deeper parts of the *haor* are usually called *beel/bil* (marsh). As the rainy season returns and the run-off from the Meghalaya arrives with renewed vigor, the *beel* return to full capacity, connecting with the neighboring *beel*, and the *haor* is reborn (Figure 7.3). At the turn of the twentieth century, there were 503 *beel* in Sylhet district spread in five subdistricts, which likely became thirty-five large *haor* during the monsoon (Allen 1905, 17–20). The current total extent of the *haor*

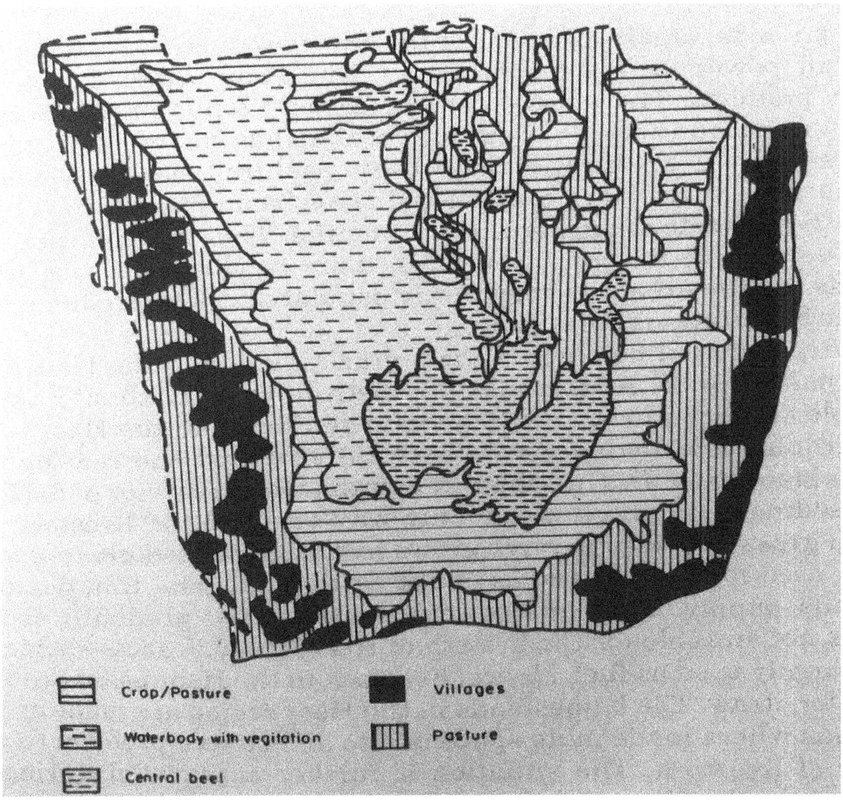

FIGURE 7.3 Transect of a typical *haor*.
Source: Syed Iqbal Ali, "Haor Basin Eco-System," in *Environmental Aspects of Surface Water Systems of Bangladesh*, edited by A Atiq Rahman, Saleemul Huq and Gordon R Conway, (Dhaka: University Press Limited), 83.

in Bangladesh, including those in Sylhet and the neighboring Mymensingh region, is about 8,100 km^2 (Ali 2000, 77) (Figure 7.3).

A Felt Habitat?

Like hunting, foraging, and pastoralism, agriculture is a primordial connector between humans and nature. The cultivation of a rich range of agricultural products of the *haor* was instrumented by inanimate elements from deep geological time. Beside atmospheric factors, silt-laden waters from the Meghalaya carried fertilizing breccia (clastic sedimentary rocks) and oxide among other, which originated from the volcanic eruptions that took place during the formation of the Meghalaya ranges (Fisher 1840, 815). Among traditional products, rice (*Oryza sativa*) and different kinds of pulses have been prominent, with their sub-varieties being seemingly "infinite," as Fisher (1840, 820) noted. Decades later, in the early twentieth century, at least thirty-nine varieties of rice of three principal cultivars (Aman, Aush, and Boro) were noticed by government officials. Whereas these three principal cultivars grew on the *haor* itself, another principal kind, Shail, was sown on the higher ground known as *tilla,* which was the only place inside the *haor* that was not submerged during the monsoon. Rice that was produced met most of the demand of the local population and surplus rice was occasionally exported upland, for example, to famine-stricken Silchar (Allen 1905, 114–119).

The "infinite" variety was not "mere fanciful distinction, but made with reference to well-marked peculiarities, either of quality or fruition." Fisher (1840, 820) elaborated:

> ...among the Sayl [Shail], which grows on the high lands, there are grains which come to maturity in the short space of six weeks, while there are other, as the Burwa [Boro], which can be raised on the Autumn lands in the winter. It may not really attract attention, but the careful inquirer will, I think, find it no small advantage, that there are so many grains whose times of coming to perfection are unequal, as they afford, under proper management, a sure resource against the loss of crops of more ample, but more slow growth.

Traditionally, crop production depended on cattle, especially cows and buffalo, whose food was in abundance in the *haor.* The receding water left a large swath of grazing feed including *doob* (tender grass), reeds, and elephant grass, which grew ten to twenty feet in height. In the dry season, cows, buffaloes, and ponies were let loose on the pasture and they came not only from the *haor* region, but beyond, such as Kachar, located to the northwest more than 100 km away (Allen 1905, 4; Evans 1883, 37). Some reports noted the difficulties of obtaining grasses for the cattle in the middle of the rainy season when the *haor* flooded everything, although rice hay was stored for use during times of shortage.

The *haor* ecology, which was plugged into both the human and the more-than-human worlds, was not just related to the question of livelihood. Northeastern Bengal's subsistence and commercial produce and the communication channels that the *haor* and its connected rivers provided, put them at the crossroads of considerable human mobility spanning the sub-Himalayan highland, deltaic plainland, and the Bay of Bengal. In this broader context, Ludden (2003, 5080; 2019, 75–100) notes that the city of Sylhet, surrounded by the *haor,* exposed the inability of national maps to account for nature's social geography.

Human bonding with the *haor* was not established only with agrarian economic activities and the *haor*'s role in facilitating mobility. It was the cultural practices embedded in the *haor* ecology that secured deeper connections between the human and the nonhuman world, where cultural ethos similar to that of central and western parts of the Bengal Delta developed. In the latter, large rivers and their tributaries promoted a life of relative ease, spiritual questing, and a universalist ethos as reflected in the syncretistic Vaishnava tradition and the songs of Lalon Shah. The fragility of life represented by the erosion of river banks, and the confluence of rivers that symbolized the meaninglessness of barriers of race, color, and religion were the hallmarks of these traditions. In the *haor* region, culturally embedded dialectics between forces of nature became more prominent because, unlike in a large riverine environment, *haor* represented a complex clime made up of both the wild and the ordinary, where the universal met with the local in harmony but also in contestation. This dialectic was reflected in popular folk tales featuring Kenaram, an infamous robber in the *haor* area. Kenaram, an orphan raised by his uncle and eventually sold, found himself in the *haor* jungle leading a gang of robbers who lived on forest products such as the milk of wild buffaloes, and who established a reign of terror. He would shed human blood as easily as a swamp tiger and represented the ruthless side of nature. He was considered a legendary bogyman or a beast, the "wildest of them in human shape," and, like the Archangel of whom Milton wrote, was as tall as "Tenriffe or Atlas, unremoved: His stature reached the sky." The Bengali oral folk tradition has the exact equivalent of this description: "আসমান জমীনে ঠেকে যখন হয় খাড়া." (Sen 1928, xxvii).

The dramatic transformation of Kenaram, as far as poetic narratives go, must also be traced to the felt habitat of the *haor* which produced bards and singers who composed in tune with nature. A poet of the region who was encountered by Kenaram and was about to be killed by him found an opportunity to sing his last song to Kenaram. When the singer, Bangshi Das, sang with his fullest devotion and perfect melody drawing even the birds to listen, Kenaram threw down his sword, broke into tears, and became a disciple of Bangshi Das, taking to singing in praise of Manasa (the snake goddess) for the rest of his life. When Kenaram sang, "the very trees became spell-bound and shed tears in the shape of dew." (Sen 1923, 286, 290). This was narrated by the poetess Chandra, the daughter of Bangshi Das, who was rumored to have been present during the encounters between her father and Kenaram. These events, according to oral tradition, are popularly believed to have taken place sometime in the sixteenth century.

Kenaram's life and behavioral transformation speak to the beauty and the beast that exist both in human and nonhuman worlds. The society that abandoned him was punished with his brutality, but the transformation was also caused by the same society in conversations with nature that touched his inner core. Kenaram represents the multidimensionality of the felt habitat of the *haor* where there was only a thin boundary between human and nonhuman world. To understand Kenaram's transformation, therefore, it is important to look at the more-than-human aspect of the *haor* as well. In the character of Kenaram we see a duality of nature—harshness, bestiality and ferocity on the one hand and melody, endowment and wellness on the other—although the multinatural unity prevailed at the end, a process that saw the allegory and active presence of humans, birds, tigers, and buffaloes, among others.

Historically, *haor* ecology invited a live part of the biotic universe in which the humans were but one among many. The *haor* looked at its best in its deep-ecological clime, being a sanctuary of a wide range of quadrupeds and avifauna. Simson, a colonial official stationed in the region, called the *haor* area the "home of the swamp-deer." He particularly mentioned the unique features of the horns of these deer, which had a regular number of tines, generally twelve, although he caught one with twenty tines. These were locally called *barah singha* or twelve-horned deer. He noted that from their home in the *haor* of Sylhet, the *barah singhas* extended their presence in Assam to the base of the Himalayas, and were also found as far northwest as Purnea, and to the Sundarbans in the south (Simson 1886, 209–211). Elephants, pigs, black leopard, jackals and wild chicken, among others, abounded. At the beginning of the twentieth century, "a large population of tigers" walked on the *haor* fringes of Sylhet, some of them being spotted as late as the 1960s (Husain 1974 22). Likewise, wild buffaloes and rhinoceroses were in good number in the early twentieth century.

The species of birds found in the *haor* are local, regional, as well as migratory visitors from higher in the Himalayas and beyond. Allan Octavian Hume noted that most of the birds scattered across the whole of eastern Bengal in the rainy season retreated to Sylhet in the drier winter months (Hume 1881, 109). *Embezilla pusilla* arrived in the region all the way from Siberia (Murray 1888). Punta Mukhi (Bengali)/*Spatula clypeata*, Sarah/Shareil abounded and *Dendrocygna javanica* came in "flocks in thousands." Birds, such as Khyah partridges, were so many in number and flew in such enormous flock that, as Simpson noted, one could catch three birds at one go with a stick from the back of an elephant. This easiness involving the hunting of birds led Simpson to think that catching them could no longer be considered sport and he withdrew from his hunting trip in the *haor*.

In the more-than-human context, such trans-local and transregional mobility put the *haor* literally at the center of connections with as disparate places as Shillong and Siberia. But the felt habitat reached its climax for about three months once in a year in the small mound areas above the flood level, the *tilla*. In the rainy season when the *haor* resembled the "scriptural deluge," in the *tilla*—about 400 to 500 feet in height with easy slopes, where human settlement had taken place—animals of all kind would gather like in a Noah's Ark. But they would gather as guests rather than refugees, retreating along with the receding water only to come back with the next monsoon season. This was the time of the year when the intimacy of humans and more-than-humans was renewed like in a seasonal fair. Simson (1886, 163) provides a vivid picture of this:

> I have sailed across these places when the human inhabitants only existed on artificial mounds crowded with huts, filled to overflowing with human beings huddled together for the short time, waiting for a fall of twelve or eighteen inches in the flood. To each hut you would see a canoe or two fastened. In some places on the top of similar mounds a few trees would be seen, their bases standing in the water and their branches waving over it; these branches were covered with an extraordinary collection of living creatures—snakes and lizards on every branch, rats and animals of the cat tribe, squirrels and wading or short-winged birds. Curious unknown insects, turtles and crab-like creatures unknown to me. As the boat came near, the snakes and the rats would leave the trees and come swimming towards it.

From Felt Habitat to Natural Resources

In his famous 1837 lecture, Ralph Waldo Emerson made a scathing criticism of the division of labor for its culpability in metamorphosing humans into "a thing, into many things" ignoring humans' holistic potential as a natural being. Emerson's colleague Henry David Thoreau elaborated on this critique of the division of labor and noted how it left fewer people attached to the "spirit" of nature. Thoreau suggested that a professor who was eating his huckleberry pudding while writing a book in his library was distanced from the field of the berry by three layers of division—the picker, his hired labor aided by patented machine, and the cook preparing the pudding. Thoreau advised that the professor ought to divide himself only between the huckleberry field and the library (Thoreau 1887, 36, 37). In suggesting this, Thoreau foreshadowed what Lynn Margulis (1998) called "a symbiotic planet" which represents the chemistry of physical contacts between everything and everyone. Karl Marx from an entirely different intellectual tradition also criticized the division of labor because it killed enthusiasm and alienated workers from the production process. Marx considered man and nature as being the source of the two "original agencies" contributing to the creation of wealth (Foster et al. 2010, 63), although such recognition falls short of salvaging Marxism from its obsession with human's industrialization-driven foray into nature.

The deeper engagement between humans and nature that defines a felt habitat—where livelihood activities occur without priority for production of surplus capital out of nature—did not agree with the tradition of classical economics and the Victorian idea of progress, particularly in Britain's imperial frontiers. About the time Emerson, Thoreau, and Marx debated the idea of the division of labor on both sides of the Atlantic, in the Sylhet *haor* Fisher (1840, 822) observed an agrarian society where:

> the first steps of economical improvement have hardly been taken, the advantages of the division of labor not having yet been appreciated, or rather the introduction of that principle having been prevented, by the want of accumulated capital, to meet the expense and delay that must precede the more ample returns which it ensures.

Fisher lamented that no cultivator ever followed agriculture in the *haor* as an exercise in speculation, or ventured to cultivate land beyond meeting the need for his family and government tax, and that fulfilling those requirements was considered a success. Although the division of labor did not take place in the Bengal delta as expected due to the lack of industrialization and urbanization, the colonial state's drive for commercial agricultural expansion, revenue generation, and the accumulation of private capital in the *haor* region made imperious inroads into the ecosystem. For example, beginning in 1878, some of the *beel* areas were drained for tea gardens (Allen 1905, 139).

In the late nineteenth century, the British colonial government offered liberal reward for capturing tigers such terms that fifty to sixty tigers were being eliminated in the *haor* annually (Ward 1884, 169, 170). Simson, with extensive knowledge of the region, was candid about it, noting that deer and other animals were "gradually being exterminated in these inundations" as they were killed by the dozen rather than the traditional two or three taken at a time in the pre-gun era. This was due to the "stupid government arrangements" of selling muskets and guns to the public, Simpson noted (1886, 211).

If the wildlife stood between the state and the *haor*, the communal organization of the *haor*-centric livelihoods presented hurdles to the colonial state's drive to control and accumulate capital. For example from the fisheries in the *haor* areas, in the last few decades of the colonial era, benign desires to secure the interests of authentic fishermen in the *haor* was coupled with the idea that the "middlemen" should be replaced by "leaseholders"—reflecting on the emergence of a sort of division of labor. A middleman was usually someone within the local community with social acceptance and some financial resources. A leaseholder was someone who was either a local or an outsider with greater financial security, influence, and a relationship with the state structure at a higher level, one who could loyally satisfy the needs of revenue and market demand. The state's intention sometimes looked genuine in that its regulations recommended leaseholders from local *haor* area rather than from outside, and they promoted the protection of brood-fish and fry. But to legitimize the lease-holding schemes it was necessary to argue in favor of larger lease areas for a longer-period of at least three years. This involved greater amount of deposit money than fishermen or riparian-land occupants were able to afford. *Bathan*, the grazing areas for buffalo, cattle, and elephants, also came under such lease-holding systems (See Figure 7.4). Cutting fodder grass without the permission of the permit-holder, another example of an agrarian form of division of labor as sanctioned by the state, became punishable crime in the *haor* area (Government of Assam 1946).

On the top of the "resourcification" of the *haor* in the colonial period, decolonization unfolded additional challenges. The terrestrial ecosystem linked to the Shillong Plateau-Bay of Bengal hydrological cycle was divided into northeastern India and East Pakistan (Bangladesh since 1971). Over the decades, two different ecological profiles have emerged for the Meghalaya raindom and for the Bangladesh *haor* zones. Some literature draws attention to the irony of the world's rainiest place, Sohra, faced with a shortage of drinking water due to a climate-induced reduction of rainfall (Dasgupta 2021). A futuristic projection, on the other hand, shows that in the case of expected average rise of temperature by 3°, precipitation would increase in Meghalaya in the next five decades (Sharma 2017). At the crossroads of uncertain climate trajectories, the Meghalaya State Cabinet recently approved a draft water policy that included a proposal to construct "check-dams" with a mechanism that would stop rainwater within an hour after the rainfall, preventing it from flowing into Bangladesh (Laitphlang 2019). The rationale of this policy is that Meghalaya state needs to develop its "common resource pool" for all resident of the state in terms of the various

PERMIT FOR GRAZING

(*Words in italics apply to 'notified areas' only.*)

District________ Book No.________ Permit No.______

Name________________________________

Residence________________________________

1	2	3	4	5	6						7
Locality	Date of expiry of permit	Description of animal over two years of age	Number of animals	Date of payment	Fees paid						Remarks
					1st *kist*			2nd *kist*			
					Rs.	as.	p.	Rs.	as.	p.	
		1. Buffaloes.									
		2. Cattle.									
		(*a*) *Cows.*									
		(*b*) *Castrated males.*									
		(*c*) *Uncastrated males (other than breeding bulls).*									
		(*d*) *Breeding bulls.*									
		3. Elephants.									
		4. Elephant calves up to the age of two years.									

*Date of issue*________ ________________________

Signature and designation of officer issuing permit.

FIGURE 7.4 Grazing permit ledger. Source: *Government of Assam, Assam Land Revenue Manual* (vol. 1, 6th edition, Shillong: Assam Government Press, 1946), 249.

uses of water. The construction of a new common resource pool through the provincialization of the terrestrial ecology, as seen in other regions of India, sits uncomfortably with villagers inhabiting the sites of such dams and has met with strong protest (Special Correspondent 2021). On the other hand, the scarcity of surface water in Meghalaya is caused by, among other things, the draining of surface water into the gaping pits and "ratholes" of abandoned coal mines, which are privately owned (Rajendran 2019).

In Bangladesh, the *haor* region is now the flagship of the national natural resource regime. National pride was much enhanced by the recognition of the Tanguar *haor* (Figure 7.2) as a Ramsar site by the UNESCO in 2000. With its picturesque watery landscape that becomes an inland sea in the monsoon season, and the wide range of birds, domestic tourism has flourished. Concerns for wetland ecology expressed by the government and civil society invoke the *haor* as a trope in the conversations about natural resources of the nation. Yet, in the *haor* areas, there are often instances of local community fishermen clashing with fishing leaseholders who are connected to the established power structure (Haque and Kazal 2008). Wildlife and biodiversity began to decline in the late nineteenth century and continue to this day. Many species of birds survived mass extinction partly because a large percentage of their population arrives as "tourists" for a few months and returns with refreshed stock, but the gradual decline of their numbers in recent decades is remarkable. In the winter of 2007, for example, twenty deeper portions of Hakaluki *haor* hosted 51,521 birds of forty-two species. Of these, 18,555 birds representing twenty-nine species were migratory birds while the remaining thirteen species were local or "permanent residents." (Goswami 2007). In the winter of 2021, the total number of birds in this *haor* was recorded as 24,551, showing a reduction by more than 50% in fifteen years (Correspondent 2021).

The villagers of Meghalaya and of the *haor*, thus, face different sets of challenges, one from multipurpose dams and mines and the other from limited access to their habitual zones of livelihood. Yet both groups find themselves bedeviled by the post-colonial state's hegemonic penetration of the multinatural world of the *haor* in the name of managing "natural resources." Such hegemony has been synced with the state-building efforts that privilege visibility and legitimization often at the expense of ecosystems and the livelihood that depend on them. The felt habitat has now been translated into "natural resources" that must be exploited to aid economic growth. The unnervingly wild nature in which Kenaram and Bangshi Das had their formative encounter has now been replaced by roads, bridges, and tourist boats. In the summer of 2022, the Meghalaya-Sylhet region received 2,500 mm of rainfall, the highest in the past one hundred years, leaving nine million people homeless and destroying more than one hundred thousand hectares of crops. At a time when *haor* could be the best natural option for retaining and draining the unusually heavy rain water coming from Meghalaya especially due to climate change, a $100 million "all-weather" embanked highway running right through the *haor* actually exacerbated the flooding situation by way of water-logging along a 124-km passage for water from the mountains to the *haor* (Das, 2022). These instances, as Wouters (2020) and Iqbal (2022) notes, hinges on neoliberal advances at the expenses of larger ecological commons that dominated the Himalayan-Tibetan waterscapes in the not so distant a past.

Conclusion

Spread out between the Brahmaputra and Meghna River systems, the *haor* resemble much of the rest of the Bengal Delta plains, but geomorphologically they represent a distinct clime. They are not, strictly speaking, connected to the glacial regime or major rivers of the mainland Himalaya, nor to their sediment load, and they represented a relatively stable bonding between human and more-than-human worlds compared to the larger deltaic river system which often saw the dismantling of settlements with the erosion of the river banks. Studies of *haor*, therefore, complement the recent explorations of the Himalayan rivers and their climes (Saikia 2019; Amrith 2018; Iqbal 2010; Schendel 2009).

The first set of these nuanced multinatural bonds reflect the way the *haor* are connected with the process of co-creating the world's heaviest rainfall regime. The Bay of Bengal-Meghalayan ranges-*Haor*—each of these spaces play their part in the reception, drainage, evaporation, condensation, and precipitation of water. The *haor*'s role is even more significant as the retainer of both the drained water and the accompanying nourishing silt from the Meghalaya and other surrounding ranges. In this way, the *haor* offer a window to the long-term human quest to understand the hydrologic cycle and its endowment (Duffy 2017).

Once the inanimate world of hydrology and geology prepared the ground for the biotic community, a truly vibrant terrestrial clime emerged with infinite possibility for both humans and more-than-human beings, as this chapter has shown. Over the millennia the combined geological and hydrological forces ushered in a place-based pluriverse, to take note of Larsen and Johnson (2017, 1–22), that embodies a sense of belonging and togetherness amid the fraught questions of existence. It is in the spirit of this felt habitat that, despite the colossal loss of biodiversity in the past two centuries, the *haor* still harbor 143 types of fish, 300 types of plant, 9 types of amphibians, 40 types of reptiles, 257 types of birds, and 29 types of mammals including more than 20 million humans (Prokriti o Jibon 2019).

Yet, as this chapter has shown, the *haor* have hit the last limit of collective resilience due to a series of historically specific events and ideologies, such as the ecologically insensitive partition of India and expanding neoliberal projects. With the climate change seen in hindsight, it is no longer tenable to look at the changes in rainfall pattern in Meghalaya or challenges facing the Bangladesh wetlands from a national perspective. Emerging challenges must take us back to the larger relatable commons that hold the Bay of Bengal, the lesser Himalayan ranges, and the *haor* and the deltaic rivers together. A historical understanding of the *haor* would bring environmental humanities into conversation with developmentalist projects and territorially sensitive national protocols at a time when the precarity of the changing climate demands trajectories beyond nation-states.

Like my fellow authors in this book I have used the concept of "clime" to respond to the growing challenge of fragmentation of locally distinctive ecologies, which played crucial role in the past in sustaining life and livelihood of humans and more-than-human entities. The racial and imperial persuasion around the abstract understanding of climate by some scholars of the medieval world and Enlightenment-era Europe have long lost appeal; now the concept of climate continues to be heavily informed by scientific instrumentality that are often embedded in the new metanarrative of "climate change." In response, recent works, including this book, seek to bring climate home—a home that we agree to call "clime." Heidegger's fourfold relations of "dwelling" between sky and earth and among mortals and the divine—as mediated by the ontopoetics of existence—closely mirror this understanding of clime. *Haor* is an understudied example of such a clime located on the monsoon pathways between the Himalaya and the Bay of Bengal, where an assemblage of human and more-than-human actors co-exist in what I have termed the "felt habitat." Through the case of the *haor* I have suggested that this earthly clime, with all its density of dwelling and dwellers, is the soul of this felt habitat.

References

Ali, Syed Iqbal. 2000. "Haor Basin Eco-System," in *Environmental Aspects of Surface Water Systems of Bangladesh*, edited by A. Atiq Rahman, Saleemul Huq, and Gordon R Conway. Dhaka: The University Press Limited.

Allen, B. C. 1905. *Assam District Gazetteers Vol. II: Sylhet*. Calcutta: Caledonian Steam Printing Works.

Amrith, Sunil. 2018. *The Unruly Waters Unruly Waters: How Rains, Rivers, Coasts, and Seas Have Shaped Asia's History*. New York: Basic Books.

Carey, Mark, and Philip Garone. 2014. "Forum Introduction." In *Forum: Climate Change and Environmental History*, edited by Mark Carey, Philip Garone, Adrian Howkins, Georgina H. Endfield, Lawrence Culver, Sam White, Sherry Johnson, James Rodger Fleming, Philip Garone, and Mark Carey. *Environmental History*, 19(2): 282–293, https://doi.org/10.1093/envhis/emu004.

Chakrabarti, Dilip K. 1992. *Ancient Bangladesh. A Study of Archeological Sources*. Dhaka: University Press Limited.

Correspondent. 2021. "Hakaluki Haor Sees Decline in Migratory Birds." *The Financial Express*, February 28, 2021. https://thefinancialexpress.com.bd/national/country/hakaluki-haor-sees-decline-in-migratory-birds-1614487268. Accessed 24 September 2022.

Das, Sadhan Kumar. 2022. "The Worst Flood in the Region in the Last 122 Years," *International Politics and Society*. https://www.ips-journal.eu/interviews/the-worst-flood-in-the-region-in-the-last-122-years-6044/ Accessed 9 September 2022.

Dasgupta, Debarshi. 2021. "Even the Wettest Place on Earth is Starting to Run Dry." *The Straits Times*, August 5, 2021. Accessed January 26, 2022. https://www.straitstimes.com/asia/south-asia/even-the-wettest-place-on-earth-is-starting-to-run-dry.

Duffy, Christopher J. 2017. "The Terrestrial Hydrologic Cycle: An Historical Sense of Balance." *WIREs Water* 4. doi: 10.1002/wat2.1216.

Evans, J. Griffith. 1883. "Anthrax in Kachar and Manipur, India." *The Veterinary Journal and Annals of Comparative Pathology*, 16(1): 99–100.

Fisher, Thomas. 1840. "Memoir of Sylhet, Kachar, & the adjacent Districts." *Journal of the Asiatic Society of Bengal*, 9(2): 808–843.

Foster, John Bellamy, Brett Clark, and Richard York. 2010. *The Ecological Rift. Capitalism's War on the Earth*. New York: Monthly Review Press.

Frieze, Henry S. 1862.*Virgil's Aeneid: With Explanatory Notes*. New York: D. Appleton and Company.

Goswami, Rajat Kanti. 2007. "Migratory Birds Rise in Hakaluki Haor." *The Daily Star*, 10 December 2007. Accessed 26 January 2022. https://www.thedailystar.net/news-detail-14882.

Government of Assam. 1879. *Glossary of Vernacular Terms in the Province of Assam*. Shillong: Assam Secretariat Press.
Government of Assam. 1946. *The Assam Land Revenue Manual*. Vol. 1. 6th edition. Shillong: Assam Government Press.
Government of Bangladesh. 2012. *Master Plan of Haor Areas, Main Report*. Vol. II. Dhaka: Center for Environmental and Geographic Information Services.
Government of Eastern Bengal and Assam. 1907. *Report on the Administration of Eastern Bengal and Assam, 1905–1906*. Shillong: Eastern Bengal and Assam Secretariat Press.
Government of India. 1911. *Census of India* 1911. Vol. 3, Assam. Part 1.
Gupta, Prabhat Kumar. 1982. "Haor." *Sera Sandes, 1368–1387*. Calcutta: Ananda Publishers, 178–180.
Haque, A. K. Enamul and Mizanul Hoque Kazal. 2008. *Rich Resources, Poor People. The Paradox of Living in Tanguar Haor.*
Hume, Allan Octavian. 1881. *The Game Birds of India, Burmah, and Ceylon*. Vol. 3. Calcutta: A. O. Hume and C. H. T. Marshall.
Hunter, William Wilson. 1875. *Statistical Account of Assam*. Vol. 2. London: Trubner and Co.
Husain, Kazi Zaker. 1974. *An Introduction to the Wildlife of Bangladesh*. Dhaka: F. Ahmed.
Iqbal, Iftekhar. 2010. *The Bengal Delta. Ecology, State and Social Change 1840–1943.* Basingstoke: Palgrave Macmillan.
Iqbal, Iftekhar. 2022. "Rivers of Mobility. Multi-Ethnic Societies and Ecological Commons in a Fluvial Asia." In *Yunnan-Burma-Bengal Corridor Geographies: Protean Edging of Habitats and Empires*, edited by Dan Smyer Yü and Karin Dean. London and New York, 105–121.
Laitphlang, David. 2019. "Meghalaya Cabinet approves draft water policy." *Hindustan Times*, July 14, 2019. Accessed 26 January 2022. https://www.hindustantimes.com/india-news/meghalaya-becomes-first-state-to-come-up-with-water-policy/story-eORJdawl5L7wtTLJTinQmI.html.
Lamb, H. H. 1996. *Climate, History and the Modern World*. London and New York: Routledge.
Larsen, Soren C. and Jay T. Johnson. 2017. *Being Together in Place: Indigenous Coexistence in a More Than Human World*. Minneapolis: University of Minnesota Press.
Lorimer, Jamie. 2012. "Multinatural Geographies for the Anthropocene." *Progress in Human Geography*, 36(5): 593–612.
Ludden, David. 2003. "Investing in Nature around Sylhet: An Excursion into Geographical History." *Economic and Political Weekly*, November 29.
Ludden, David. 2019. "Cowry Country: Mobile Space and Imperial Territory." In *Asia Inside Out—Itinerant People*, edited by Eric Tagliacozzo. Cambridge, MA: Harvard University Press.
Macdonald, H. F. 1958. "Virgil's Migrating Birds." *Greece & Rome*, 5(2): 185–185.
Mei, Rui and M Ashfaq. 2014. "Moisture Sources for South Asian Monsoon Precipitation from CMIP5 Simulations." Paper presented at American Meteorological Society Conference, Atlanta. Accessed January 26, 2022. https://ams.confex.com/ams/94Annual/webprogram/Paper237264.html.
McAdie, Alexander. 1922. "Monsoon and Trade Winds as Rain Makers and Desert Makers." *Geographical Review*, 12(3): 412–419.
Margulis, Lynn. 1998. *Symbiotic Planet. A New Look at Evolution*. New York: Basic Books.
Murray, James A. 1888. *The Avifauna of British India and Its Dependencies: A Systematic Account*. Vol. 2.

Nag, Sajal. 2015. "Rain, Rain, Come Again: Cherrapunji, the Rainiest Spot on Earth." In *RCC Perspectives: Unruly Environments*, 3: 25–29, edited by Siddhartha Krishnan, Christopher L. Pastore, and Samuel Temple.

Olsson, J. T. 2014. "The World in Arab Eyes: A Reassessment of the Climes in Medieval Islamic Scholarship." *Bulletin of SOAS*, 77(3): 487–508.

Paerregaard, Karsten. 2020. "Communicating the Inevitable: Climate Awareness, Climate Discord, and Climate Research in Peru's Highland Communities." *Environmental Communication*, 14(1): 112–125.

Prokriti o Jibon. 2019. হাওর ব্যবস্থাপনা [Haor Management] Episode 334. December 15. Accessed 26 January 2022. https://www.youtube.com/watch?v=YB93INEF1vI.

Rajendran, C. P. 2019. "The Meghalayan Plateau Is a Geological Marvel Under Ecological Threat." *The Wire*, April 14. Accessed 26 January 2022. https://thewire.in/the-sciences/the-meghalayan-plateau-is-a-geological-marvel-under-ecological-threat.

Saikia, Arupjyoti. 2019. *The Unquiet River: A Biography of the Brahmaputra*. Delhi: Oxford University Press.

Scott, David. 1828. "Additional Observation on the Great Cavern of Boobon." In *The Edinburgh Journal of Science*. Vol. IX, April-October, 1828. 55. (For a note on marine remains in Sylhet from a "comparatively recent" geological period, see Hunter above).

Sen, Dineschandra. 1923. *Eastern Bengal Ballads, Mymensingh*. Calcutta: University of Calcutta.

Sen, Dineschandra. 1928. *Eastern Bengal Ballads*. Vol. III, part I. Calcutta: University of Calcutta.

Sharma, Dinesh C. 2017. "Why Are We Sweating in Shillong?" *DownToEarth*, July 4. Accessed 22 January 2022. https://www.downtoearth.org.in/news/climate-change/why-are-we-sweating-in-shillong--58212.

Simson, Frank B. 1886. *Letters on Sport in Eastern Bengal*. London: R.H. Porter.

Smyer Yü, Dan. 2021. "Situating Environmental Humanities in the New Himalayas: An Introduction." In *Environmental Humanities in the New Hiamalays: Symbiotic Indigeneity, Commoning, Sustainability*, edited by Dan Smyer Yü and Erik de Maaker. London and New York: Routledge.

Special Correspondent, "Residents of 12 Meghalaya villages oppose dam on India's clearest river," The Hindu, April 11, 2021. https://www.thehindu.com/news/national/other-states/meghalaya-locals-oppose-dam-on-indias-clearest-river/article34291199.ece.

Thoreau, Henry David. 1887. *The Writings of Henry David Thoreau, vol 8: Winter*, edited by H.G.O Blake. Boston: Houghton Mifflin Company.

Van Schendel, Willem. 2009. *A History of Bangladesh*. Cambridge: Cambridge University Press.

Ward, Susan R., ed. 1884. *A Glimpse of Assam*. Calcutta: S.R. Ward, Thomas S. Smith City Press.

Wouters, Jelle J. P. 2020. "Neoliberal Capitalism and Ethno-Territoriality in Highland Northeast India: Resource-Extraction, Capitalist Desires and Ethnic Closure." *Geopolitics*, 25: 1–23, doi:10.1080/14650045.2020.1812581.

PART III

Bridging Disciplines and Teaching Clime Changes

8

STORYLINING CLIMES[1]

Theodore G. Shepherd and Chi Huyen Truong

Introduction

Climate change is recognized as one of the key challenges of our time. From the international perspective, which is to say within the United Nations Framework Convention on Climate Change (UNFCCC), the understanding of climate change and its implications is represented in the many reports from the Intergovernmental Panel on Climate Change (IPCC). The IPCC is divided into three working groups representing physical science, impacts and adaptation, and mitigation. Despite the broad scope of the problem, there has long been criticism of the fact that the experts are mainly drawn from only a subset of the social sciences (with a heavy focus on economics), and hardly at all from the humanities (Bjurström and Polk 2011; Corbera et al. 2016). It is then perhaps no surprise that the international community's collective response to the climate crisis reflects this technocratic perspective (Hulme 2009). The UNFCCC Paris Agreement of 2015 is widely seen as a crucial turning point in that response, but the novelist-scholar Amitav Ghosh's interpretation is otherwise (Ghosh 2016, 156): "...the [Paris] Agreement's rhetoric serves to clarify much that it leaves unsaid: namely, that its intention, and the essence of what it has achieved, is to create yet another neo-liberal frontier where corporations, entrepreneurs, and public officials will be able to join forces in enriching each other." Ghosh is suggesting not only that climate change is a consequence of colonialism (both past and ongoing; see also Ghosh 2021), but that the international response to climate change may further entrench it.

Clime as a dictionary term is sometimes seen as synonymous with climate but has a more affective implication. Within the environmental humanities and anthropology, the word has been adopted as referring to the cultural experience of climate, as discussed by Hulme (2009) and by Bristow and Ford (2016).

DOI: 10.4324/9781003347026-12

This might seem like a world apart from the technocratic nature of climate science represented by the IPCC. However, there is an older historical tradition in climate science which is more readily aligned with the concept of clime, and a recent development of storyline approaches to climate science. Together these offer a meeting point. Despite differences of perception, there is a shared physical reality out there, which all forms of being are experiencing, albeit in different ways. Only by finding ways to bridge between these different ways of knowing (Bristow and Ford 2016) and between climate change and daily lives (Dilling and Moser 2007) will it be possible to take up the challenge raised by Ghosh of resisting and indeed fighting back against a technocratic approach to climate-change "solutions" (Hulme 2009). The call for interdisciplinarity in this endeavor is hardly new; recent references are Coen (2021) and Schipper, Dubash, and Mulugetta (2021). The purpose of this particular piece—co-authored by a climate scientist and an anthropologist—is to contribute to this wider enterprise by drawing on the various scholarly traditions and concepts in climate science and in anthropology within the context of storylining climes. The piece is thus "multipolar" from an epistemic perspective.

Why consider high altitudes and high latitudes together? From the perspective of climate science, there are interesting parallels between the two kinds of regions. This follows from the simple fact that atmospheric temperature decreases both with increasing altitude and with increasing latitude. Although these temperature decreases occur for different reasons, they have similar consequences, because lower temperatures imply drier and more stable air, and the possibility of creating and maintaining frozen forms of water. These features in turn make the regions particularly sensitive to climate change. Basic physics implies that, all else being equal, the warming from increased greenhouse gases is greatest in the coldest parts of the climate system. (The Antarctic is an exception because the Antarctic ice sheets are so massive that they keep the continent cool.) Moreover, the melting or thawing of frozen forms of water which occurs with warming can change the very landscape of a region. As part of these landscapes of vastness and intricate interconnectedness, diverse human communities have demonstrated, through daily practice and oral tradition, the finest of their wisdom in adapting to their challenging environments and their changes over time. In this piece, we, therefore, choose case studies from both the Himalaya and the Arctic to illustrate the connections between altitudinal and latitudinal highlands and the complexity and challenges that local communities in Nepal and Alaska are facing while navigating unprecedented and accelerated changes. The piece is thus also "multipolar" from a geographical perspective.

We (a word we use exclusively to describe ourselves, as authors) start by reviewing relevant recent developments in both climate science and anthropology, to lay the table for our subsequent discussion. We then discuss our two case studies, which represent stories of loss, intertwining the climate-science and anthropology aspects. We do not pretend to be shedding any light on those two stories; our knowledge is entirely second-hand, obtained through published accounts.

Their purpose within our piece is to demonstrate how naturally the different perspectives of climate science and anthropology can be weaved together in particular contexts. Our final section provides a synthesis, where we identify several conceptual threads which, in our view, provide promising opportunities for creative dialogue between the social sciences, humanities, and natural sciences, within the context of climate change.

Storylining in Climate Science

As with many of the natural sciences, climate science has two distinct historical traditions. One tradition may be said to be descriptive, and is generally called "climatology." In disciplinary terms, climatology falls within physical geography but is also often linked with agricultural sciences because of the importance of climate for agriculture. This tradition is anchored in observational data and there is a strong focus on statistics and spatial mapping. The second tradition may be said to be explanatory. In this case, the focus is on general (or abstract) principles anchored in the laws of physics, and the mathematics involved is that of analysis rather than statistics. The latter distinction is significant because analysis and statistics have been largely separate mathematical disciplines, often housed in completely separate academic departments. While the word "explanatory" suggests a role for observations (as the thing to be explained), it is understood that for a natural (rather than an experimentally controlled) system, theoretical predictions will only be loose approximations of reality. As a result, there is a large element of subjectivity within the explanatory tradition in what is to be explained, and in what constitutes an explanation. Yet it has generally been the explanatory tradition which has claimed the mantle of objectivity (Hulme 2013b).

Both traditions seek to make sense of climate, but in very different ways: the descriptive tradition by a detailed understanding of the essential properties of different locations, and the explanatory tradition by an appeal to abstract physical laws. Both traditions have surely existed in some form for millennia. Within the West, the descriptive tradition particularly flourished from roughly the middle of the nineteenth century to the middle of the twentieth century, enabled by advances in meteorological instrumentation and in the field of statistics. It was also very much enabled by empire (Mahony and Endfield 2018). For example, Coen (2018) describes how the Habsburg (or Austro-Hungarian) Empire used climatology as a way of unifying its diverse political components, separated by history and language and with considerable local autonomy. It did this by showing how the climate of the different regions could reflect local differences while at the same time being interconnected (and interdependent) across spatial scales. The British Raj in South Asia had a different agenda, using climatology to support agricultural development and to predict year-to-year variations in the monsoon circulation, whose failure could lead to drought and famine (Amrith 2020).

The modern era of the explanatory tradition in the West can be traced back somewhat earlier, to the rise of mathematical physics beginning with Newton's laws in the late seventeenth century (Lorenz 1967). The perspective at that time was very much global and again was linked with empire insofar as a key explanatory target was the pattern of the trade winds, so important for imperial commerce in the days of sailing ships. The basic structure of the large-scale atmospheric circulation in the tropics, inferred from physical principles by George Hadley in 1735, bears his name and remains a central theoretical paradigm even today. However, the application of the laws of physics to fluid dynamics can only be performed by hand for steady (time-independent) flows, which severely restricted their applicability to climate for several centuries (in contrast with, for example, the motion of the planets). In the interim, the descriptive tradition reigned supreme. But with the advent of the digital computer in the middle of the twentieth century, the field of climate modeling was born. That development, together with the emergence of the pressing scientific question of anthropogenic climate change around the same time, rapidly changed the face of climate science. Martin-Nielsen (2015) chronicles this transition in the UK, as reflected in the experience of Hubert Lamb, a leading expositor of the descriptive approach to climate and its intimate link to history (Lamb 1995).

The descriptive tradition is far from dead: the annual "State of the Climate" report produced by the American Meteorological Society in collaboration with leading climate scientists around the world, which is entirely descriptive, runs to nearly 500 pages. However, the questions around anthropogenic climate change involve causation, and climate scientists agree that in order to establish causation, it is necessary to perform counter-factual, or "what-if-things-were-different?," calculations. And that requires a mathematical theory expressing the relevant cause-effect relationships, which climate models solve numerically, and which can be manipulated to perform counter-factual calculations. This is the main source of climate models' authority in the public sphere (Hulme 2013b). An accessible description of climate modeling is provided by Saravanan (2022), covering both its historical development and its strengths and limitations. The challenge for climate modeling is fundamentally two-fold. First, the equations can only be solved very approximately (because they only represent the relevant physical processes in a coarse-grained way). Second, the solutions provide single realizations of the simulated climate system (analogous to the single realization we experience of the real climate system), and due to a phenomenon known as chaos (Lorenz 1995), these simulated realizations do not track the real system and can only be meaningfully analyzed statistically, in terms of their aggregate properties (Shepherd and Sobel 2020).

The conundrum is that despite the replacement of a descriptive approach to climate by an ostensibly explanatory approach, one is thrown back on description. This is because the laws of physics solved by the models govern nearest-neighbor interactions (in both space and time), such as conservation of energy and momentum, yet the aspects of climate to be explained are emergent properties (for example, the South Asian monsoon) which are extended in both space and time, and without any direct connection to the nearest-neighbor interactions.

In particular, it is very difficult to understand why different climate models (which implement these nearest-neighbor interactions in slightly different ways) produce different statistical descriptions of climate. Some climate scientists try to understand these cross-scale connections using phenomenological theories, but all such theories are subjective and controversial. For the most part, climate scientists just abandon such attempts at understanding and combine ensembles of simulations from different climate models into a grand ensemble, for example, in the IPCC assessment reports. In the language of Hastrup (2013), the mode of configuration of knowledge has gone from narration to simulation to counting. Along the way, a thick description has been replaced by a very thin one. This procedure is presented as the climate science community's best answer to the relevant counter-factual questions regarding anthropogenic climate change, but is very ad hoc and is widely accepted to be fundamentally flawed. It only persists as the community standard because of the lack of any consensus on an alternative, and because it has the pretense of objectivity (colloquially referred to as "one model, one vote").

If the virtual climates simulated by the models were close to the observed climate, and their responses to anthropogenic climate change were similar, then this pragmatic approach might be acceptable. However, that is not the case (Shepherd 2014). The aspects of climate known as "thermodynamic" are robust: increasing greenhouse gases warm the climate system, which melts ice and causes sea level to rise. Those general consequences are enough to justify taking action to limit climate change. But the aspects of climate known as "dynamic" (namely those to do with winds and circulation systems) are neither particularly well simulated nor do they respond robustly to increasing greenhouse gases. This leads to considerable uncertainty in regional aspects of climate change, including extreme events (Shepherd 2019). While the climate models are slowly improving, the progress is very slow indeed, and most of the increased confidence in statements from the IPCC assessment reports over the years has come from a lengthened and deepened observational record, rather than from the climate models. But this raises the question of what is considered to be an observational record, which leaves many gaps around the world (Figure 8.1).

One of the reactions to this rather embarrassing situation has been to argue that in the face of such scientific uncertainty, and with such high stakes, climate science cannot be left to the scientists alone. The situation has been described as "post-normal" science (Funtowicz and Ravetz 1993), and so far the discussion has largely been carried out on the margins of climate science (Krauss, Schäfer and von Storch 2012; Hulme 2013a). This perspective aligns with the growing view among philosophers of science that purely epistemic values are not determinative of what constitutes good scientific practice, and that contextual values can and should enter in (Kuhn 1977; Douglas 2009). Acknowledging different contextual values is key to understanding why people disagree about climate change (Hulme 2009, 92): "Detached from its cultural anchors, climate change becomes a malleable envoy enlisted in support of too many rulers." The fact that Working Group

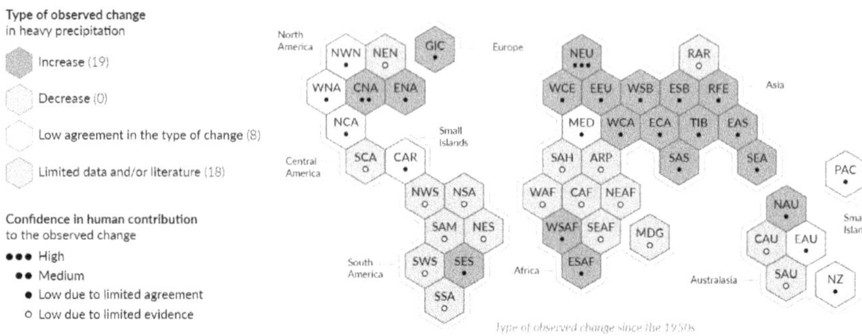

FIGURE 8.1 Synthesis of assessment of observed change in heavy precipitation and confidence in human contribution to the observed changes in the world's regions. Each hexagon corresponds to one of the IPCC AR6 WGI reference regions. Note there are no instances of Decrease.

Source: From Figure SPM.3 of IPCC (2021b).

I of the IPCC (representing physical climate science) has recently acknowledged this view creates a potential opening, although the "value-free ideal" provides useful cover for climate scientists and the working practice of climate science has yet to respond to this challenge (Pulkkinen et al. 2022).

Another reaction has come from within the climate science community, through the development of so-called "storyline" approaches to the representation of climate information (Shepherd et al. 2018). Such approaches navigate the uncertainty landscape by providing conditional explanations, either of the past or of a plausible future, through the device of *contingency*. By contingency is meant that while certain causal factors may not be predictable a priori, they can either be known (from observation) or hypothesized (from some kind of mechanistic theory); and that once these factors are prescribed, one can make sense of the situation of interest. The storyline approach—which we call "storylining"—represents a paradigm shift from the physics-based explanatory tradition in climate science, which aims for unconditional explanations (or, for the future, predictions) based on spatio-temporally invariant laws of nature (Shepherd and Sobel 2020). Yet it has a strong parallel with narrative-based or historical approaches that play a prominent role in other branches of the natural sciences, e.g., evolutionary biology (Shepherd and Lloyd 2021). Through the device of contingency, storylining provides a bridge between the descriptive and explanatory traditions in climate science discussed earlier. The term "physical climate storyline" is now part of the IPCC Working Group I Glossary (IPCC 2021a, 2250), which means it can be considered mainstream.

Storylining also aligns well with a decision-making perspective. Stirling (2010) has argued that when knowledge is uncertain, experts should resist the pressure (or temptation) to over-simplify the situation by framing their information in a "single, definitive" form. Instead, they should frame their information in a "plural, conditional" form, in order to adequately reflect the complexity of the situation. Such a framing also addresses any concern that a narrative approach to science might lead to "just-so stories," as multiple possibilities are articulated and can each be assessed as to their plausibility (Shepherd and Lloyd 2021). Finally, storylines engage an emotional element which has been shown to be a necessary ingredient for human decision-making (Damasio 1994). Storylines thereby provide a connection "between what we know and how we feel, between our affective, discursive and epistemological selves" (Bristow and Ford 2016, 6), which is essential for meaningful communication about climate change (Dilling and Moser 2007).

Enlivening and Worlding in Anthropology

Description, or detailed descriptive accounts, has always been part and parcel of anthropological craft since anthropology became an independent field of inquiry in the 1920s. Anthropologists of a large part of the twentieth century were traditionally concerned with human cultures of diverse societies, most oftentimes non-Western and studied as if these cultures or societies had not been altered by the colonial expansion from the West. Operating within the school of thought within which an anthropologist was trained, s/he would conduct a year-long ethnographic field research in a relatively small-sized community and produce an ethnographic monograph documenting its ways of life, showing how that society worked or how its culture would make sense to its members. However, the postmodern critique in the 1980s, about the same time as the advent of chaos theory and post-normal science, shook up social sciences, including anthropology (Marcus and Fischer 1986). The ways in which indigenous cultures had been altered, corrupted, and distorted by European imperial and colonial power, and how indigenous populations saw themselves through the eyes of the colonial Others, or "worlding," became central concerns of social-cultural anthropologists. The latter thus joined the postcolonial scholarship in giving voice to colonial "subjects," or the subaltern (Spivak 1988). Anthropologists thereby became more self-reflexive, with an increased awareness of how both the context of investigation and the context of explanation shape their anthropological accounts. Under the influence of post-structuralism, this recognition of the contextualization of knowledge laid the foundation for subsequent anthropological analyses of science, policy, and public discourse.

Over the last four decades, there are two streams in anthropological inquiry that are relevant for the purpose of this chapter. One is a movement toward an extension of difference beyond human societies and cultures. In this ontological turn, difference is not just within and between various societies and cultures of humanity, but also between worlds of beings—human and other-than-human beings (O'Gorman and Gaynor 2020). In this world of many worlds, not only nonhuman living beings such as plants, animals, and multitudes of species but also other beings such as water, rivers, mountains, etc., have their own part to contribute to the interaction and intra-action, regardless of human observations.

To this end, ethnography, or "lively ethography" as van Dooren and Rose propose, is seen as an appropriate concept-making genre that would enliven other-than-human actors and give voice to their accounts which would not otherwise be heard (Kirksey and Helmreich 2010; van Dooren and Rose 2016). A sociological concept of agency, understood as the capacity of individuals, human and other-than-human alike, to make their own choices and decisions and take action, is brought to the fore with a newly acquired intellectual advocacy.

The other stream is a development from an earlier branch in anthropology called "human ecology," looking at social and political systems from a materialist perspective, and encompassing technology and ecology. Seeking to maintain the relevance of their field, anthropologists have kept abreast of contemporary issues and issues of change: from development to globalization, from environmental degradation to climate change. Discovering interconnections, interdependencies, and possible causal linkages at the trans-local or global scale that underlie these changes marks the move away from descriptive documentation of local observations toward counter-factual explanatory accounts in anthropology. With the signature advantage of ethnographic field research at the local scale, anthropological studies that take up climate change phenomena as topics of inquiry demonstrate how locally embedded knowledge would be of significant scientific and political relevance (Dove 2014). Scholars of political economy and political ecology offer a unique contribution from their critical perspectives on power relations and inequity at the global scale as these are experienced by local communities (Pettenger 2007; Oliver-Smith 2012). The focus on power and politics helps anchor the ontological turn and make it an opening: instead of turning to difference-for-the-sake-of-difference, ethnographic storying as becoming-witness invites both response-ability and responsibility (Rose and van Dooren 2017). Humanities scholars under this influence further the postcolonial concept of "worlding" to unmake the colonial (human beings') universe and move toward conceptualizing a pluriverse, a world of many worlds, where all beings not only have their agency, intentionality, and sentience (Blaser and de la Cadena 2018), but also co-constitute these complex, cross-cutting, and intertwining worlds (O'Gorman and Gaynor 2020).

Ethnographic insights, historical perspective, and a holistic view are three distinctive contributions of anthropology to the study of climate change (Barnes et al. 2013). Since the millennium, anthropology of climate change has become a fast-growing branch of the field (Strauss and Orlove 2003; Crate and Nuttall 2016). A productive conversation between anthropology and climate science has started with a thorough examination of the social construction of climate change and climate change knowledge (Pettenger 2007; Hastrup 2013). The call for interdisciplinarity as a necessary, if not mandatory, approach for global change research has also been put forth among climate change anthropologists (Castree et al. 2014). However, a multitude of disciplines differs significantly from a built-in integration across disciplines or a genuine dialogue between disciplines, of which examples remain rare. An innovative duography—a pair of studies that juxtapose two complementary perspectives on climate change—by Boyer (2019) and Howe (2019) is a welcome experiment in reconciling global versus local and human versus nonhuman dichotomies and hints toward a multi-scalar and interdisciplinary endeavor. However, so far, the level of participation by indigenous communities in climate fieldwork has been patchy (Figure 8.2).

One way to make the global-local linkages productive for the anthropological inquiry of climate change is a recently refreshed take of "clime" by environmental historians and scholars of environmental humanities (Carey and Garone 2014). As climate serves to stabilize the perception of weather (Hulme 2015), the notion of clime points to the way in which climate is as much a socio-cultural construction as it is a scientific construction. Scholars working in this line seek to uncover the agency of climate through place-based lived experience and memories of extreme weather events or disasters. This enlivening ethnographic approach bears witness to clime, letting it speak for itself and thereby storying clime.

There would seem to be an opportune moment of conjuncture between the physical climate storyline approach and the recent ontological opening in social sciences and humanities, in particular the distinctively anthropological writing genre—thick-description ethnography, in the tradition of Geertz (1973)—when it is applied to place-based clime. Climate science storylining and enlivening ethnography converge in their questioning of objectivity based on a single and constant source (truth) and in their conviction of the contingency or conditionality of knowledge. This outlook goes back to Thomas Bayes for the former, and Friedrich Nietzsche for the latter. The storylining approach bridges between the descriptive and explanatory traditions in climate science, employing a combination of quantitative and qualitative elements in interpreting and giving meaning to data. Enlivening ethnography connects multiple agencies, human and other-than-human alike, within a co-constituted spatio-temporal platform where voices and actions can be heard and visualized. Storylining returns single, definitive forms of knowledge to their plural, conditional framing; enlivening

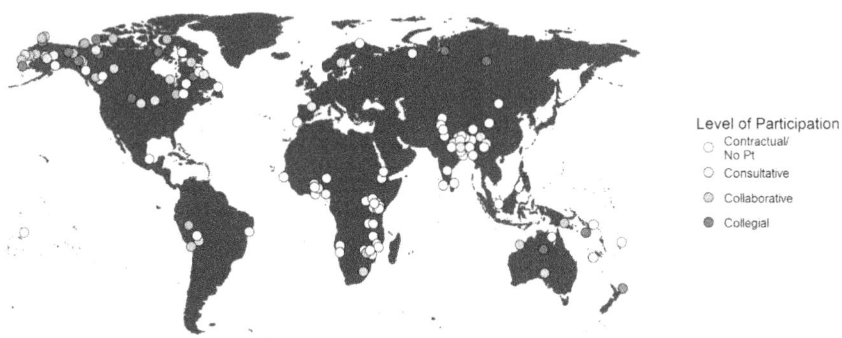

FIGURE 8.2 Patterns across space: global distribution of field sites classified by levels of indigenous community participation.

Source: From Figure 4 of David-Chavez and Gavin (2018).

ethnography, through the device of storying clime, allows multiple possibilities to emerge as imaginable if not plausible, through narrative. Storylining works well with past events pivoting toward plausible future pathways; storying clime focuses on the here and now of the experience of as many involved agencies as possible. Storylining seems to readily lend itself to facilitating engagement in conversation between diverse communities of scientists, social scientists, and humanities scholars, as well as practitioners. Storying clime may find storylining its desirable companion in the realization of response-ability and responsibility.

A Story of Loss: The Melamchi Flood Disaster

Administratively, Melamchi is only nine years old, since it was made into a municipality in 2014. It is located in the southwest of Sindhupalchowk District in the Bagmati Province of Nepal, about two hours' drive, or 69 km, from Kathmandu. In 2017, as part of the administrative restructuring under the new federal constitution of 2015, four villages and a part of Shivapuri Nagarjun National Park were added to the original seven villages that made up the municipality, covering an area of about 160 km^2 (A-NOT 2018) and inhabited by approximately 53,000 people. The name comes from the Melamchi Bazaar, a centuries-old trading center serving the ethnically diverse population of the district.

The Sindhupalchowk District is spread between mid-hills and high-hills/mountain terrain, from well below 1,000 m to Loenpo Gang peak at 6,979 m above mean sea level. This vast elevation difference results in diverse flora and fauna reflecting tropical, sub-tropical, and temperate climates. The average annual rainfall in the Melamchi basin is approximately 2,800 mm, concentrated during the monsoon season from mid-June to mid-September, and varying considerably by location within the valley, with higher rainfall amounts at the higher elevations (Office of Melamchi Municipality 2019). Melamchi River, a tributary of the Indrawati, originates from the Jugal Himal at 5,875 m elevation. It runs 41 km southward and meets up with the Indrawati at Melamchi Bazaar. The Melamchi Water Supply project, which was started in 1998 and completed in 2021, was designed to divert 170 million liters of water per day to supply the capital city of Kathmandu.

Melamchi inhabitants are made up of various ethnic and caste groups: 34% are Tamang, 24% Bahun, and 16% Chhetri. Hindu culture, practiced by the Bahun and the Chhetri, dominates the lowland settlements of the valley, while Buddhist culture, practiced by the Tamang and other ethnic groups, is prevalent in the mid-hills and high-hills villages. Among caste-ethnic groups, the Dalit—a marginalized caste previously known as "untouchable"—remain the most socio-economically disadvantaged. Agriculture accounts for 94% of the economically active populations, with cultivated land occupying about 65% of the total watershed area (Neupane and Rai 2018). Paddies are cultivated by the Bahun and the Chhetri cultivate irrigated fields in the lowland, while millet and maize are major crops of the Tamang in the mid- and high-hills. Agricultural land is the major physical asset for the lowland groups and agriculture is traditionally viewed as a superior occupation in local culture. Vegetable farming, livestock and poultry, fisheries, wage labor, and services supplement livelihoods. However, with the increase in the cash economy, out-migration, and international remittances, the caste-occupation-based hierarchy that followed elevation has started changing: the poorer upstream groups have tended to migrate more and thus earn greater remittances, which are displayed in local consumption patterns. As a result, the Tamang, who used to depend on the Bahun or Chhetri for cash or wage labor, have started redefining their identity (Pokharel 2010).

In 2015, a major earthquake, referred to as the Gorkha earthquake, hit Nepal. Sindhupalchowk was one of the districts worst-affected, recording the highest magnitude aftershock of 6.7. The casualties in Sindhupalchowk (3,057) account for one-third of the Nepal total (8,964). Research is yet to confirm whether the 2015 earthquake made the district's landscape more prone to landslides (Kincey et al. 2021; Marc et al. 2019), but more frequent landslides were reported during the three years following the earthquake and local communities seem to have been aware of increased risks. On 14 August 2020, a landslide claimed eleven lives with twenty-seven reported missing in the Jugal Rural Municipality of Sindhupalchowk. A local resident recalled that "There were foreboding signs for weeks, and small landslides had been breaking out above the village" (Bushal 2021). The memorandum the community submitted to the District Disaster Management Community requesting help a week before the landslide struck was not acted upon (Tiwari 2020).

And then came the June 2021 flood event. (We draw on Maharjan et al. 2021 for the following details.) The monsoon of 2021 started around 10 June, with heavy rainfall in the headwaters of the Melamchi river. It had been preceded by heavy snowfall from two cyclones, leading to snowmelt and a rain-on-snow situation, which is conducive to heavy streamflow. On 16 June, a flash flood from the two converging rivers—the Melamchi and Indrawati—hit Melamchi Bazaar, claiming the lives of five people with another twenty missing, and bringing heavy damage to the Melamchi Water Supply project as well as to numerous individual homes and assets. Two days later, a landslide occurred on the slope of the left bank of the Melamchi River, setting in motion a chain of events which buried the farmland in the valley with debris (Figure 8.3).

From a geophysical perspective, the Melamchi flood event is an example of what is called a compound event, with cascading hazards. The temporal sequence of events is not known precisely, and may not be ever known, because of the patchwork nature of the physical evidence: human reporting (including photographs), digital records from a small number of weather stations, and images from satellite instruments (but with large gaps in time). Nevertheless, an overall narrative can be put together. The spring snow cover was a precondition for heavy streamflow once the monsoon commenced. Earlier landslides had created natural dams, which were then overtopped by the heavy streamflow. This discharged a high-speed flow of water within the narrow valley, hyper-concentrated with

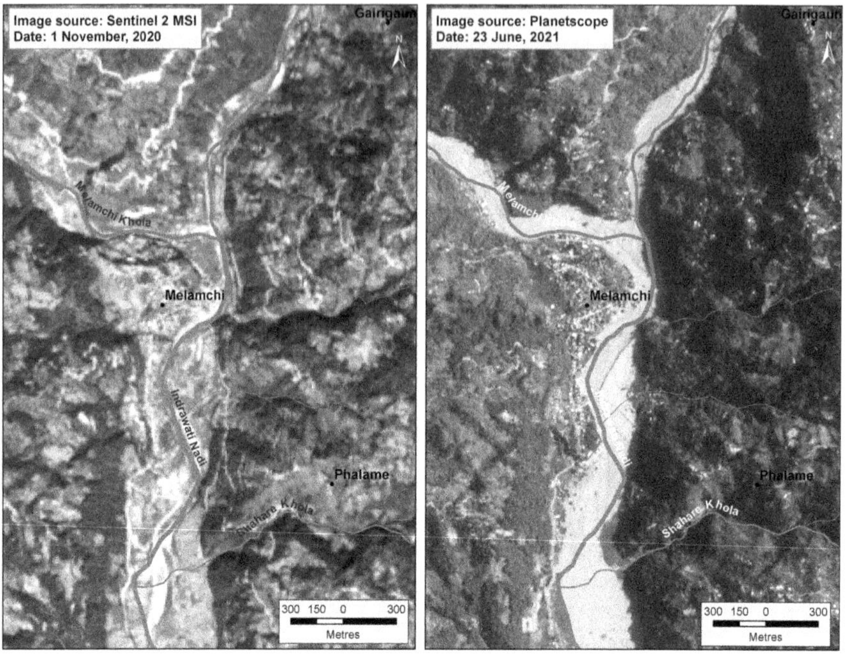

FIGURE 8.3 Debris deposition as seen from satellite images taken before (left) and after the Melamchi event (right).

Source: From Figure 14 of Maharjan et al. (2021).

debris, leading to numerous riverbank collapses and further landslides and ultimately depositing a massive amount of sediment downstream, inundating the farmland in the valley. The immense power of these cascading processes came from the vertical drop, which is a source of potential energy. But it was also a compound event from a human perspective. Upstream communities had no warning and suffered heavy casualties where multiple small landslides occurred during the event. Downstream communities were warned by the upstream communities and were able to take sufficient precautions; families along the riverbank in Melamchi Municipality had evacuated before the flood reached them, and thus no life was lost in these communities. However, the permanent loss of irrigated farmland made it more challenging for the downstream communities, whose livelihoods relied predominantly on agriculture, to rebound following the event.

Cascading events of a similar scale are not rare in the Hindu Kush Himalaya. Earlier the same year, in February 2021, a massive rock and ice avalanche from an elevation of 6,063 m rapidly transformed into an extraordinarily large and mobile debris flow descending the Ronti Gad, Rishiganga, and Dhauliganga Valleys in Chamoli, Uttarakhand, India (Shugar et al. 2021). The cascade of events caused more than 200 deaths or missing persons and damaged infrastructure, including two hydropower projects. Unfortunately, local people on site had little to no warning.

There is much agency in how the Melamchi event unfolded, and much contingency. Landslides and debris flows are natural phenomena in this region, but are expected to become more frequent with climate change because of destabilization of landscapes from warming, heavier precipitation, and more rain-on-snow events (Krishnan et al. 2019). The evolution of any particular event depends on the history of previous events, through legacies in the form of natural dams from past landslides, and moraine dams. In this case, the 2015 Gorkha earthquake may also have played a role in further destabilizing the landscape. The Melamchi event gained the attention it did in part because of the fact that it affected the Kathmandu water supply, and that the impact was so visually dramatic. The history of ethnic-caste hierarchies in the region meant that the wealthier downstream communities suffered fewer casualties because of the warning from the poorer upstream communities, but were more affected in the long term. Indeed, the landscape change for the downstream communities is essentially irreversible, and has completely altered their future by depriving them of their home and livelihood. All of this is explainable, but none of it was predictable from abstract theory, nor did it have to happen.

A Story of Loss: Usteq

The curves of Kuskokwim River, filled with sediment-rich dark brown slushy water, contour the boreal forests, swamps, and tundra in multiple shades of green, dotted by icy blue lakes of the Yukon-Kuskokwim Delta in southwest Alaska (Figure 8.4). The expansive, windswept subarctic Delta, one of the largest deltas in the world (500,000 sq. miles), has an almost triangular shape, with the Yukon River as its northern border, Kuskokwim to the southeast, and the Bering Sea to the southwest. It is the home of approximately 25,000 people, 85% of whom are Alaskan Natives, the Yup'ik, and the Athabaskan.

The Yup'ik live in small villages with less than a thousand inhabitants, and practice a subsistence-based lifestyle, that they term *yuuyaraq* (our way of life), consisting of hunting, gathering and fishing, with employment and a small cash economy limited to urban areas. Yup'ik folklore records the experience of change, both in their environment and in society. Indeed the Yup'ik name for the city of Quinhagak, *Kuinerraq*, means "new river channel" and the Yukon River itself, as the collective memory recalls, was narrower hundreds of years ago than it is today. Periodic relocation was part of community experience in search of peace from tribal wars during the pre-contact period (Lim et al. 2021). Relocations due to changes in water and land also took place in the past, and memories of such relocations can still be recalled. The elderly share a long-held popular belief of the Quinhagak community having to relocate five times before finding a permanent settlement. A combination of remote sensing and ethnographic inquiry reveals settlement sites of a recent past, with some of the relocations due to changes in nature such as a lake becoming dried up (Lim et al. 2021).

The Arctic is the most rapidly warming part of the world, and Alaska is warming particularly rapidly (Taylor et al. 2017). In contrast to mountainous regions

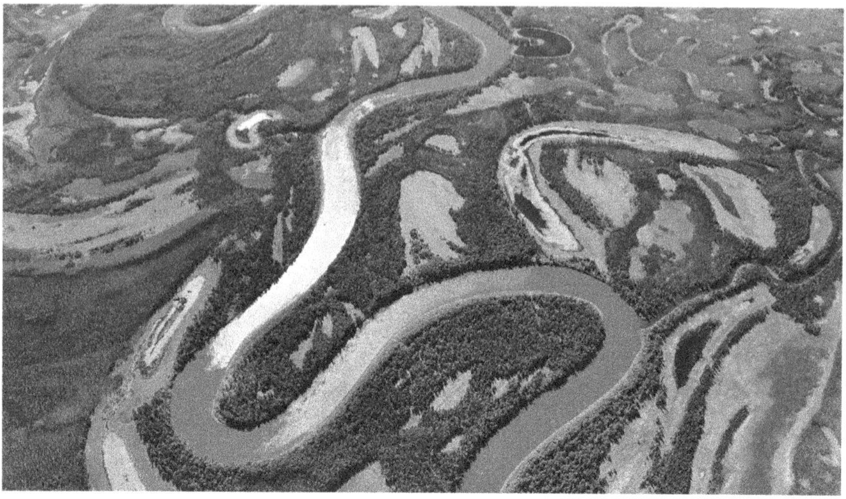

FIGURE 8.4 Kuskokwim River, in southern Alaska. This aerial photo was taken by scientists on a NASA DC-8 "flying laboratory" in their mission to study how thawing permafrost affects hydrology in the landscape by measuring the elevation of rivers and lakes.

Credit: Peter Griffith, NASA. https://climate.nasa.gov/climate_resources/159/kuskokwim-river/, accessed August 12, 2022.

like the Hindu Kush Himalaya, Arctic inhabitants live in direct proximity with the frozen components of the climate system, and indeed rely on them either for the stability of the land under their feet (permafrost) or for transportation and hunting (perennial lake and sea ice) (Singh, also Andersen, this book). As the climate warms, the permafrost thaws, which leads to riverbank erosion and landform collapse. Moreover, the ice-free summer season becomes longer, which not only adversely affects transportation and hunting, but prevents the formation of land-fast ice in coastal regions, thereby exposing coastlines to storm surges. The rapidity of Arctic warming means that these changes are occurring at unprecedented rates, most dramatically when storm surges combine with permafrost thaw in coastal regions. As a result, landscapes are changing before people's eyes, with meters of land collapsing into the sea in a single night (Climate Justice Resilience Fund 2019). The fact that the Yup'ik had to create a new word to describe the phenomenon, *usteq* (roughly translated as "surface caves in"), documents its unprecedented nature.

In a US government report published nearly twenty years ago (GAO 2003), the danger posed by *usteq* was estimated to have affected at least 86% of 213 Alaska Native communities as evidenced by 190 disaster emergencies recorded in the last quarter of the previous century. A number of subsequent reports and assessments—the Alaska Climate Impact Assessment (2008) and the Alaska Baseline Erosion Assessment (2009)—documented the lack of progress in either the state of knowledge or action to assist affected communities (GAO 2009).

Ten years after the 2009 GAO Report, the Denali Commission Report of 2019 surveyed 187 of Alaska's rural communities and evaluated individual threats to public infrastructure associated with erosion, flooding, thawing permafrost, and the compound threats imposed by interactions between the three. It proposed an analytical process in scoring and ranking communities' risks. The Denali Report identifies thirty-eight communities facing immediate and serious threat from flooding, and seventy-nine under immediate serious threat from compound vulnerabilities. The Report also classifies scales of "time to damage"—with the most urgent damage possibly occurring in the next five years, compared to erosion which occurs slowly over twenty years. This evaluation of risks by a panel of multiscientific experts, however, does not characterize the actual status of risk faced by the concerned communities, nor does it capture the perceptions of risks by the local people. The Report acknowledges that uncertainty is directly correlated with the quality and quantity of available data and the rankings are only to be used to identify communities requiring additional investigation (University of Alaska Fairbanks Institute of Northern Engineering 2019).

In response to this challenge, some communities are considering co-relocation, a process in which the population of an entire community relocates to a new site where residents can continue to practice their subsistence lifeways (Ristroph 2017). As early as the late 1990s, residents from three of the most-at-risk rural communities, Newtok, Kivalina, and Shishmaref, took the initiative in planning to co-relocate their entire community with the assistance of the government. They met with many difficulties, including expense, delay, lack of leadership among state and federal agencies in facilitating relocation, and confusion between the laws and institutions related to relocation and climate change. It took twenty-five years for Newtok, a village of 375 residents located by the Ninglick River in Southwest Alaska, to relocate to a new site, Mertarvik, 14 km away on higher, volcanic ground. With an average rate of 20 m of shoreline loss every year, Newtok faced a slow-moving disaster (Cole 2018). Lack of investment in public works due to the prolonged uncertainty during the more than two decades of negotiation and planning left residents without plumbing and sanitation. In exchange for the costs of relocation, the people of Newtok will eventually relinquish their lands to the Yukon Delta National Wildlife Refuge (Welch 2019). In the meantime, many other communities who remain in between protect-in-place and relocation continue to experience multiple forms of uncertainties. In Quinhagak, which is on the coast and at the mouth of a river, there has been contamination from the sewage lagoon, inaccessibility of the only functional dock, and a housing crisis due to subsidence and subsequent mold and rot. In Nunapitchuk, which is about 12 km inland from the coast and built on three riverbanks, there has also been contamination from the sewage lagoon and dumpsites, the failure of foundations of important buildings, and the removal of the power plant to the nearby village of Kasigluk (Bronen et al. 2020).

In the absence of a methodological and governance framework to evaluate climate change impacts, there have been noteworthy efforts to develop monitoring and assessment tools that are led by indigenous researchers and communities (Bronen et al. 2020). This approach engages community members as equal partners in the design and implementation of research that incorporates indigenous knowledge and addresses their needs. Such an adaptive governance framework, as a result of multi-year work with the two communities, Nunapitchuk and Qingagak, has integrated indigenous science and Western science in the co-production of knowledge and planning for the course of action that makes sense to the people themselves.

However, the process to prepare and implement community adaptation plans takes place against the backdrop of complex legal context and changing socio-economic conditions in Alaska. Climate-change-induced environmental changes are not the only challenges Native Alaskan—the Yup'ik, Athabaskan, and Inupiat—are facing: a socio-economic system that is no longer sustainable, lack of ownership and control over land or associated natural resources that could be used for adaptation, the high costs of living and limited access to jobs and health care due to distance from urban centers, the decline of population in some areas, youth unemployment, substance abuse, and domestic violence (Ristroph 2017). While there is no one-size-fits-all package of solutions for any combination of these issues, three keys to sustainable development have been identified: practical self-rule, capable governing institutions, and cultural match (Cornell and Kalt 2003). Self-rule is found to promote citizen engagement, put the development agenda in indigenous hands, and link decisions and their consequences. Effective institutions of community self-governance, institutional integrity, and practical action may pave the way for productive economic development. Lastly, institutions have to resonate with and build upon indigenous political conceptions if they are going to deliver the common goods.

Sweetgrass-Braiding

The two stories of loss—one of high altitude, one of high latitude; one extremely fast-paced on a human timescale, one relatively slow-paced, yet still occurring within living memory—give us windows into the two climes to be storylined (Figure 8.5). In Melamchi, the instability of landforms, and the man-made infrastructure, local media, government, and the scientific community, all had or could have had some role in the unfolding of the events. In Alaska, the pace at which *usteq* occurs, the existing experience in regard to relocation of each of the at-risk communities, their internal coherence and relationship with external agencies, all may shape that experience quite differently from one community to another. Singh (2021) argues that to develop a justice-centered pedagogy of climate change, it is helpful to view climate change through three transdisciplinary meta-concepts: balance-imbalance, planetary boundaries and limits, and complexity. All three are exemplified in our case studies. Climate change arises from imbalance in the Earth's energy budget, and leads to imbalance in cryospherically affected landforms. Limits are breached when these landforms collapse.

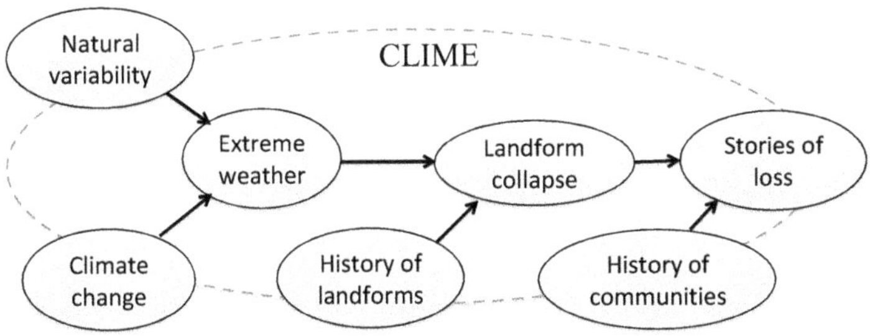

FIGURE 8.5 Storylines of our two case studies, represented together in the form of a causal network (Shepherd and Lloyd 2021) depicting the various actors in the events.

Complexity is evident in the intertwined history of weather, communities, and landforms, together representing clime.

No story ends just when or where the story-teller stops narrating. Six months after the Melamchi flood, Dalit communities from the upstream regions had still not received any relief support from the government. "[T]elling our stories of pain is useless," said a community member, expressing his frustration to a journalist and telling the visitor to come back in the next monsoon season to see if he were still alive (Bushal 2021). In Alaska, Newtok was the first community that made its move; two or three other villages are planning theirs. The way in which each community can tell its story in the face of climate change-induced and other socio-economic challenges will vary from one community to another and will depend on the extent to which their knowledge and perceptions are incorporated in the adaptation and resilience-building process.

Our two stories of loss illustrate through concrete example how climes can be storylined. We recognize that these stories could have been told quite differently. While some portion of the vignettes might have been left out without our intention, we have brought together strands of experiences and understandings that in another account might have been overlooked. We now step back to reflect on how the current moment in both anthropology and climate science provides an opportunity for productive and creative engagement between the humanities, social sciences, and natural sciences. In this respect, we can already identify several potential threads for this dialogue.

The first is the role of *time*. In climate science (as in physics), time is a continuum, and there is no fundamental distinction between past and future. Time only acquires meaning through physical imbalances in the climate system, but for the most part these imbalances are time-reversible. The present is essentially meaningless because it is only a snapshot in time, while climate by definition consists of long-term statistics (the classical definition in climatology is thirty years). Conventional anthropology is essentially the opposite; it deals with the here and now. Climate science is challenged by the here and now and by contingency, and anthropology is challenged by imbalance and temporal change;

in both cases, the ontologies are stretched. History deals with the past, with the present projected remotely as a mental canvas for the moral of the story to be inscribed. Sub-branches within anthropology and humanities have cross-fertilized and experimented to make the temporal dimension more productive. Anthropologists of climate change have started familiarizing themselves with anticipation and anticipating nature (Strauss and Orlove 2003; Hastrup 2013; Salazar 2017); Whyte (2021) has argued that indigenous peoples have represented climate-change time in terms of changing kinship relationships. Andersen (this book) describes how in the tundra of West Greenland, climate and social change have led to a responsive and relational pluriverse of "deep, accelerated, and troubled times." Our stories of loss show how in the world-as-lived, the past, present, and imagined future come together through intertwining compatible concepts in both climate science and anthropology.

The second thread is the nature of *agency and intentionality*. From a climate science perspective, agency is closely linked to causality and counter-factual reasoning, which is in turn linked to imbalance, as an intervention in the normal functioning of a system. In conventional anthropology, agency resides in the human domain and there has been a reluctance to consider the agency of climate lest it lead to the pernicious concept of "climate determinism," which is an instrument of colonialism. Yet a factor can be influential without being determinative. Enlivened anthropology has substantially widened the cast of actors with speaking parts in the drama. By providing voice to the South Asian monsoon, Amrith (2020) similarly provides an "enlivened history." Ghosh (2016) argues that literature needed to be enlivened by allowing climate change to have agency within the narrative, and further argues (2021) for its intentionality, suggesting that Earth is "fighting back" against what humans are doing to it. The latter appeals to the "Gaia hypothesis," broadly understood as the notion that the Earth is a living entity, which was first articulated by the atmospheric scientist Lovelock (1972). Although the Gaia hypothesis has been highly controversial among climate scientists, the objections would seem to rely on an overly anthropomorphic view of intentionality. One can accept the Gaia hypothesis while remaining within a conventional scientific paradigm, provided one accepts that emergent phenomena can arise in ways that are not immediately apparent from reductionist mechanisms (the nearest-neighbor interactions discussed earlier). In this respect, Karen Barad's relational ontological notion of "agency [as] an enactment, a matter of possibilities for reconfiguring entanglements," offers a way of thinking about causality and response-ability in complex systems (Dolphijn and van der Tuin 2012, 54, 55).

The third thread is *chaos*. During the last two decades of the twentieth century, anthropologists working within structuralist or post-structuralist traditions started drawing on chaos theory for inspiration (Wagner 1991; Strathern 1995; Appadurai 1996). Most of chaos theory (Lorenz 1995) describes *temporal chaos*, namely that even in a deterministic system, small differences in the state of a system grow in time until the two states are completely different. But there is also the concept of *spatial chaos* (Nowak and May 1992), namely that complex quasi-steady patterns can arise in a system, which are unpredictable yet which exhibit a coherent structure. Traditional climate science averages over all the chaos (Shepherd and Sobel 2020), but there are two problems with this. The first is that in a chaotic system, even the average behavior is highly sensitive to the specification of the system, which is why modeled climates can differ in significant ways from the observed climate. The second is that the observed realization of climate is subject to both temporal and spatial chaos. History can be thought of as describing a process of temporal chaos. Anthropology can be thought of as describing a process of spatial chaos. Both depend on contingency for their explanations. This extends to climate science, where storylining is a way to make sense of the observed record, and of the differences between climate models, through the device of contingency. Contemporary anthropology is attracted to chaos theory's notion of contingency precisely because of its power of contextualization (Mosko 2005). This helps close the loop at the meeting point with the ontological opening where emphasis is placed on the relations and intra-actions of human and other-than-human beings instead of beings as entities-unto-themselves.

The fourth thread is the *two ways of knowing* which probably permeate any scholarly discipline. What we have called the "descriptive" tradition in both climate science and anthropology would seem to be well aligned with attributes such as concrete, differentiated, ritual, and spatial. And what we have called the "explanatory" tradition would seem to be well aligned with attributes such as (respectively) abstract, aggregated, narrative, and temporal. One way may dominate the other at any given time, but there is invariably an ebb and flow between them within any discipline, and between disciplines. If they are seen as a strict dichotomy then it is a false dichotomy, since both ways are needed for balance. Other dichotomies are more ambiguously placed. In climate science, abstraction is perceived as more objective than description, but in anthropology it would be the opposite; this reflects Karen Barad's observation that "objectivity is a matter of responsibility and not a matter of distancing" (Dolphijn and van der Tuin 2012, 57). And where episodic knowledge (stories) and semantic knowledge (facts) fit within this framework will also differ between disciplines. Similar to the relation between storying and clime, storylines represent a "concrete abstraction"—a form of thick-description ethnography—that can be put to work by bridging between the two strands, or "climing" in the usage of Paerregaard (this book). In *Braiding Sweetgrass*, Kimmerer (2013) describes how indigenous Native American knowledge of plants can be combined with the Western science of botany to yield a productive synergy. Related to the theme of water, she points out (55) that the Ojibwe word for a bay, *wiikwegamaa*, is a verb rather than a noun, indicating that water is "being" a bay at that moment in time. As with Amrith's (2020) depiction of the South Asian monsoon (see also Amrith and Smyer Yü, this book), this more dynamic, or enlivened, representation of water is well aligned with the climate-science concept of the water cycle (Dorigo et al. 2021). We adopt sweetgrass-braiding as a metaphor for how different ways of knowing can be brought together by storylining climes (Figure 8.6).

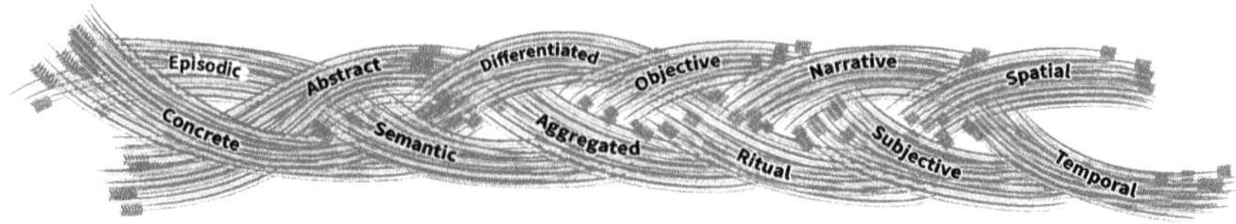

FIGURE 8.6 Sweetgrass-braiding of epistemic dichotomies, in the spirit of Kimmerer (2013). One strand reflects "descriptive" ways of knowing: concrete, differentiated, ritual, spatial. Another strand reflects the contrasting "explanatory" ways of knowing: (respectively) abstract, aggregated, narrative, temporal. A third strand contains more ambiguous dichotomies: episodic vs. semantic, and objective vs. subjective.

The novelist Wendell Berry describes the following dichotomy (Berry 2017, 182, 184):

> The Rational Mind is motivated by the fear of being misled, of being wrong. Its purpose is to exclude everything that cannot empirically or experimentally be proven to be a fact.
>
> The Sympathetic Mind is motivated by fear of error of a very different kind: the error of carelessness, of being unloving. Its purpose is to be considerate of whatever is present, to leave nothing out.
>
> The definitive practical aim of the Sympathetic Mind is to adapt local economies to local landscapes. This is necessarily the work of local cultures.

This is yet another expression of our two ways of knowing, and one can easily see on which side Berry is leaning. Yet even this dichotomy has a scientific interpretation: in that language, the Rational Mind is afraid of so-called Type 1 errors (a false alarm), while the Sympathetic Mind is afraid of so-called Type 2 errors (a missed warning). Climate science has tended to prioritize the former, but storyline approaches prioritize the latter (Lloyd and Oreskes 2018).

We are reminded of Lévi-Strauss's (2021, orig. 1962) "pensée sauvage"—an untamed, intuitive state of mind as distinct from a purposeful mind, the latter cultivated with an aim to yield a return. In a circumstantial world where a universal package of solutions is a guarantee of failure, scientists and social scientists, policy makers, and communities of a particular clime may find a bricoleur's way sensible: rendering knowledge useful and usable through joint interpretation of it. Thus, it is not one or the other: sweetgrass-braiding is possible and bricolage is necessary. In this process, all scholarly disciplines are challenged. Rodrigues and Shepherd (2022) challenge climate science by repurposing the economist E.F. Schumacher's *Small is Beautiful* (1973) to ask how climate-change science might look if it were structured "as if people mattered." In their multilayered unpacking of the phrase "climate change is all about us," Bristow and Ford (2016, 7) similarly challenge the humanities: "Like lichens and coral reefs, the disciplines of the humanities may also be considered as climatically sensitive forms. Climate change is all about us not least as a cascade of collapsing and outmoded philosophical and scholarly categories."

Berry's evocation to "be considerate of whatever is present, to leave nothing out" is nothing other than a plea for enlivening, and is echoed by Ghosh (2021, 76) who discusses the "loss of meaning that is produced by the vision of world-as-resource." Kimmerer, too, emphasizes that the rational needs to serve the sympathetic, rather than the other way around (as, for example, in the use of traditional knowledge to provide commodities). Ghosh sees anthropogenic climate change as the manifestation of colonialism, both historical and present-day, in which only a small subset of humanity is seen to have a legitimate voice. Against this, enlivening ethnography would allow one to stay tuned to the voices that would not have been heard or the existence that would not have presented of various beings, be they a river or its banks, steep slopes, or valleys. The Anthropocene Himalayan and Arctic land-and-waterscapes have already been transformed by the local and the non-local, including by the scientific community itself, to the point of no return. Enlivening ethnography by storylining clime would open ways for human and other-than-human beings to narrate their experience of worlding and of bearing witness of others in their co-constituted pluriverse. In the words of Ezra Pound (1993, orig. 1948),

> Pull down thy vanity, it is not man
> Made courage, or made order, or made grace,
> Pull down thy vanity, I say pull down.
> Learn of the green world what can be thy place...

Note

1 This chapter is one of the outcomes from The European Union's Horizon 2020 Research and Innovation Programme under Grant agreement No. 820712, through the RECEIPT project.

References

Alaska Baseline Erosion Assessment. 2009. Elmendorf Air Force Base, Alaska: U.S. Army Corps of Engineers.

Alaska Climate Impact Assessment Commission. 2008. *Final Commission Report, Alaska Climate Assessment Commission*. Juneau, AK: Alaska State Legislature.

Amrith, Sunil. 2020. *Unruly Waters*. London: Penguin Books.

A-NOT Architecture and Architects. 2018. *Land Pooling Feasibility Study of Bahunepati Area*. Melamchi Municipality, Office of the Municipal Executive, Melamchi, Sindhupalchowk.

Appadurai, Arjun. 1996. *Modernity at Large: Cultural Dimensions of Globalization*. Minneapolis: University of Minnesota Press.

Barnes, Jessica, Michael Dove, Myanna Lahsen, Andrew Mathews, Pamela McElwee, Roderick McIntosh, Frances Moore, Jessica O'Reilly, Ben Orlove, Rajindra Puri, Harvey Weiss, and Karina Yager. 2013. "Contribution of Anthropology to the Study of Climate Change." *Nature Climate Change* 3: 541–544.

Berry, Wendell. 2017. *The World-Ending Fire*. London: Penguin Books.

Bjurström, Andreas, and Merritt Polk. 2011. "Physical and Economic Bias in Climate Change Research: A Scientometric Study of IPCC Third Assessment Report." *Climatic Change* 108: 1–22.

Blaser, Mario, and Marisol de la Cadena. 2018. "Proposals for a World of Many Worlds." In *A World of Many Worlds*, edited by M. de la Cadena and M. Blaser, 1–22. Durham and London: Duke University Press.

Boyer, Dominic. 2019. *Energopolitics: Wind and Power in the Anthropocene*. Durham, NC: Duke University Press.

Bristow, Tom, and Thomas H. Ford. 2016. "Climates of History, Cultures of Climate." In *A Cultural History of Climate Change*, edited by T. Bristow and T. H. Ford, 1–14. London: Routledge.

Bronen, Robin, Denise Pollock, Jacquelyn Overbeck, DeAnne Stevens, Susan Natali, and Chris Maio. 2020. "Usteq: Integrating Indigenous Knowledge and Social and Physical Sciences to Coproduce Knowledge and Support Community-Based Adaptation." *Polar Geography* 43 (2, 3): 188–205.

Bushal, Ramesh. 2021. "Six Months after Melamchi Floods in Nepal, No Respite for Displaced People." *The Third Pole*, December 7, 2021, accessed January 26, 2022. https://www.thethirdpole.net/en/livelihoods/melamchi-floods-in-nepal-six-months-on/

Carey, Mark, and Philip Garone. 2014: "Forum Introduction." *Environmental History* 19: 282–293.

Castree, Noel, William M. Adams, John Barry, Daniel Brockington, Bram Büscher, Esteve Corbera, David Demeritt, Rosaleen Duffy, Ulrike Felt, Katja Neves, Peter Newell, Luigi Pellizzoni, Kate Rigby, Paul Robbins, Libby Robin, Deborah Bird Rose, Andrew Ross, David Schlosberg, Sverker Sörlin, Paige West, Mark Whitehead, and Brian Wynne. 2014. "Changing the Intellectual Climate." *Nature Climate Change* 4: 763–768.

Climate Justice Resilience Fund. 2019. "There Is No Coming Back from Disappearing Coastlines." *Climate Home News*, July 23, 2019, downloaded January 26, 2022. https://www.climatechangenews.com/2019/07/23/no-coming-back-disappearing-coastlines/

Coen, Deborah R. 2018. *Climate in Motion: Science, Empire, and the Problem of Scale*. Chicago and London: University of Chicago Press.

Coen, Deborah R. 2021. "A Brief History of Usable Climate Science." *Climatic Change* 167: 51.

Cole, Dermot. 2018. "How Newtok's Relocation Efforts Could Be an Example for Other Climate-Threatened Northern Communities." *Arctic Today*, April 29, 2018, accessed January 25, 2022. https://www.arctictoday.com/newtoks-relocation-efforts-example-climate-threatened-northern-communities/

Corbera, Esteve, Laura Calvet-Mir, Hannah Hughes, and Matthew Paterson. 2016. "Patterns of Authorship in the IPCC Working Group III Report." *Nature Climate Change* 6: 94–99.

Cornell, Stephen, and Joseph P. Kalt. 2003. *Alaska Native Self-Government and Service Delivery: What Works?* Cambridge, MA: The Harvard Project on American Indian Economic Development.

Crate, Susan A., and Mark Nuttall. 2016. "Introduction: Anthropology and Climate Change." In *Anthropology and Climate Change: From Actions to Transformations*, edited by S. A. Crate and M. Nuttall, 11–34. New York: Routledge.

Damasio, Antonio. 1994. *Descartes' Error*. New York: Putnam.

David-Chavez, Dominique M., and Michael C. Gavin. 2018. "A Global Assessment of Indigenous Community Engagement in Climate Research." *Environmental Research Letters* 13: 123005.

Dilling, Lisa, and Susanne C. Moser. 2007. "Introduction." In *Creating a Climate for Change: Communicating Climate Change and Facilitating Social Change*, edited by S. C. Moser and L. Dilling, 1–27. Cambridge: Cambridge University Press.

Dolphijn, Rick, and Iris van der Tuin. 2012. "'Matter Feels, Converses, Suffers, Desires, Yearns and Remembers': Interview with Karen Barad." In *New Materialism: Interviews and Cartographies*, 48–70. Ann Arbor, MI: Open Humanities Press.

Dorigo, Wouter, Stephan Dietrich, Filipe Aires, Luca Brocca, Sarah Carter, Jean-François Cretaux, David Dunkerley, Hiroyuki Enomoto, René Forsberg, Andreas Güntner, Michaela I. Hegglin, Rainer Hollmann, Dale F. Hurst, Johnny A. Johannessen, Christian Kummerow, Tong Lee, Kari Luojus, Ulrich Looser, Diego G. Miralles, Victor Pellet, Thomas Recknagel, Claudia Ruz Vargas, Udo Schneider, Philippe Schoeneich, Marc Schröder, Nigel Tapper, Valery Vuglinsky, Wolfgang Wagner, Lisan Yu, Luca Zappa, Michael Zemp, and Valentin Aich. 2021. "Closing the Water Cycle from Observations across Scales: Where Do We Stand?" *Bulletin of the American Meteorological Society* 102: E1897–E1935.

Douglas, Heather E. 2009. *Science, Policy, and the Value-Free Ideal.* Pittsburgh, PA: University of Pittsburgh Press.

Dove, Michael. 2014. "Introduction: The Anthropology of Climate Change. Six Millennia of Study of the Relationship between Climate and Society." In *The Anthropology of Climate Change: An Historical Reader*, edited by M. Dove, 1–36. Malden, MA: Wiley Blackwell.

Funtowicz, Silvio O., and Jerome R. Ravetz. 1993. "Science for the Post-Normal Age." *Futures* 25: 739–755.

Geertz, Clifford. 1973. "Thick Description: Toward an Interpretive Theory of Culture." In *The Interpretations of Cultures*, edited by C. Geertz, 3–30. New York: Basic Books.

Ghosh, Amitav. 2016. *The Great Derangement.* Chicago and London: University of Chicago Press.

Ghosh, Amitav. 2021. *The Nutmeg's Curse.* Chicago and London: University of Chicago Press.

Government Accountability Office (GAO), U.S. 2003. "Alaska Native Villages: Most Are Affected by Flooding and Erosion, but Few Qualify for Federal Assistance." Washington DC: U.S. Government Accountability Office.

Government Accountability Office (GAO), U.S. 2009. "Alaska Native Villages: Limited Progress Has Been Made on Relocating Villages Threatened by Flooding and Erosion." Washington DC: U.S. Government Accountability Office.

Hastrup, Kirsten. 2013. "Anticipating Nature: The Productive Uncertainty of Climate Models." In *The Social Life of Climate Change Models: Anticipating Nature*, edited by K. Hastrup and M. Skrydstrup, 1–29. London: Routledge.

Howe, Cymene. 2019. *Ecologics: Wind and Power in the Anthropocene.* Durham, NC: Duke University Press.

Hulme, Mike. 2009. *Why We Disagree About Climate Change: Understanding Controversy, Inaction and Opportunity.* Cambridge: Cambridge University Press.

Hulme, Mike. 2013a. *Exploring Climate Change through Science and in Society.* London: Routledge.

Hulme, Mike. 2013b. "How Climate Models Gain and Exercise Authority." In *The Social Life of Climate Change Models: Anticipating Nature*, edited by K. Hastrup and M. Skrydstrup, 30–44. London: Routledge.

Hulme, Mike. 2015. "Climate." In *Living Lexicon for the Environmental Humanities*, Vol. 6. Durham, NC: Duke University Press.

IPCC. 2021a. "Annex VII: Glossary." In *Climate Change 2021: The Physical Science Basis. Contribution of Working Group I to the Sixth Assessment Report of the Intergovernmental Panel on Climate Change*, edited by V. Masson-Delmotte, P. Zhai, A. Pirani, S. L. Connors, C. Péan, S. Berger, N. Caud, Y. Chen, L. Goldfarb, M. I. Gomis, M. Huang, K. Leitzell, E. Lonnoy, J. B. R. Matthews, T. K. Maycock, T. Waterfield, O. Yelekçi, R. Yu, and B. Zhou, 2215–2256, doi:10.1017/9781009157896.022. Cambridge and New York: Cambridge University Press.

IPCC. 2021b. "Summary for Policymakers." In *Climate Change 2021: The Physical Science Basis. Contribution of Working Group I to the Sixth Assessment Report of the Intergovernmental Panel on Climate Change*, edited by V. Masson-Delmotte, P. Zhai, A. Pirani, S. L. Connors, C. Péan, S. Berger, N. Caud, Y. Chen, L. Goldfarb, M. I. Gomis, M. Huang, K. Leitzell, E. Lonnoy, J. B. R. Matthews, T. K. Maycock, T. Waterfield, O. Yelekçi, R. Yu, and B. Zhou, 3–32, doi:10.1017/9781009157896.001. Cambridge and New York: Cambridge University Press.

Kimmerer, Robin Wall. 2013. *Braiding Sweetgrass*. London: Penguin Books.

Kincey, Mark E., Nick J. Rosser, Tom R. Robinson, Alexander L. Densmore, Ram Shrestha, Dammar Singh Pujara, Katie J. Oven, Jack G. Williams, and Zuzanna M. Swirad. 2021. "Evolution of Coseismic and Post-Seismic Landsliding after the 2015 Mw 7.8 Gorkha Earthquake, Nepal." *Journal of Geophysical Research: Earth Surface* 126: e2020JF005803.

Kirksey, Eben S., and Stefan Helmreich. 2010. "The Emergence of Multispecies Ethnography." *Cultural Anthropology* 25 (4): 545–576.

Krauss, Werner, Mike S. Schäfer, and Hans von Storch. 2012. "Introduction: Post-Normal Climate Science." *Nature and Culture* 7 (2): 121–132.

Krishnan, Raghavan, Arun B. Shrestha, Guoyu Ren, Rupak Rajbhandari, Sajjad Saeed, and Jayanarayanan Sanjay. 2019. "Unraveling Climate Change in the Hindu Kush Himalaya: Rapid Warming in the Mountains and Increasing Extremes." In *The Hindu Kush Himalaya Assessment—Mountains, Climate Change, Sustainability and People*, edited by W. Philippus, A. Mishra, A. Mukherji, and A. B. Shrestha. Cham, CH: Springer Nature, Switzerland AG.

Kuhn, Thomas S. 1977. "Objectivity, Value Judgment, and Theory Choice." In *The Essential Tension: Selected Studies in Scientific Tradition and Change*, edited by T. Kuhn, 320–329. Chicago: University of Chicago Press.

Lamb, Hubert H. 1995. *Climate, History and the Modern World*, 2nd ed. London: Routledge.

Lévi-Strauss, Claude. 2021. *Wild Thought: A New Translation of La Pensée Sauvage*, translated by Jeffrey Mehlman and John Levitt. Chicago: University of Chicago Press.

Lim, Jonathan S., Sean Gleason, Warren Jones, and Willard Church. 2021. "Nuna Nalluyuituq (The Land Remembers): Remembering Landscapes and Refining Methodologies through Community-Based Remote Sensing in the Yukon-Kuskokwim Delta, Southwest Alaska." *Archeological Prospection* 28: 339–355.

Lloyd, Elisabeth A., and Naomi Oreskes. 2018. "Climate Change Attribution: When Is It Appropriate to Accept New Methods?" *Earth's Future* 6: 311–325.

Lorenz, Edward N. 1967. *The Nature and Theory of the General Circulation of the Atmosphere*. Geneva, CH: World Meteorological Organization.

Lorenz, Edward N. 1995. *The Essence of Chaos*. Seattle, WA: University of Washington Press.

Lovelock, James E. 1972. "Gaia as Seen through the Atmosphere." *Atmospheric Environment* 6 (8): 579–580.

Maharjan, Bikash Sudan, Jakob Friedrich Steiner, Arun Bhakta Shrestha, Amina Maharjan, Santosh Nepal, Mandira Singh Shrestha, Birendra Bajracharya, Ghulam Rasul, Maxim Shrestha, Miriam Jackson, and Nishikant Gupta. 2021. *The Melamchi Flood Disaster: Cascading Hazard and the Need for Multihazard Risk Management*. International Centre for Integrated Mountain Development, Kathmandu, Nepal.

Mahony, Martin, and Georgina Endfield. 2018. "Climate and Colonialism." *WIREs Climate Change* 9: e510.

Marc, Odin, Robert Behling, Christoff Andermann, Jens M. Turowski, Luc Illien, Sigrid Roessner, and Niels Hovius. 2019. "Long-Term Erosion of the Nepal Himalayas by Bedrock Landsliding: The Role of Monsoons, Earthquakes and Giant Landslides." *Earth Surface Dynamics* 7: 107–128.

Marcus, George E., and Michael M. J. Fischer. 1986. *Anthropology as Cultural Critique*. Chicago: University of Chicago Press.

Martin-Nielsen, Janet. 2015. "Ways of Knowing Climate: Hubert H. Lamb and Climate Research in the UK." *WIREs Climate Change* 6: 465–477.

Mosko, Mark S. 2005. "Introduction: A (Re)turn to Chaos: Chaos Theory, the Sciences, and Social Anthropological Theory." In *On the Order of Chaos: Social Anthropology and the Science of Chaos*, edited by M. S. Mosko and F. H. Damon, 1–46. New York: Berghahn Books.

Neupane, Madhu Krishna, and Dhyanendra Bahadur Rai. 2018. "Diversification of Livelihood Strategies in Melamchi Valley, Sindhupalchok District, Nepal." *The Geographical Journal of Nepal* 11: 113–126.

Nowak, Martin A., and Robert M. May. 1992. "Evolutionary Games and Spatial Chaos." *Nature* 359: 826–829.

O'Gorman, Emily, and Andrea Gaynor. 2020: "More-Than-Human Histories." *Environmental History* 25: 711–735.

Office of Melamchi Municipality. 2019. *Preparation of Municipal Transport Master Plan of Melamchi Municipality*. Melamchi, Sindhupalchowk, Nepal.

Oliver-Smith, Anthony. 2012. "Disaster Risk Reduction and Climate Change Adaptation: The View from Applied Anthropology." *Human Organization*, 72 (4): 275–282.

Pettenger, Mary E. 2007. *The Social Construction of Climate Change: Power, Knowledge, Norms, Discourses*. Aldershot, UK: Ashgate.

Pokharel, Binod. 2010. "Changing Relations between High Castes and Tamang in Melamchi Valley." *Dhaulagiri Journal of Sociology and Anthropology* 4: 65–84.

Pound, Ezra. 1993. "Canto LXXXI." *Poetry Foundation,* accessed August 12, 2022. https://www.poetryfoundation.org/poems/54320/canto-lxxxi

Pulkkinen, Karoliina, Sabine Undorf, Frida Bender, Per Wikman-Svahn, Francisco Doblas-Reyes, Clare Flynn, Gabriele C. Hegerl, Aiden Jönsson, Gah-Kai Leung, Joe Roussos, Theodore G. Shepherd, and Erica Thompson. 2022. "The Value of Values in Climate Science." *Nature Climate Change* 12: 4–6.

Ristroph, Elizaveta Barrett. 2017. "When Climate Takes a Village: Legal Pathways Toward the Relocation of Alaska Native Village." *Climate Law* 7 (4): 259–289.

Rodrigues, Regina R., and Theodore G. Shepherd. 2022. "Small is Beautiful: Climate-Change Science as if People Mattered." *PNAS Nexus* 1: pgac009.

Rose, Deborah Bird, and Thom van Dooren. 2017. "Encountering a More-Than-Human World: Ethos and the Arts of Witness." In *The Routledge Companion to the Environmental Humanities*, edited by U. Heise, J. Christensen, and M. Niemann, 120–128. London: Routledge.

Salazar, Juan Francisco. 2017. "Speculative Fabulation: Researching Worlds to Come in Antarctica." In *Anthropologies and Futures: Researching Emerging and Uncertain Worlds*, edited by J. F. Salazar, S. Pink, A. Irving, and J. Sjorberg, 151–170. London: Routledge.

Saravanan, R. (Ramalingam) 2022. *The Climate Demon*. Cambridge and New York: Cambridge University Press.
Schipper, E. Lisa F., Navroz K. Dubash, and Yacob Mulugetta. 2021. "Climate Change Research and the Search for Solutions: Rethinking Interdisciplinarity." *Climatic Change* 168: 18.
Schumacher, Ernst F. 1973. *Small is Beautiful: A Study of Economics as if People Mattered*. London: Blond & Briggs.
Shepherd, Theodore G. 2014. "Atmospheric Circulation as a Source of Uncertainty in Climate Change Projections." *Nature Geoscience* 7: 703–708.
Shepherd, Theodore G. 2019. "Storyline Approach to the Construction of Regional Climate Change Information." *Proceedings of the Royal Society A* 475: 20190013.
Shepherd, Theodore G., and Adam H. Sobel. 2020. "Localness in Climate Change." *Comparative Studies of South Asia, Africa and the Middle East* 40 (1): 7–16.
Shepherd, Theodore G., and Elisabeth A. Lloyd. 2021. "Meaningful Climate Science." *Climatic Change* 169: 17.
Shepherd, Theodore G., Emily Boyd, Raphael A. Calel, Sandra C. Chapman, Suraje Dessai, Ioana M. Dima-West, Hayley J. Fowler, Rachel James, Douglas Maraun, Olivia Martius, Catherine A. Senior, Adam H. Sobel, David A. Stainforth, Simon F. B. Tett, Kevin E. Trenberth, Bart J. J. M. van den Hurk, Nicholas W. Watkins, Robert L. Wilby, and Dimitri A. Zenghelis. 2018. "Storylines: An Alternative Approach to Representing Uncertainty in Physical Aspects of Climate Change." *Climatic Change* 151: 555–571.
Shugar, Dan H., et al. 2021. "A Massive Rock and Ice Avalanche Caused the 2021 Disaster at Chamoli, Indian Himalaya." *Science* 373: 300–306.
Singh, Vandana. 2021. "Toward a Transdisciplinary, Justice-Centered Pedagogy of Climate Change." In *Curriculum and Learning for Climate Action: Toward an SDG 4.7 Pathway for Systemic Change*, edited by R. Iyengar and C. T. Kwauk, 169–187. UNESCO-IBE: Brill.
Spivak, Gayatri Chakravorty. 1988. "Can the Subaltern Speak?" In *Marxism and the Interpretation of Culture*, edited by C. Nelson and L. Grossberg, 271–313. Urbana: University of Illinois Press.
Stirling, Andy. 2010. "Keep It Complex." *Nature* 468: 1029–1031.
Strathern, Marilyn. 1995. *The Relation: Issues in Complexity and Scale*. Cambridge: Prickly Pear Press.
Strauss, Sarah, and Ben Orlove. 2003. "Up in the Air: The Anthropology of Weather and Climate." In *Weather, Climate, Culture*, edited by S. Strauss and B. Orlove, 3–14. New York: Berg.
Taylor, Patrick C., Wieslaw Maslowski, Judith Perlwitz, and Donald J. Wuebbles. 2017. "Arctic Changes and Their Effects on Alaska and the Rest of the United States." In *Climate Science Special Report: Fourth National Climate Assessment, Volume I*, edited by D. J. Wuebbles, D. W. Fahey, K. A. Hibbard, D. J. Dokken, B. C. Stewart, and T. K. Maycock, 303–332. Washington, DC: U.S. Global Change Research Program. https://doi.org/10.7930/J00863GK
Tiwari, Anish. 2020. "We Had Long Been Asking the Authorities to Move Us to Safer Location." *Kathmandu Post*, August 15, 2020, accessed January 25, 2022. https://kathmandupost.com/province-no-3/2020/08/15/friday-s-landslide-a-preventable-tragedy-sindhupalchok-villagers-say
University of Alaska Institute of Northern Engineering. 2019. Statewide Threat Assessment: Identification of Threats from Erosion, Flooding, and Thawing Permafrost in

Remote Alaska Communities. Report Prepared for the Denali Commission. Report #INE 19.03

Van Dooren, Thom, and Deborah Bird Rose. 2016: "Lively Ethography: Storying Animist Worlds." *Environmental Humanities* 8 (1): 78–94.

Wagner, Roy. 1991. "The Fractal Person." In *Big Men and Great Men: Personification of Power in Melanesia*, edited by M. Godelier and M. Strathern, 159–173. Cambridge: Cambridge University Press.

Welch, Craig. 2019. "Climate Change Has Finally Caught Up to This Alaska Village." *National Geographic*, October 22, 2019, accessed January 25, 2022. https://www.nationalgeographic.com/science/article/climate-change-finally-caught-up-to-this-alaska-village

Whyte, Kyle. 2021. "Time as Kinship." In *The Cambridge Companion to Environmental Humanities*, edited by J. J. Cohen and S. Foote, 39–55. Cambridge: Cambridge University Press.

9

NOT JUST THE SCIENCE

A Transdisciplinary Pedagogy for Cryospheric Climes

Vandana Singh

Introduction

How might we think about teaching climate change, especially with regard to the dramatic changes that are occurring in the cryosphere? I begin with two stories from the North Slope of Alaska, where I visited in 2014. The first tale I owe to the traditions of the Iñupiaq peoples of the region, as retold by Edna Ahgeak MacLean, Elder and linguist (Chance 2002, 13):

> One day an *avingaq* [mouse] decided to venture outside his hole and assess the rest of the world. When he stood up on his hind legs, lo and behold, to his surprise, he was able to reach the heavens. When he reached down, he felt the ground. When he reached in all directions, he was able to touch the limits of the world. He concluded that he was the largest person on the face of the earth.
>
> In reality, the poor mouse had surfaced from his hole in the ground into an old Iñupiaq boot sole turned upside down. The top of his heaven was the sole of the *atungak* and the outer limits of his world were the sides of the *atungak*.

My second story is a real-life-based fictionalized account I created for the classroom, based on my conversations with scientists and Iñupiaq community members as well as the work of scholars documenting Indigenous knowledge about sea ice (Eicken 2010; Gearheard et al. 2013).

> In the days when sea ice was still thick in the springtime, a Scientist went to Utqiagvik with a team of fellow scientists to study sea ice. He knew it well: its physics and chemistry, its seasonal cycles, but he wanted to expand

DOI: 10.4324/9781003347026-13

> his understanding through field measurements. It was a perfect day, as the group walked from the shore over the ice. Clear sky, and nothing in the weather forecast to cause concern. Their footsteps crunched solidly, giving no inkling that beneath the surface layer of ice was frigid liquid water. It was the sea ice thickness, among other properties, that they intended to measure. The ice glittered where the low light of the sun caught the pressure ridges that were formed when the liquid water underneath pushed chunks of ice over each other. They walked to the edge where the sea ice met the liquid ocean, and proceeded to set up camp.
>
> As they were doing so, the Iñupiaq Elder who was their guide suddenly said, without preamble, that they had to get off the ice right now! They were skeptical, because the day was perfect, and there were no indications that anything might go wrong. But the urgency in his voice compelled them to obey, and they packed up. As they walked back toward the shore, there was a deafening crack behind them, and the region of sea ice on which they had set up camp broke off and floated into the Arctic Ocean.
>
> In those days, with no coastguard presence, being borne off into the harsh Arctic environment at the mercy of unpredictable currents was close to a death sentence. The Elder had saved their lives.
>
> The Scientist had thought there was only one way to know sea ice, and it was with this purpose that he had gone out to its far edge with his instruments. But the Elder knew sea ice in a completely different way. The ice had spoken to him, but not to the Scientist. The Scientist never forgot the experience.

The North Slope of Alaska has been home to the Iñupiat for thousands of years. Here, the tundra meets the frigid waters of the Arctic Ocean, and in winter, late fall, and early spring, the top surface of the ocean freezes and extends from the shore over liquid water, for a few miles. This is sea ice: not land, not sea, a liminal zone.

On a relatively warm April day (about −12°) I stood at the edge of sea ice near the town of Utqiagvik, looking at polar bear tracks that my kindly Iñupiaq hosts pointed out to me. The pressure ridges glittered in the sunlight, and the ice field stretched toward the horizon, the distant liquid water boundary lost to sight. Despite the snow goggles, I squinted in the dazzling light. The astronomical accident of the Earth's tilt, which deprives the Earth's northern extremity of the sun's light for nearly half the year, along with the tectonic movements that gave rise to this nearly enclosed ocean, create an environment where ice—and sea ice in particular—is both protagonist and storyteller (Figure 9.1).

My visit to Utqiagvik was the result of a failure and a fascination. The failure was in the classroom. As a theoretical physicist and educator concerned about the environment, I had started introducing basic climate science into my general physics courses. My hope was that in doing so I would be equipping students with the knowledge they needed to prepare for change, and perhaps to become

FIGURE 9.1 Shore of the Arctic Ocean near Utqiagvik, Alaska, with sea ice in the distance.

change-makers. These naïve expectations were quickly dashed when students' reactions revealed both a cognitive and emotional disengagement. Climate topics like the greenhouse effect, the carbon cycle, the absorption spectrum of carbon dioxide, and current and future impacts, were scattered through the semester, and key details were easily forgotten; meanwhile, emotional reactions ranged from grief, anger, and apathy, to paralyzing despair. "I'm just one person, what can I do," or "it doesn't affect me, so I'm not going to worry about it" were common refrains. The experience drove home to me the fact that *teaching just the science was not enough, and could even become counter-productive.* Clearly, I needed to become a student again. I decided to begin with the Arctic, which is "ground zero for climate change," with temperature rise now nearly four times the global average (Rantanen et al. 2022).

When I finally stood on that remote shore on a cloudy April day, the crisis of climate change that I had been studying and teaching for the last few years became viscerally real. Breathing the frigid air, with the sea ice glittering a few feet ahead of me, the "acute disjunction" which climate creates between our "affective, discursive and epistemological selves" (Bristow and Ford 2016, 6) seemed to dissolve in the undifferentiated whiteness of sea, land, and sky. The ice roads of Utqiagvik and the complete absence of vegetation on the snowy tundra indicated an entirely different way of being a human animal. During my short stay, I interviewed Iñupiaq town officials, a whaler, an Elder and educator, and a multitude of scientists, learning from them and from various scholarly works, and through them, from the ice. It became clear to me that reducing the climate crisis to a solely scientific-technological issue washed out "the thick textures of lived experience" (Bristow and Ford 2016, 6) that are necessary for a meaningful engagement with the crisis. This was the start of a journey in search of vehicles of meaning-making that could illuminate the transdisciplinarity of the climate crisis. One of the most powerful such vehicles is the story.

In the physics classroom, stories like the ones I have told above give rise to multiple questions. What does the mouse's story mean? How did the Iñupiaq elder know the ice was going to break off? Why were the scientists studying sea ice? Further, these stories are grounded in the local—the particularity of Iñupiaq history and culture, the specifics of changing social-environmental conditions along the northern Alaskan coast. We might consider these stories emerging from a particular cryospheric clime. Following the notion of a clime as the link between weather and place (Carey et al. 2014; Fleming 2014) the cryosphere might usefully be considered to be a mosaic of climes that share certain common features, but that are also distinct. Thus, the North Slope of Alaska is distinctly different, in social-environmental-historical terms, than, for example, the Yukon region to its southeast in Canada, or the Greenland ice sheet, or the highest peaks of the Himalayas—yet all are part of the cryosphere. These clime-specific stories, where the social and biophysical, the real and imaginary seem to be irrevocably intertwined, appear in stark contrast to the language of climate science: the graphs showing temperature change or carbon dioxide concentration or sea ice extent, or the journal papers estimating climate sensitivity or attempting to quantify the ice-albedo effect. In addition, physical scientists tend to seek the universal, and the local context only appears when universal laws are applied to a specific situation—whereas human cultural stories are firmly grounded in localness, in place.

The question arises: what use are stories? Why would anyone, much less a physicist, introduce them as part of a pedagogy of climate change? How might they inform a useful and inspired understanding—affective and cognitive—of climate science? And might the reverse be also possible—that one could discover stories in the detached, formal narratives of climate science? As a physicist, it is more than obvious to me that matter is active in the universe and has stories to tell, stories generally ignored in modernity's obsession with the exclusively human. As an educator, I also know that stories of the nonhuman seen through the lens of science can be made accessible, meaningful, and relevant to the non-scientist. But how—and why—should we bridge the divide between disciplines, between stories like those quoted above and the narratives of science, between human and nonhuman, local and planetary?

What I learned and continue to learn from my teachers, human and nonhuman, has led me to a fundamental reconceptualization of climate pedagogy, which foregrounds its transdisciplinary nature and makes issues of justice and ethics central. It proposes four dimensions of an effective pedagogy:

the scientific-technological, the transdisciplinary, the epistemological, and the psychosocial-action dimensions. Rather than reductively exploring each aspect, these dimensions are manifested relationally through stories and via three interdisciplinary meta-concepts (Singh 2021a). My primary impetus for developing such a pedagogy is this: when we are confronted with a phenomenon such as the climate crisis, which defies our current paradigms, we must go to it with a radical openness and humility, divesting ourselves (to the extent possible) of preconceived frameworks. In other words, we must acknowledge the phenomenon itself as teacher. In this chapter I begin with an invocation to sea ice in Alaska, focusing on its physical nature, seasonal cycles, and decline. I acknowledge the teachings of the sea ice and those of the climate problem as a whole. I then briefly summarize my pedagogical framework, including three key transdisciplinary meta-concepts. Thereafter I return to the Alaskan Arctic to illustrate these ideas, bringing out the connections between clime and climate, local and planetary, social and scientific, and human and nonhuman. I elaborate on the notion of cryospheric clime, and show how stories can help animate the concept.

Sea Ice: A Scientific Invocation

> It can be stiff and silent but also blasting and crushing with terrible noise, and it can advance and retreat as if a living being. Every year sea ice extends across 14–16 million km^2 in the Arctic (and 17–20 million km^2 in the Antarctic Southern Ocean) and then it shrinks and withers by the end of the summer to a fraction of its winter might. As it forms, persists, advances, and melts, it changes the ocean circulation, regulates global climate, and affects life in the polar regions.
>
> *(Krupnik et al. 2010, 2)*

In the late fall, the Arctic Ocean water starts to freeze. Tiny ice crystals, called frazil, begin to form. Depending on whether the sea is calm or rough, these crystals coalesce in different ways to build sea ice. This new ice, called first-year ice, may be anywhere from a few centimeters to a couple of meters thick. Sea ice is not as solid as it appears. During its formation, the concentrated salt water that is unable to freeze forms tiny pockets and passages, called brine channels, within the body of the sea ice. Sea ice, through its formation, reaches across scales, from the millimeter dimension of the frazil and the brine channels to the millions of square kilometers of the Arctic ice cap.

The shiny surface of the sea ice reflects the sun's energy back into space without much absorption; the cryosphere as a whole, rejects 30% of the sun's energy due to its shininess or albedo, thus limiting how much is absorbed by the Earth's system. In the summer, the sea ice begins to melt, and as it does, the brine channels expand and connect, forming a labyrinthine maze, releasing the concentrated salt water into the liquid ocean below. This highly dense salt water sinks toward the ocean floor, providing part of the impetus for the great system of ocean currents, the Global Conveyor Belt. As the summer progresses, some of the first-year ice, if thick enough, will survive the seasonal melt, helping to thicken the ice the next winter. This survivor ice, called multi-year ice, can be up to 20 m thick. Because it partly melts in the summer and allows the brine to run out, multi-year ice is a source of fresh water for Iñupiaq hunters and polar expeditions. Meanwhile, the brine channels provide space for ice algae to grow, which form the base of the sea ice ecosystem, a web of relationships that starts with the ice algae and continues all the way to polar bears, bowhead whales, and humans. For the Iñupiat on Alaska's North Slope, the sea ice provides much needed sustenance in the harsh winter months through the fresh meat of animals that spend time on it (seals and polar bears) and by making it possible to walk to where bowhead whales can be hunted. Thus the crucial importance of Arctic sea ice is simultaneously situated in the local (ecosystem well-being and Iñupiaq subsistence and culture) and the global (ocean currents and albedo). In fact, the paleohistory of the Earth's climate indicates the central role of Arctic sea ice in determining the onset and completion of the past million years of glacial and interglacial periods. Sea ice inhabits multiple temporal scales, from its winter/summer expansion and contraction to its paleohistoric past and its current rapid decline, exhibiting temporal multiplicities—"deep, accelerated, and troubled," like the tundra it borders (Andersen, this book).

From the shore near Utqiagvik, the sea ice appears stationary. But, as the Scientist and Elder story implies, ocean currents and winds are always moving it around. The Beaufort Sea just east of Utqiagvik happens to have a circular current that keeps sea ice in the area long enough to form multi-year ice. Eventually, the ice floats out of the Arctic Ocean via the Fram Strait on the East coast of Greenland, and melts in the warmer waters of the Atlantic Ocean.

Thus the extent of sea ice depends on new formation in winter versus the summer melt, and also on the amount of multi-year ice. When sea ice formation happens at the same speed as sea ice loss, the average amount of sea ice remains constant (a condition of dynamic balance), apart from seasonal fluctuations. But global heating is causing sea ice to melt at an alarming rate. Not only is there less ice formation and more summer melt, but multi-year ice has decreased significantly. The Beaufort gyre, which was once a sea ice nursery, is turning into a graveyard (NASA 2018). With summer melt occurring faster than winter formation, and with the disappearance of multi-year ice, it is not surprising that sea ice extent is dropping rapidly, into a state of imbalance. This has ramifications for the local ecosystems, human populations, and the entire planet—culminating in what Sverker Sörlin calls a cryohistorical moment, signifying human hegemony in the vanishing of a crucial element of the climate (Sörlin 2015, 327). Thus sea ice is going from balance to imbalance, and because it exists in intimate relation with the local and the global, it simultaneously affects and is affected by planetary-scale imbalances in carbon dioxide concentrations and energy (Loeb et al. 2021).

Sea ice belies the expectations of those who are not nurtured in its vicinity. Neither land nor ocean, it can be as stationary and solid as the tundra, or it can break or move in sudden and dangerous ways. In pictures, it looks as quiet as a snow-covered landscape in lower latitudes, but it is not silent—it creaks, groans, whines, and thunders.

The graph below is a compressed representation of the sea ice's story (Figure 9.2).

Teachings of the Sea Ice

What might we learn from the sea ice that is pedagogically useful for understanding clime and climate? Sea ice is a necessary aspect of coastal cryospheric climes of the Arctic and Antarctic, but its importance crosses boundaries of geography and time. To summarize, it tells us that the *local and global are intimately connected; it spans large scales of space and time.* In addition, it warrants a *transdisciplinary understanding*, since the meaning of sea ice, the implications of its decline, and how we might mitigate and adapt to its loss involve multiple disciplines, including the natural and social sciences, Indigenous culture, history and epistemology, and considerations of economics and justice. From my scientific invocation arise three additional "meta-concepts" that form the basis of my pedagogical framework. One is the idea of *balance or steady state,* and its opposite, *the state of change or imbalance.* As the dynamics of the sea ice change due to global heating, the speed

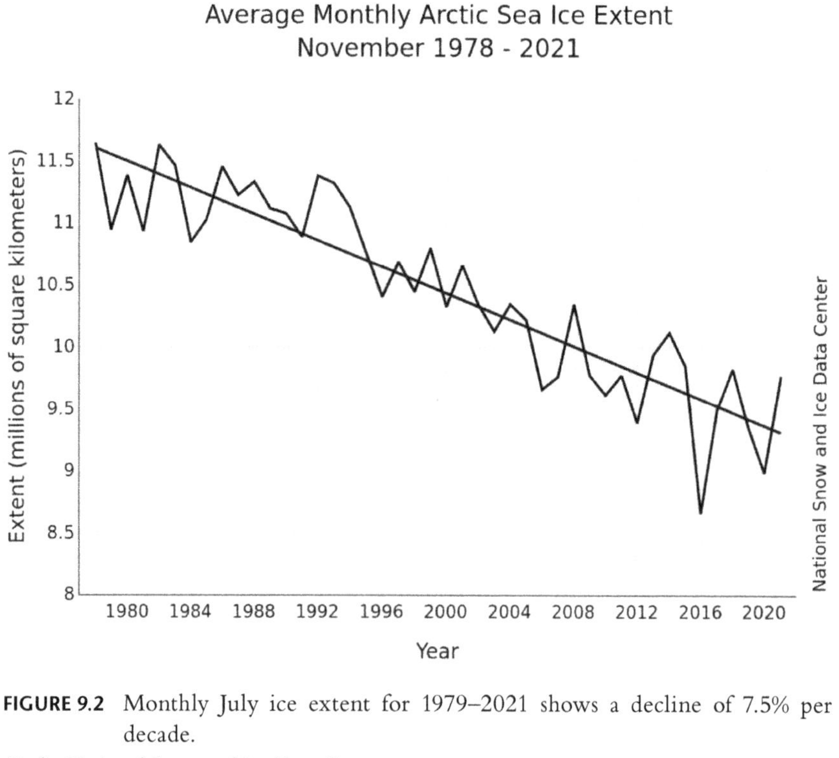

FIGURE 9.2 Monthly July ice extent for 1979–2021 shows a decline of 7.5% per decade.

Credit: National Snow and Ice Data Center.

of melt is gaining over the speed of formation. Scientists predict that Arctic summer sea ice will be completely gone by midcentury, and possibly even by 2035 (Guarino et al. 2020). This is unprecedented in human history, and represents a limit, boundary, or threshold with serious implications. Is it a threshold of no return? Such *critical limits, boundaries, and thresholds* crop up in multiple climatic phenomena. Sea ice also demonstrates a richness of relationships—between and within human and biophysical systems that extend from the local to the planetary. The nature of these relationships is *complex*, in a specific way that I will demonstrate with another story from the Arctic.

Larry Merculieff of the Alaskan Aleut Peoples tells the story (Landis and Fries-Gaither 2009) of a scientist who visited the Native Elders seeking information about moose populations in the region. Instead of answering him directly, the Elders asked him if he knew how many beaver dams were in the area. The scientist was puzzled—he was interested in moose, not beavers. The Elders pointed out that beaver dams affect the foraging areas of the moose. The scientist was looking at the moose in isolation from other species, but the Elders had a more holistic outlook, focused on *relationships*. Systems in which relationships are as important as the parts, so much so that the relationships can change the nature and function of the parts, are generally categorized as complex systems. Unlike simple systems, where causal relations are straightforward and often linear, the relationships in complex systems tend to be nonlinear and involve multiple kinds of causality. Ecosystems, social networks, the human endocrine system, and the global climate system are all examples of complex systems. Thus *complex entanglements* or interconnections abound in human-biophysical systems, including the climate. (Here I distinguish between quantum entanglement and complex entanglement, where the latter refers to connections and relationships at our scale that are a result of inherent relationality). Real-world systems with complex entanglements transcend dichotomies of human/nonhuman or social/ecological. What do beavers have to do with moose populations? How did the Elder know the ice was going to break? These questions are about relationships that recognize complexity in an embodied way. Survival and well-being in various environments require the ability to read the biophysical surroundings through direct bodily immersion, to know it as a participant, from within, through multiple and multispecies *relationships*.

Complex systems are not fully amenable to reductionism. The metaphor for a simple system is a clock: understand the springs and gears, and you've understood the clock. A spring is always a spring, and a gear always a gear. Contrast that with the "parts" of the global climate system, where the oceans, or the Amazon rainforest, can flip from net carbon absorbers to net carbon emitters. Additionally, the subsystems of complex systems are often themselves complex. Complexity in the climate system entails cascading and domino effects that may start out small and local, but that can quickly spread to other regions and even precipitate systemic changes in the whole.

Listening to the sea ice through the stories of the Iñupiat, it becomes clear also that the Iñupiat, despite a complicated relationship with the oil industry, have not contributed significantly to global climate change, and yet are subject to its worst impacts. The thinning ice is dangerous to walk on, making it hard to set up trails and camps for whaling, and increasing the risk of accidents; coastal erosion is worsened by both sea level rise and the lack of protective sea ice; whales are changing their routes in response to warming waters, and while polar bear populations are not yet endangered, they are declining in some areas because the seals are disappearing with the sea ice. Polar bears have been seen in Utqiagvik, apparently looking for food. No matter what the clime, neither the attribution nor the impacts of climate change are evenly distributed. Thus sea ice loss in particular, and climate change in general, are inherently problems of multispecies justice.

The teachings of the sea ice echo the lessons of the climate as a whole. The three meta-concepts of balance/ imbalance, limits, boundaries and thresholds, and complex entanglements can serve as boundary objects. Boundary objects (Star and Griesemer, 1989) are concepts flexible enough to have meaning across different social worlds, yet robust enough to retain their identity; they are useful even when there is a lack of consensus. Later in this chapter, I will show how these meta-concepts travel across the realms of the human/ biophysical, local/ global, scientific/social, and Indigenous/Scientific worlds through the help of another boundary object: the (appropriately chosen and contextualized) story.

An Effective Pedagogy of Climate Change

Like the sea ice, the larger phenomenon of climate change exists over enormous scales of space and time, through complex entanglements between and within human-biophysical systems. It is inherently transdisciplinary, transcending disciplinary boundaries; further, issues of justice and power are central to any serious consideration of climate change. Yet modern globalized industrial culture prioritizes short-term, spatially disconnected thinking, ignores complex interconnections and multiple kinds of causal relations in favor of simple, linear causality, separates disciplines into watertight compartments that rarely interact, and engenders a hierarchical social system that favors those with power and privilege. It is not surprising, then, that the mainstream education system reproduces rather than challenges the status quo. The lack of radical visions for climate education is one of five macro-level roadblocks that have been identified by scholars (Kwauk 2020). Climate change is rarely taught (or for that matter, acted upon) in a manner commensurate with its essential teachings. When it is taught, it is usually in a science class as a narrow scientific-technical subject, emphasizing a view of climate as "an objective reality to be manipulated through material intervention," and implying an "an unambiguous separation between Nature and culture" (Hulme 2008, 12). In a compartmentalized education system, an alternative option is to teach it in every class, through every discipline, by adopting an approach that may start in one disciplinary context, but reaches toward transdisciplinarity, paying attention to complex interconnections across space, time, and species, embracing localness as well as planetary-scale phenomena without privileging the latter, and foregrounding justice.

My only option currently has been to teach climate change in this manner in all my general physics undergraduate classes. This has also required changing classroom power dynamics, coming up with ways to avoid piecemeal learning, and taking inspiration from best practices, all of which are necessary to support an effective pedagogy of climate change. This I describe elsewhere (Singh 2021a) as one that foregrounds the transdisciplinary aspect of the climate problem, entangling the scientific with the social-cultural, so that students can develop a cognitive, affective, and integrated understanding that inspires them to engage in useful ways with the problem. The three transdisciplinary meta-concepts mentioned above provide the scaffolding for the essentials of climate science but also go beyond science to wider concerns. Here I focus on how these meta-concepts serve as boundary objects, and the role of stories in this context.

Across Space, Time, and Discipline: The Meta-Concepts

The sea-ice-mass imbalance currently underway in the Arctic is causally related to planetary-scale imbalances. Carbon dioxide concentration in the atmosphere is rising every year, a consequence of the imbalance in the carbon cycle: carbon dioxide is being put into the atmosphere faster than it is being removed. Through the greenhouse effect, this imbalance has an impact.

Thus we see that **balance and imbalance** manifest across scales, allowing us to connect the local to the planetary, and clime to climate. Depending on the degree and timescale of imbalance, the system at each scale may cross thresholds or limits that commit that system to change in one direction, changes that may not be reversible on human timescales, such as the changes in Arctic sea ice ecosystems, or the melting of the great ice sheets of Greenland and Antarctica. Interestingly, the causal flow in these nested phenomena is not necessarily only from the planetary to the local; as we know from the crucial role of Arctic sea ice, its continued melt means that more of the sun's energy is retained in the Earth system, exacerbating the greenhouse effect through vicious cycles, including, but not limited to the ice-albedo effect (Goosse et al. 2018). Arctic sea ice is considered a tipping element (Wunderling 2021) of the Earth's climate system. Tipping elements are Earth-system components susceptible to tipping points, that is, when they are subject to an external perturbation, they can switch to a qualitatively different state relatively quickly, and since tipping elements interact with each other, when one crosses a tipping point, this may instigate the others to do so as well; when the changes are all in the same direction, this tipping cascade may shift the entire Earth system to a new state, crossing multiple planetary boundaries.

Planetary or Earth-system boundaries are the planetary-scale manifestations of the second meta-concept of **limits, boundaries, and thresholds**; they are set by natural cycles and processes that define stable environmental conditions within which humans and the rest of the biosphere can thrive. Of the nine planetary boundaries proposed by Steffen and others (Steffen et al. 2015), humankind is in violation of five, including climate change and biodiversity loss, which represent, broadly speaking, a shift from conditions of relative balance to imbalance, toward a new and unknown balance. Tipping elements, therefore, connect regional phenomena like Arctic sea-ice loss to the planetary scale. The transition from balance to imbalance and the crossing of limits and boundaries at multiple scales is a result of **complex entanglements** between elements of Earth's biophysical system. At local scales, ecosystems may demonstrate relatively steady states, as well as shifts that are sometimes unidirectional, leading to the collapse of the old ecosystem and the establishment of a new one (for example, the classic case of kelp forests giving way to urchin barrens) because of such complex interconnections. Meanwhile, complex interconnections between tipping elements are only now being identified; for example, the possibility of Amazon rainforest dieback due to Arctic sea ice melt affecting ocean currents. This "domino causality" is one of many kinds of causal relations (apart from simple linear causality) that occur in complex systems (Grotzer 2012, 41).

Thus far I have attempted to demonstrate that the three meta-concepts travel across spatiotemporal scales, and are therefore useful for connecting the local and the global, clime and climate. But do these meta-concepts also travel across social worlds, an essential feature of a boundary object? And what is the role of story?

To explore this, let us return to the story of the Scientist and the Elder, and the question I posed earlier: how did the Elder know that the ice was going to break? The story allows us to explore more deeply the human components of the cryospheric clime of the Alaskan north shore, through the history and onto-epistemology of the Iñupiat. The Iñupiat have lived with the ice for millennia, knowing its seasons, moods, and voices through an immersive, embodied experience in which the body itself becomes a delicate environmental sensor. The sea ice has allowed the Iñupiat to live in one of the harshest environments on the planet by providing fresh food in the form of seals and bowhead whales. This is especially true in the winter, a half-year of semi-darkness, when the north shore is disconnected from the rest of the world by severe weather and killing temperatures, and the only available food is meat. Subsistence hunting is still necessary for survival, despite modernity and capitalism having transformed the lives of the Iñupiat. To hunt the whale, teams cut trails across the sea ice every fall and spring, carrying skin boats and supplies on snowmobiles. The ice allows the teams to get close to where the bowheads swim. The slightest change in conditions, learned through oral transmission of knowledge by Elders as well as immersive experience, warns and directs the hunter. The whale hunt is a sacred ritual, a key part of the identity of the Iñupiat. The animal is thanked for the sacrifice of its life, and the meat is never sold; it can only be shared. The belief is that the whale gives itself to the hunter. The entire community takes part in the hunt, the dragging of the carcass over the ice, and the butchering of the meat. Stored in outdoor ice cellars, the meat can feed multiple families over the long, dark winter.

The Iñupiat think of the sea ice as "home," a place that makes them homesick if they are away from it for too long. It is a place for travel, leading them to where the bowhead whales swim. It is where young people play sports, where the children learn to be Iñupiat. As Elder Nancy Neakok Leavitt says, sea ice is crucial for the very existence of the Iñupiat, allowing a cashless, subsistence way of life, providing their bodies with the kind of food needed to survive in the cold. Other Elders describe the sea ice as being "like a garden" (Gearheard et al. 2013, 63–69, 146). The vast repertoire of sounds that the ice makes, its variety of shapes and types are reflected in the complexity of the language, Iñupiaq. Such a detailed lexicon has survival value, but it also demonstrates how the sea ice has shaped a people, a culture, and a way of knowing.

This relationship with the ice and its creatures, nurtured over millennia, has resulted in a concept that is shared by many Indigenous cultures: "collaborative reciprocity" (Sakakibara 2016, 163), which speaks to the relationships between humans and nonhumans (including the physical landscape). It acknowledges that if an animal provides its body for one's survival, one must give back to the animal in some way. Far more than mere "give and take," this reciprocity is centered on the belief that "humans and animals physically and spiritually constitute each other, that the soul, thoughts, and behaviors of animals and people interpenetrate in a collaboration of life" (Sakakibara 2016, 163, 164). This is consistent with the notion that "we become with and through diverse others, ancestors and contemporaries, human and nonhuman" (Rose and Van Dooren 2017, 122). Perhaps it is in this sense that a whaler, in a video (PBS 2007) about the Iñupiaq whaling tradition, reportedly has the experience of "dying with the whale" when the whale is killed. The whale's gift demands an answering respect that is expressed through the centrality of whaling in Iñupiaq life and culture, from origin stories (the land itself being the body of a whale) to the annual festivities and ceremonies that are set by the whaling cycle.

Collaborative reciprocity implies the notion of **balance** in at least two senses: one, through the equality of human and nonhuman persons, and two, in mutual give and take beyond mere exchange. It acknowledges **limits, boundaries, and thresholds**, through the idea that you take only what you need, and also with regard to sharing, such as in the case of salmon-fishing timeshares between humans and grizzlies in parts of the Yukon (Clark and Slocombe 2009). When a whale is killed, every part of it is used. To waste any part is to disrespect its sacrifice. This notion of limits is consistent with the concept of societal boundaries (Brand et al. 2021) recently proposed by social-science scholars in response to the planetary boundary concept—self-limitations that societies may impose on themselves in order for humans and nonhumans to thrive. All this has material ecological consequences. Environmental history tells us that while some Indigenous cultures have inadvertently caused extinctions of species, their track record is far better than that of modern industrial civilization. Holism is a central aspect of many Indigenous cultures; the emphasis on relationships acknowledges **complex entanglements** between the human and the nonhuman. Thus the three meta-concepts that describe key features of the global climate problem are mirrored in crucial aspects of Indigenous epistemologies, even though their specific meanings are distinct (but recognizably similar) from those in the scientific context.

Stories and Climes

These stories demonstrate the importance of Indigenous paradigms. While we cannot regard Indigenous societies as "pure" cultures hermetically sealed in space and time from the rest of humankind (Smyer Yü 2021), Indigenous societies around the world are historically among the most successful human cultures. They have shaped and been shaped by their climes for millennia, without—for the most part—destroying their environment. Modern industrial civilization has the dubious distinction of threatening the survival of humankind and the biosphere in a mere 200 years or so. Collaborative reciprocity is a common feature of many Indigenous traditions, and it recognizes that the world is a priori complex. Science and modernity are still hampered by the reductionism and atomism of the Newtonian paradigm, which, despite its power, has a limited domain of validity.

In the Arctic, the vanishing of the sea ice also threatens the diminishing importance of Elders' knowledge in Iñupiaq communities. The accumulated knowledge of past generations includes an understanding of biophysical change, but under rapidly changing new conditions, these knowledge systems may not have time to change with the environmental changes. The loss of these knowledge systems, whether due to colonialism or modernity or rapidly changing environmental conditions, is an example of epistemicide (Santos 2014).

This point leads us back to the story of the mouse. This is at one level a teaching story about human hubris and foolishness. But we can also interpret it as a story about paradigms and epistemologies. Our preconceived notions about the world shape our reality. They limit our experiences and imaginations, resulting in what I call *paradigm blindness*. Elsewhere I discuss (Singh 2021b) how the Newtonian paradigm can be seen to undergird modern industrial civilization. We are constrained by this construction of reality, which habituates us to short-term, spatially disconnected, simple linear thinking, and limits our empathic reach and our relationships to a few humans like ourselves. Perhaps we are deluded into feeling larger than we are, rather like the poor mouse in the overturned shoe.

Indigenous cultures use stories to transmit values, traditions, and complex ecological information across generations. Scholars have pointed out the use of stories in strengthening community and building cultural resilience in the face of unprecedented social-environmental change, for example in the Iñupiaq community of Point Hope, Alaska (Sakakibara 2016) and among the Viliui Sakha people in northeastern Siberia (Crate 2008), where Indigenous people are creating new stories in response to these changes. Scholarly work on stories (Dillon and Craig 2022) points to their multiple functions: enabling multiple points of view, creating and consolidating collective identity, and providing a narrative model that enables explanation and understanding, including an anticipatory narrative model to deal with possible futures.

In the classroom as well, stories—whether mythical/ metaphoric, real-life, or fictional—can function in multiple ways. In the context of a pedagogy based on transformational learning theory (Mezirow and Taylor 2009), shared stories that are read, enacted, and interpreted together can help build a sense of community and enable an epistemic shift in the classroom and beyond. By breaking down walls between disciplines, and inviting the whole student to engage, they can facilitate both the cognitive and affective aspects of learning. In their transdisciplinarity, and by revealing the centrality of relationships, stories mirror the teachings of the climate itself. In this chapter, I have shown how three scientific meta-concepts can act as boundary objects across scales and social worlds. Carefully curated stories can also be boundary objects, enabling these meta-concepts to travel between worlds. Stories originating in the human dimension can take us, through these meta-concepts, to the scientific realm, and from the local to the global.

What about stories coming from the opposite direction, that is, from matter, the subject of scientific exploration, to the social worlds of humans? It has been suggested that climate scientists need to become better storytellers, because the language of science does not translate well to the public sphere. Instead, an approach that builds trust, contextualizes the content to the local community, uses the complexity and thick description inherent in a good story, and includes good listening as much as good storytelling has been recommended (Harris 2019). In the classroom as well, attention to student reception of a story is crucially important.

A fine example of a story crossing the boundary from the science side is the notion of storylining (Shepherd and Lloyd 2021), where the power of narrative is invoked to connect local-scale environmental change (for example, a single ecological event) to the bigger picture (statistics-based climate science), thus making causal sense of the local-global connection. A storyline is also a boundary object, because it can inhabit the local and global simultaneously, allowing the navigation of multiple social worlds, with implications for climate action, and law and policy. It can enable a transdisciplinary understanding of climate change (Shepherd and Truong, this book). As a means of storifying science in the classroom, it can also be pedagogically useful. If we allow nonhumans to be agents, then scientific phenomena can be rendered as stories. I have used approaches similar to storylining—storified renderings of causal relations in scientific phenomena, for instance, stories in which the sea ice speaks—in the classroom. I have also used embodied learning, such as the dramatized enactment of physical processes like the greenhouse effect, to help students transcend the subject-object separation in science toward a more "participant-observer" approach to learning.

This allows me to extend the concept of clime, thus: clime is an enactment or a performance in which the actors are human, nonhuman living beings, as well as elements of the local landscape, wind, and weather patterns. Through their multiple mutual relationships, cultural, ecological and geophysical distinctiveness is co-produced, such that a "sense of place" comes into being, what one might call an "ethos of place," to extend the notion of ethos as "what makes a group or kind distinct" (Van Dooren and Rose 2016, 80). Ecological classifications like bioregions (biomes) and ecoregions, generally don't take into account human settlements, cultures, or histories, whereas a clime—which may span ecoregions and fall within bioregions—entangles all of these. Cryospheric climes all share the physical characteristic of high albedo or shininess due to ice and snow, but are made distinct from each other by particularities of place, weather and climatic patterns, and ecosystems and species, including humans. The cryospheric clime of the north shore of Alaska, where the Iñupiat live, is defined by sea ice and tundra, the seasonal migration of the bowhead whale, and the history, customs, politics, and epistemology of the Iñupiat, which make it distinct from, for example, the cryospheric clime of Ladakh in the Himalayas. Because the three meta-concepts manifest in specific ways in particular climes, they help to particularize climes in the context of the climate crisis. Thus sea-ice-mass imbalance in the Alaskan Arctic impacts subsistence hunting and polar bear health (Whiteman, John P. 2018), while glacial melt in the Himalayas results in spring floods and summer drought that affect agricultural productivity. These meta-concepts are therefore useful in describing the current and future projected changes to "ethos of place."

In this chapter, I have described how the acknowledgment of climate as teacher allows for the emergence of a radically transdisciplinary pedagogy that – through varieties of selected stories, help make evident three meta-concepts that serve as boundary objects, traveling between climate and clime as well as between disciplines. Crucially, these stories center considerations of justice, including epistemic justice. The acknowledgment of nonhuman actors, including the elements of weather and climate, leads to an expansion of the concept of clime as an enactment with multiple human and nonhuman actors. Thus the cryospheric clime of the Alaskan North Slope is co-produced through the performances of sea ice, wind and ocean currents, polar bears and bowhead whales, and the history and culture of the Iñupiat, always entangled, always in mutuality.

References

Brand, Ulrich, Barbara Muraca, Éric Pineault, Marlyne Sahakian, Anke Schaffartzik, Andreas Novy, Christoph Streissler, et al. (2021) From planetary to societal boundaries: An argument for collectively defined self-limitation, *Sustainability: Science, Practice and Policy*, 17 (1): 264–291, DOI: 10.1080/15487733.2021.1940754

Bristow, Tom, and Thomas H. Ford. 2016. "Climates of History, Cultures of Climate." In *A Cultural History of Climate Change*, edited by Tom Bristow and Thomas H. Ford. Abingdon: Routledge.

Carey, Mark, Philip Garone, Adrian Howkins, Georgina Endfield, Lawrence Culver, and Sherry Johnson. 2014. "Forum: Climate Change and Environmental History." *Environmental History* 19 (2): 281–364. https://doi.org/10.1093/envhis/emu004.

Chance, Norman A. 2002. *The Iñupiat and Arctic Alaska: An Ethnography of Development*. Mason: Cengage Learning.

Clark, Douglas A., and D. Scott Slocombe. 2009. "Respect for Grizzly Bears: An Aboriginal Approach for Co-Existence and Resilience." *Ecology and Society* 42. Accessed January 2022. URL: http://www.ecologyandsociety.org/vol14/iss1/art42/.

Crate, Susan A. 2008. "Gone the Bull of Winter?: Grappling with the Cultural Implications of and Anthropology's Role(s) in Global Climate Change." *Current Anthropology* 49 (4): 569–595. https://doi.org/10.1086/529543.

Dillon, Sarah, and Claire Craig. 2022. *Storylistening: Narrative Evidence and Public Reasoning*. Abingdon: Routledge.

Eicken, Hajo. 2010. "Indigenous Knowledge and Sea Ice Science: What Can We Learn from Indigenous Ice Users?" In *SIKU: Knowing Our Ice: Documenting Inuit Sea Ice Knowledge and Use*, edited by Shari Fox Gearheard, Lene Kielsen Holm, Henry Huntington, and Andrew R. Mahoney, 357–376. Dordrecht: Springer Netherlands.

Fleming, James R. 2014. "Climate Physicians and Surgeons." *Environmental History* 19: 338–345.

Gearheard, Shari F., Lene K. Holm, Henry P. Huntington, Joe M. Leavitt, Andrew R. Mahoney, Margaret Opie, Toku Oshima, and Joelie Sanguya. 2013. *The Meaning of Ice: People and Sea Ice in Three Arctic Communities*. Hanover, New Hampshire: International Polar Institute Press.

Goosse, Hugues, Jennifer E. Kay, Kyle C. Armour, Alejandro Bodas-Salcedo, Helene Chepfer, David Docquier, Alexandra Jonko, et al. 2018. "Quantifying Climate Feedbacks in Polar Regions." *Nature Communications* 9. https://doi.org/10.1038/s41467-018-04173-0.

Grotzer, Tina. 2012. *Learning Causaility in a Complex World: Understandings of Consequence*. Lanham, MD: Rowman & Littlefield.

Guarino, Maria-Vittoria, Louise C. Sime, David Schröeder, Irene Malmierca-Vallet, Erica Rosenblum, Mark Ringer, Jeff Ridley, et al. 2020. "Sea-Ice-Free Arctic during the Last Interglacial Supports Fast Future Loss." *Nature Climate Change* 10: 928–932. Accessed January 10, 2022. https://doi.org/10.1038/s41558-020-0865-2.

Harris, Dylan M. 2019. "Telling Stories About Climate Change." *The Professional Geographer*. https://doi.org/10.1080/00330124.2019.1686996.

Hulme, Mike. 2008. "The Conquering of Climate: Discourses of Fear and Their Dissolution." *The Geographical Journal* 174: 5–16. https://doi.org/10.1111/j.1475-4959.2008.00266.x.

Krupnik, Igor, Claudio Aporta, and Gita J. Laidler. 2010. "SIKU: International Polar Year Project #166 (An Overview). In *SIKU: Knowing Sea Ice - Documenting Inuit Sea Ice Knowledge and Use*, edited by Igor Krupnik, Claudio Aporta, Shari Gearheard, Gita J. Laidler, and Lene Kielsen Holm, 2. Dordrecht: Springer.

Kwauk, Christina. 2020. *Roadmaps to Quality Education in a Time of Climate Change*. Brookings Institute. Accessed January 10, 2022. https://www.brookings.edu/wp-content/uploads/2020/02/Roadblocks-to-quality-education-in-a-time-of-climate-change-FINAL.pdf.

Landis, Carol, and Jessica Fries-Gaither. 2009. "Fruitful Collaboration: Western Science and Native Ways of Knowing." Beyond Penguins and Polar Bears website. Accessed August 8, 2022. https://beyondpenguins.ehe.osu.edu/issue/peoples-of-the-arctic/fruitful-collaboration-western-science-and-native-ways-of-knowing.

Loeb, Norman, Gregory C. Johnson, Tyler J. Thorsen, John M. Lyman, Fred G. Rose, and Seiji Kato. 2021. "Satellite and Ocean Data Reveal Marked Increase in Earth's Heating Rate." *Geophysical Research Letters* 48. https://doi.org/10.1029/2021GL093047.

Mezirow, Jack, and Edward W. Taylor, ed. 2009. *Tranformative Learning in Practice: Insights from Community, Workplace, and Higher Education*. Hoboken: Wiley.

NASA. 2018. Disappearing Arctic Sea Ice. March 15. Accessed September 15, 2021. https://www.youtube.com/watch?v=hlVXOC6a3ME.

PBS. 2007. "Iñupiaq Whale Hunt." PBS Learning Media Website. Accessed August 28. 2022. https://mass.pbslearningmedia.org/resource/echo07.sci.life.coast.eskimo/Iñupiaq-whale-hunt/.

Rantanen, Mika, Alexey Yu Karpechko, Antti Lipponen, Kalle Nordling, Otto Hyvärinen, Kimmo Ruosteenoja, Timo Vihma, and Ari Laaksonen. 2022. "The Arctic has warmed nearly four times faster than the globe since 1979." *Commun Earth Environ* 3 (1): 168. https://doi.org/10.1038/s43247-022-00498-3.

Rose, Deborah Bird, and Thom Van Dooren. 2017. "Encountering a More-Than-Human World: Ethos and the Arts of Witness." In *The Routledge Companion to the Environmental Humanities*, edited by U. Heise, J. Christensen and M. Niemann. London: Routledge. https://doi.org/10.4324/9781315766355.

Sakakibara, Chie. 2016. "People of the Whales: Climate Change and Cultural Resilience Among Iñupiat of Arctic Alaska." *Geographical Review*, 107: 159–184.

Santos, Boaventura de Sousa. 2014. *Epistemologies of the South: Justice Against Epistemicide.* Abingdon: Routledge.

Shepherd, Theodore G., and Elisabeth A. Lloyd. 2021. "Meaningful Climate Science." *Climatic Change* 169: 17. https://doi.org/10.1007/s10584-021-03246-2.

Singh, Vandana. 2021a. "Toward a Transdisciplinary, Justice Centered Pedagogy of Climate Change." In *Curriculum and Learning for Climate Action: Toward an SDG 4.7 Pathway for Systemic Change*, edited by Christina Kwauk and Radhika Iyengar, 169–187. UNESCO-IBE.

Singh, Vandana. 2021b. "Toward an Effective Pedagogy of Climate Change: Lessons from a Physics Classroom." Preprint. https://arxiv.org/abs/2008.00281.

Smyer Yü, Dan. 2021. "Symbiotic Indigeneity and Commoning in the Anthropogenic Himalayas." In *Environmental Humanities in the New Himalayas: Symbiotic Indigeneity, Commoning, Sustainability*, edited by Dan Smyer Yü and Erik de Maaker, 239–260. Abingdon: Routledge.

Sörlin, Sverker. 2015. "Cryo-History: Narratives of Ice and the Emerging Arctic Humanities." In *The New Arctic*, edited by Joan Nymand Larsen, Birgitta Evengård, and Øyvind Paasche, 327–339. New York: Springer.

Star, Susan Leigh, and James R. Griesemer. 1989. "Institutional Ecology, 'Translations' and Boundary Objects: Amateurs and Professionals in Berkeley's Museum of Vertebrate Zoology, 1907–1939." *Social Studies of Science* 19 (3): 387–420. https://doi.org/10.1177/030631289019003001.

Steffen, Will, Katherine Richardson, Johan Rockström, Sarah E. Cornell, Ingo Fetzer, Elena M. Bennett, Reinette Biggs, et al. 2015. "Planetary Boundaries: Guiding Human Development on a Changing Planet." *Science* 347 (6223). https://doi.org/10.1126/science.1259855.

Van Dooren, Thom, and Deborah Bird Rose. 2016. "Lively Ethography: Storying Animist Worlds." *Environmental Humanities* 8 (1): 77–94.

Whiteman, John P. 2018. "Out of Balance in the Arctic." *Science* 359: 514, 515.

Wunderling, Nico, Jonathan F. Donges, Jürgen Kurths, and Ricarda Winkelmann. 2021. "Interacting Tipping Elements Increase Risk of Climate Domino Effects under Global Warming." *Earth System Dynamics* 12 (2): 601–619.

PART IV

Multispecies Clime Change from the Little Ice Age to the Anthropocene

10

CLIMING THE GREAT GORGE

The Many Discoveries of the Yarlung Tsangpo Gorge[1]

Ruth Gamble

Still-growing mountains and ever-eroding Yarlung Tsangpo River have created a singular landscape at the eastern end of the Himalaya. As the river descends through the hundred-kilometer horseshoe-shaped gorge around Namcha Barwa (7,782 m), it tumbles over multiple waterfalls and rapids, dropping in altitude from around 3,000 to 600 m. Within the gorge, landslides, outbreak floods, and earthquakes regularly move large boulders and reshape its precipitous riverine landscapes. The only thing that mutes the river's roar is the pounding rain, carried up the gorge by the South Asian Monsoon, a climate system that crosses oceans and subcontinents (Amrith and Smyer Yü, this book). At the bottom of the gorge, they are intensified by localized climate systems, becoming heavier and lasting longer, creating one of the earth's wettest regions and a tropical rainforest. As the rains travel up the gorge, they soak forests of rhododendron, conifer, and sal with seasonal rain, before coating higher altitude alpine meadows with seasonal snow and slush. On Namcha Barwa's slopes, the snow replenishes glaciers.

As the rains mingle with the ascending earth, the gorge's steepness creates a disorientating assemblage of climes, reminding us that the English word "clime" derives from the Greek word *klima*, "slope." Like other authors in this book (Amrith and Smyer Yü, Delly and Aiyadurai) and Mark Carey and Philip Garone (2014), I understand these climes to be more than just landscapes and places. They are the sites of energy exchange between climate, hydrological, and geological systems that provide the stuff of life. Over millennia, life in myriad forms has sought out niches within the gorge's spectrum of climes. As the gorge formed and reformed, microbes, fungi, plants, and animals found homes within its biomes. Many settled in one clime; others climbed between them. Most changed their new habitats to suit their needs, modifying their environments dramatically or imperceptibly. The gorge's concentrated topographical diversity

DOI: 10.4324/9781003347026-15

facilitated unusual proximity between plant and animal species that would not cross paths elsewhere. Sixteen species of ungulates, including deer, sheep, takin, and goral, found homes within it, along with macaques, langurs, red pandas, and civets, who all live between rainforest and glaciers. But the most telling marker of the gorge's diversity is its cats. Snow leopards, clouded leopards, common leopards, Asian golden cats, jungle cats, marbled cats, and tigers prowl parts of the gorge looking for prey.

Humans and their domesticated animals were the last species to enter the gorge. At first, they sought out niches to hunt and farm. But slowly, each new wave of human settlement developed divergent ways of knowing the gorge and changing it to suit their purposes. The humans described their new knowledge as discoveries, and this knowledge gave them new ways to transform themselves and the gorge; they acclimatized themselves to the gorge's climes and—as much as they could—altered its climes to suit them. As their technology improved, so did their capacity to terraform the gorge, transforming it for their purposes.

Human interactions with the gorge were never unmitigated. As a species, we do not have the sensual apparatus to understand spaces like the gorge without dividing them into smaller, comprehensible units. We know localized areas through our senses: homes, farms, and hunting and trade routes. But when we want to understand larger spaces, we focus on specific elements, routes, waterways, societies, sacralities, sovereignties, administrations, topographies, geologies, biomes, temperatures, and precipitation. We divide our world into climes through our relationships with places or our conceptualization of it. Our connections to these climes combine localized, sensual understandings with larger, abstract maps. This interwoven understanding enables us to find niches within the biome that we can shape and by which we can be shaped.

But the process of discovery not only enables an understanding of the spaces through which we move; it can also assert our authority over a place. As theorists and historians of settler states, such as Linda Tuhiwai Smith (1999, 31, 32, 59–61) and Patrick Wolfe (2006, 387–409) have explained, this process of discovery presupposes that we are the first to see things that nonhumans and other humans already know, and as such is a tool of colonization and erasure. It also supposes that the places we come to know are passive recipients of our attention when they are not.

Understanding this process as "climing," making sense of a clime, changing it, having it change us, gets closer to the reality of the process of human occupation than "discovery." As Paerregaard sets out in this book, climing is what happens when communities observe, interpret, and respond to micro-climatic changes in mountain regions with multiple ecological niches. When we look at this process within a mountainous region's environmental history, we can see how climing occurs in cycles through time and, importantly, includes responsive adaptations from the nonhuman elements of the ecological system and human communities. Humans clime the gorge, but the gorge also climes around its humans.

Few spaces offer more diversity of climes, have been "discovered" more times, and provide more insights into our complicated relationship with the globe's diverse climates than the Yarlung Tsangpo Gorge (Figure 10.1). This chapter problematizes the anthropocentric idea that the gorge was "discovered" multiple times and proposes instead that its *more-than-human* inhabitants have developed complicated relationships with its climes over time; they have climed the gorge. Outlining the gorge's socio-ecological history shows how its "discovery" has been a millennia-long climing process rather than a singularly recent event. Ungulates followed plants into the gorge. Cats followed ungulates. Human hunters followed both, tracing their tracks. Later, the hunters led pastoralists, agriculturalists, envoys of local polities, and spiritual seers from various communities—Adi, Monpa, Pemaköpa, and Tibetan. Human presence in the gorge quickened the temporalities of discovery. Humans approached the gorge from diverse entry points and developed related but distinctive understandings of it and their place within it. Their connected but discrete spiritual geographies grew symbiotically with its multiple ecologies. Most of the Adi, whose settlements were concentrated to the south, see the gorge as a zone of ancestral connection. The Pemaköpa and Tibetan peoples came to see it as the manifestation of the deity Dorjé Pakmo's body, and thus as one of the Himalaya's most sacred sites. Their presence helped make the gorge a protectorate of the nearby Po Kingdom.[2]

Nineteenth-century geopolitics further accelerated the gorge's discoveries. The British Empire sent quasi-military expeditions searching for the connection between the Yarlung Tsangpo and Brahmaputra Rivers. Their expeditions combined scientific surveys with the threat of state violence. The first Western—mainly

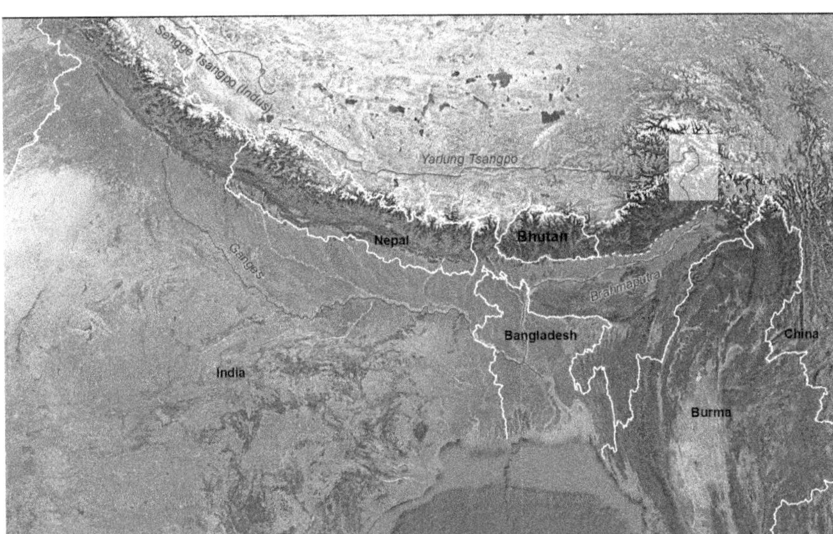

FIGURE 10.1 Yarlung Tsangpo Gorge.
Source: Adapted From Google Earth by Richard MacNeill.

British—scientists, "plant hunters" like Frank Kingdom-Ward (1885–1958), were drawn to the gorge by its biodiversity, but the British Empire's primary legacy in the region was the McMahon Line. They imagined this cartographical border with Tibet in the 1910s. It jumped from mountaintop to mountaintop, disconnected from the region's biological, hydrological, and cultural flows. Five decades later, unable to agree on the line's validity, the newly formed People's Republic of China and the Republic of India fought a war around the gorge in 1962 before securitizing and militarizing it and, in the process, "discovering" more about it. Western adventures in the gorge continued after the Chinese and Indian governments began administering the gorge. In the 1990s, a new generation of explorers, seekers, kayakers, and botanists sought out its gurus and plants, "discovering" a waterfall they named the Hidden Waterfall with the aid of local guides.

These discoveries were, in turn, contested by scientists from the Chinese Academy of Sciences (CAS), who began asserting their claims to the gorge around the same time. These scientists, headed by geomorphologist Yang Yichou (1935–2019), began studying the gorge in the 1970s. By the time the Western adventurers had made their way through it, they had photographed it from the air, established that it was the globe's deepest terrestrial gorge—"the actual Grand Canyon" (Tao 2019, 7)—ascertained that it had more hydropower potential than anywhere else on earth, and begun a ground-level survey of it, which they claimed made them the first group to walk through the gorge (Tao 2019, 7). This ground-level survey led to the discovery of more species, a greater understanding of the gorge's biodiversity, and a new contest for it between hydropower and biodiversity proponents (Harrison and Loring 2020). Both sides of this debate present it as a blank space in which they have made physical or biological "discoveries" (Tao 2019). A bifurcation between international and Chinese ecological studies has led to these "discoveries" being trumpeted within China, while the international scholarly community broadly commits itself to decolonization and the recognition of Indigenous knowledge of place.

It is possible to read this history as another example of declensionist environmental and colonial history. But the living stories of critters that cannot speak—which is to say, animals and plants—and the biographies and journals of the many humans that have documented their journeys into the gorge suggest an alternative reading. They suggest that those who stay in the gorge over generations seek the tracks of those before them and develop a sense of communion with them. It is even arguable that the gorge's multiple climes partially facilitate this diversity. At the very least, these pluralistic views of the gorge resist the gorge's hegemonic reduction to either a source of hydropower or a sealed fortress of biodiversity.

More-than-Human Rock Clim(b)ers

Before the gorge could be climed, it needed to be created. The region's geological and hydrological cycles worked together to bring this about. The gorge contains the remains of an ancient ocean that two dueling continents squeezed out and dried up as they ground into each other (Lang and Hungtington 2014). The continents buckled the earth's crust, folding and stretching it, creating weaknesses underlying magma exploited, seeping out and pushing the earth skyward. As this slow-moving geological drama continued, the river's upper reaches were pushed into a thinner atmosphere with colder temperatures. Spring ice-melt and strengthening monsoon rains began to trickle down from these heights, following the most direct route. The rising mountains forced the water to flow through the narrowing gorge, a small gap that had once been a great ocean. Sometimes, landslides blocked off even this narrow passage (Xu et al. 2020) and lakes formed. These blocks were followed by massive, breakthrough flooding (Wang et al. 2014; Hu et al. 2021). Drawn by gravity, the water followed the path of least resistance. As it flowed, it wore away at the rising earth. Occasional large flood events transported village-sized boulders and tore into cliff sides. River planes became inclines, and inclines became river planes.

As the gorge formed, life forces responded, climbing and climing it. This gorge and those adjacent to it provided a corridor through which flora and fauna from the colliding Gondwana Indian and the Eurasian continents could travel and co-evolve. The combination of the biome collision, the unsettling clime diversity caused by mountain uplift, and the steepness of mountain slopes created what Mosbrugger et al. call "species pumps" and "refugia and geographical barriers that allow(ed) for the survival of many species, as well as for allopatric (or geographically distinct) speciation" (2018, 429). Namcha Barwa, nearby Gyalha Pelri, and the Kangri Karpo Range provided rain shadows that further exaggerated the gorge's climatic diversity. On the mountains' northern sides, the snowline was higher, and only alpine meadows, rhododendron shrubs, coniferous and mixed forests grew between it and the river. On their southern sides, where the monsoon fed a localized rain cycle, semi-evergreen and evergreen broadleaf forests stretched down the slope, eventually leading to dense tropical rainforests (Peng 1997, 43–48).

Herbivores followed plants into the gorge. The gorge provided them niches and alternate biomes into which they could move when the climate or topography changed. Endemic birds, particularly rare pheasants and migratory birds, found places to bear young. Smaller mammals such as civets (masked palm, large Indian, small Indian), otters (Eurasian, small-clawed), and yellow-throated martens rustled through the brush. Red pandas, langurs, and macaques scampered across the forest canopy. Assam macaques, Rhesus monkeys, and the white-cheeked macaques found homes between altitudes of 1,395 and 2,700 m, eating fruits and leaves in the evergreen, broad-leaved forests (Li et al. 2015). Their diversity in the gorge suggests to biologists that the gorge was their evolutionary hub. They moved up and down it when climates changed; new species evolved and spread to other regions (Li et al. 2020).

Ungulates found niches in the gorge's forests too. Where they chose to live depended on their body size and food habits. Foragers, like leaf-eating gongshan muntjac, and moss-lichen-eating musk deer, sought food from multiple biomes.

Grazers, like most deer, blue sheep, goral, and serow, found grass in concentrated habitats. Takin moved throughout the gorge, eating leaves and grass.

Carnivores chased the herbivores into the forests. Although the gorge became home to two species of bear and at least one species of wild dog, its most common and diverse predators are cats. There are more cat predators here than anywhere else on the planet. The Bengal tiger prowls its lowland jungles. Snow leopards climb its highest peaks, searching for blue goats. Occasionally, these capstone predators leave their habitats at the gorge's altitudinal extremes and move into the in-between biomes that are home to common and clouded leopards (Can 2020). The gorge's medium-size cats—the jungle cat, marbled cat, and Asian golden cat—have more confined habitats but more arresting adaptations. The Asian golden cat, for example, changed its spots in the gorge, developing coat colors to suit the gorge's altitudes. High in the gorge, it is ocelot, gray, and cinnamon. At lower altitudes, it is melanistic, golden, and tightly rosette (Nijhawan et al. 2019; Wang et al. 2019). The smaller cats also eat a more comprehensive range of prey than larger cats, including rodents, snakes, lizards, birds, and muntjac deer. This diversity of cats and recent fossil finds suggest that, as for the macaques, the gorge was a site of big cat evolution (Tseng et al. 2014; Wang 2015).

The cats' and macaque's evolutionary response to the gorge was typical of a *more-than-human* climing that reacted to the gorge over extended time frames. The humans that followed them were less patient, and along with a lack of patience, they began the process of transformation and discovery that now threatens the entire region.

Climing for Refuge

Human migration into the gorge came first in trickles and then in waves. Their journey was influenced by broader peopling and migration patterns. Recent DNA analysis has suggested that the nearby plateau was peopled by Denisovan-like humans and *Homo sapiens* who interbred (Zhang et al. 2021), but the establishment of migration routes onto the Plateau and its timelines are far from settled. The gorge was one of the last places in the region to be peopled, and it was a relatively recent event sometime during the past millennium. Before this, the entrance to the gorge was repeatedly blocked by climatic and seismic fluctuations, glaciations, and landslides. Records from nearby Kongpo suggest major glaciation events to the gorge's north in 200–600 CE, 800–1150 CE, and during the Little Ice Age between 1400–1920 CE. Early glaciations probably combined with floods and seismological changes around 1000 CE (Mosbrugger et al. 2018) to block the entrance to the gorge from the north. But the last glaciation, the Little Ice Age, and concurrent political instability on the plateau (Gamble et al. 2022) may have encouraged migration into it.

Humans likely entered the gorge as hunters first, following the same game as the *panthera* predators that had preceded them: sambar, serow, goral, takin, musk deer, and several macaque species (Aiyadurai et al. 2010). Unlike the *panthera,* they also hunted other predators—the *panthera* themselves, barking dog, wild pig, and black bear—and ranged over the entire gorge, through all its biomes, looking for prey. Pastoralists followed the hunters, shepherding sheep and goats in the higher altitudes, and cows at lower altitudes. When the gorge's hydrology and seismology stabilized, agriculturalists from variant cultural groups followed, planting multiple altitude-appropriate crops. Like the cats, the humans moved up and down the gorge to deal with further environmental changes.

They entered from the Yarlung Tsangpo River Valley in the northeast, the Po region in the northwest and west, and south from the Adi lands of the Siang River Valley. The gorge was only ever lightly populated, and a large stretch of it, the most vertical space between Namcha Barwa and Gyalha Pelri, was never settled. Mountain passes from Kongpo and Po led humans into the middle and lower gorge while allowing them to avoid its steepest sections. But it is inevitable that hunters from local communities "discovered" this unsettled section before the later explorers who claimed this privilege.

Despite its light population, the gorge has hosted greater linguistic and cultural diversity than much of the already-culturally-diverse Himalaya. Adi communities were concentrated in its southern, lower altitudes, up to around 2,000 m. They introduced rice cropping, a practice they share with the broader Tani cultural group. The Tani all claim descent from Abotani, their primary ancestor, and their stories explain how he introduced *jhum* (1,000–2,500 m) and *paddy* (lower than 1,000 m) rice cultivation. The Tani grew bamboo, vegetables, and spices, and kept pigs, chickens, and mithun, a large domestic cattle endemic to the eastern Himalaya. Local traditions insist the Adi migrated down into southern Himalayan river valleys, including the Yarlung Tsangpo (which they call the Siang) River Valley, from the plateau (Aiyadurai and Lee 2017). But other evidence from comparative ethnological studies suggests they entered these valleys from the southeast (Huber 2012; Blackburn 2008, 1–12; Childs 2012; Grothmann 2012). Their material culture, focused on swidden agriculture, enabled them to undertake "micro migrations" throughout the region in response to other climatic, environmental, and political changes. This movement continued until the area was securitized in the 1960s (Huber 2012, 99). In the north, at the edge of the Tibetan Plateau in Kongpo, the Tibetans grew barley, spices, and fruit trees and kept pigs, chickens, and yaks. Kongpo residents were also famous for their forests, and supplied poplar, willow, and pine across the plateau (Bsod nam bkra shis 2011, 1–11).

With pastoralism and settlement, the gorge's humans developed taboos against killing animals at certain times and places (Janaki et al. 2021). For the most part, these taboos, along with later, broader Buddhist proscriptions against hunting, stabilized animal populations (British explorers noted this stabilization at the beginning of the twentieth century, see Kingdon Ward 1926, 110; Bailey 1957, 49, 139). They used most parts of the animal for clothing and food. Animal skins were used as weather-proof clothing (Kingdon Ward 1926, 304) and sometimes as a meat substitute (Trungpa 1985, 219). As well as taking from the gorge's wild population, all their domesticated animals became new prey for the region's predators (Naha et al. 2020). The gorge was unlike much of the Himalaya in that pack animals were not common there. Yaks were used at high altitudes and a few donkeys on other trails, but humans carried most of what was brought into the gorge (Kingdon Ward 1926, 197). Like the cats and hunters who came before them, agriculturalists and traders continued to use animals' tracks, particularly those of the roaming takin, to find accessible routes throughout the region (214, 221).

The Adi and Tibetans traded across mountain passes before the gorge was continuously populated (Blackburn 2008, 30–35). But conflict arose around the sixteenth century, when the Tibetans and their co-religionist Mönpa who lived west of the Adi, began to migrate further down the gorge. This movement was facilitated by the Kingdom of Po, which had been established in the adjacent Po Tsangpo River Valley. Its rulers, the Kanam Depa (Ka gnam sde pa), claimed descent from an early, semi-mythical Tibetan ruler, Drigum Tsenpo. They retained a form of independence until the 1920s ('Gan 'khur ba 1988, 27–43; Lazcano 2005; Schwieger 2002; Grothmann 2012), and between the sixteenth and twentieth centuries, with the aid of the Mönpa migrants, they gradually took control over the gorge.

As their influence grew, the Kanam Depa encouraged a growing movement to view the gorge as a "hidden land" (Tib. *spas yul*). The hidden lands are a series of Vajrayana Buddhist sacred sites in high-altitude south-facing Himalayan valleys in Nepal, Sikkim, Bhutan, and the eastern Himalaya. Tibetan Buddhist gurus known as "treasure revealers," *tertön* (Tib. *gter ston*), began "opening" (Tib. *sgo' byed*) these sites during the fourteenth century, claiming they were places of refuge from political and ecological strife. From this point on, whenever there was strife on the plateau, groups of Vajrayana Buddhists would migrate into these hidden lands (Samuel 2020, 51–59). The gorge lay within Pemakö (Pad ma dkod, "lotus array"), which was considered the most sacred hidden land as it was opened last. With the backing of the Kanam Depas, moving from north to south, treasure revealers gradually "opened" Pemakö to Vajrayana Buddhists. The door openers found instructions left for them by the semi-mythical Guru Rinpoche, who is revered across the Himalaya for subduing its people, deities, and spirits and, therefore, allowing Buddhism to flourish (Samuel and Oliphant 2020). Their sequential opening of the gorge was encouraged by Mongol and Qing invasions of Tibet, sectarian conflicts, and oppressive tax regimes.

The hidden-land phenomenon facilitated a complex process of refugee migrations by relying on a combination of non-dual mysticism and prejudice. Mystically, Vajrayana Buddhism insists that habitual mental states create external worlds. When a being changes its mental state, their perception of the external world also changes. From this perspective, sites that some would see as peripheral, as the home of *lopa* (Tib. *klo pa*, "barbarians"), and as harboring the threat of snakes, frogs, and jungles, could be transformed by the mind into a site of spiritual and agricultural potential. When the Kanam Depa invited the Mönpa into the middle section of the gorge in the 1800s, he claimed to be helping persecuted co-religionists migrating to escape sectarianism and high taxes (Lazcano 2005, 41–63; Grothmann 2012, 21–52). But this invitation was also a shrewd political maneuver steeped in prejudice. The Mönpa migration extended Po influence down the gorge into the area now known as Metok. The Adi who first rented the Mönpa land "covered with trees and bamboo and inhabited by demons and spirits" (Tibetan Social History Series Editing Team 1987, 21) were then displaced by further Mönpa migrations and ended up with the badlands themselves (Pad bkod 2016, 64–66). The Mönpa adapted well to this middle section of the gorge because it was similar in altitude to their home regions in Eastern Bhutan, which enabled them to grow millet and rice. They quickly became the most populous group in the central gorge and referred to themselves as Pemaköpa.

The lure of Pemakö and threats across the plateau led to a further influx of Vajrayana Buddhists during the late 1800s, this time Tibetans. This group was fleeing Qing incursions and sectarian oppression. The Kanam Depa supported the Nyingma ("Old Ones") lineage (Pad bkod 2016, 64–66), which the Lhasa-based government persecuted, as it was associated with its rivals, the Gelug (Mathes and Coura 2021, 1–17). Pemakö's newest migrants were Nyingma adherents. As Buddhists moved into the gorge, their gurus opened sites further south, culminating in settlement of the Yangsangchu-Nyerang (Tibetan and Adi, respectively) River Valley around the turn of the twentieth century. This southern movement brought them into further contact and conflict with the Adi. As the Adi were fighting among themselves, the Po made a strategic decision to back the Tangam group (Tib. *klo dkar po*, the "white southerners/barbarians") against the Simong (Tib. *klo nag po,* "black southerners/barbarians"). After the Simong Adi raided a Pemaköpa village in 1905, the Kanam Denpa sent a force that established a fort at the confluence of the Yangsangchu-Nyerang and Yarlung Tsangpo-Siang Rivers in 1908. This enabled the Tangam group to reclaim the valley, and in return, the Tangam let Tibetans, mainly Khampas from eastern Tibet, settle there (Dunbar 1917, 94).

Population increases in the gorge and its side valleys led to the spread of agriculture and more significant human-animal conflict. All the region's humans had to contend with its animals, particularly its predatory animals, in one way or another. Local lore has it that the Po rulers were successful because their gurus could tame tigers (Kunga Tsomo, personal communication). During this time, European visitors to Po monasteries noted images of tigers outside the front of each monastery, and the head guru tiger's skin seat (Bacot 1912, 187, 46).

Empires and Riddles

The Po-backed incursions into the gorge were a mild precursor for the state incursions that followed during the nineteenth and twentieth centuries. During this period, the British Empire, the Qing Empire, Republican China, Communist China, and Republican India claimed part of the gorge. To facilitate their rule, they set about producing knowledge about it, a colonizing process often

FIGURE 10.2 Yangsangchu River Valley.
Photograph by Ruth Gamble.

presented as "discoveries." The British combined military expeditions into the gorge with a quest to solve the "riddle" of the Brahmaputra River's course. These expeditions included multiple explorations by "plant hunters," like Kingdon Ward, whose work facilitated the transfer of the gorge's plants worldwide (Paterson 2000). The Qing sent in soldiers to quell local uprisings and find new sites for rice cultivation (Relyea 2015). In the second half of the twentieth century, the post-colonial Chinese and Indian states fought over and securitized an undefined border that ran through the gorge's south. Few non-military "discoveries" were allowed during this time, but a brief window of government openness in the 1990s and early 2000s allowed a new generation of Western scientists, conservations, and adventurers (including kayakers) to claim discoveries in the gorge.

The British became interested in the Tsangpo Gorge after they extended their imperial territories into Assam in the 1820s. They set up tea and opium plantations on the Assamese plains, extracted wood, and set aside large tracts as hunting reserves (Saikia 2011). They had little interest in the densely-jungled hills and, in 1873, drew an "inner line" below them and demanded a permit to cross it. Several decades later, these hills became a subject of geopolitical concern. The British were worried that the Russians—whose power they overestimated—and the Qing—whose facilities they underestimated—could attack them from Tibet (Simpson 2021, 21–69). They advocated the "science" of natural borders along watersheds and underpinned their colonial rule through knowledge acquisition.

They were also determined to establish the course of the Yarlung Tsangpo River. Several rivers joined to create the Brahmaputra in eastern Assam—the Lohit, Dibang, and Dihang. Geographers were unsure which was the Yarlung Tsangpo. At first, they tried to determine the river's course by traveling upstream into Adi (whom they knew as "Abor") territory, but this proved impossible. When they discovered the Tibetan authorities would not allow them access to survey their lands, they hired a series of Himalayan people to travel secretly into Tibet as pilgrims and traders and map it for them. These secret surveyors became known as "pundits," and their stories have been told multiple times (Stewart 2006).

In 1878, they sent a Sikkimese pundit, Nem Singh, and his assistant Kintup on the first expedition to southeast Tibet. The pair reached the gorge's northern entrance. A few years later, Kintup accompanied an unnamed Mongolian or Chinese monk on an expedition to find the course of the Yarlung Tsangpo for the British. When they arrived in the northern section of the gorge, the monk sold Kintup into slavery and ran away. Despite this and other setbacks, Kintup traveled down into the gorge's mid-section, threw marked logs into the river in the hope that he could alert the British in Assam about the river's course, and eventually returned to British India. During the translation of his post-journey debriefing, he was misquoted as saying there was a large waterfall on the river. This waterfall was marked on the Survey of India's 1889 map. In 1895, the Scottish scholar Laurence Waddell seemed to confirm the falls' existence by publishing an illustration of them procured from "a Tibetan informant." Kintup's mistranslated report and Waddell's illustration encouraged a whole generation of European geographers and explorers to imagine the "Tsangpo Gorge Falls."

Along with dreams of imperial waterfalls, the British Indian Government was interested in the river for practical reasons. First, in 1900, a large flood hurtled down the Dihang River, delivering debris, coniferous wood, and bodies into British territory (Bentnick 1913, 107). Then, in the following three years, their concern with Russian interference in Tibet rose, and this fear, along with Tibetan (and Qing) continued refusal to trade with them, led to their 1904 invasion of Tibet, the so-called "Younghusband Expedition." This expedition killed many Tibetans, occupied Lhasa, forced the thirteenth Dalai Lama, Thubten Gyatso (1876–1933), into exile in Mongolia and China, established trade envoys in Gyantse and Lhasa, and—like many British imperial invasive expeditions—gathered a good deal of scientific data about the Tibetan Plateau (Gamble 2022). Among its members was a young explorer, Frederick Bailey (1882–1967), who subsequently traveled up the Yarlung Tsangpo to Mount Kailash and began work as a trade agent and spy at Gyantse (Bailey 1957, 168).

As the Qing Empire still claimed Tibet, despite a century of decreased influence (Sperling 1976), their imperial bureaucrats reacted angrily to the invasion. Those in neighboring Sichuan called for a policy change, the adoption of the British colonizing paradigm, and the survey and settlement of Tibetan areas, particularly the lowland areas where they could grow rice (Relyea 2015). When the Khampas in Eastern Tibet revolted against this policy in 1906, the governor of Sichuan, Zhao Erfeng (1845–1911), killed many of them, moved on to Central Tibet, causing the Dalai Lama to flee to British India, and in 1910 sent several hundred soldiers into Po and nearby Dzayül. Zhao wanted control of these two regions as they provided access to Yunnan and British India. The Kanam Depa's court reacted to the Chinese invasion by fighting among themselves, giving the Chinese an advantage. The Chinese routed the Po and occupied the northern section of the gorge for several years. But when the Qing Empire fell in 1911, the Po returned fire, killing many Qing troops and forcing the rest to flee ('Gan 'khur 1988; Sperling 1976; Lazcano 2005).

While the Po ejected the Chinese from their territory, Republican soldiers replaced their Qing counterparts in neighboring Dzayül, along the Lohit River. When British Political Agent Noel Williamson saw their flags, he determined to establish their level of influence along the Dibang and Dihang Rivers and, for good measure, determine the course of the Yarlung Tsangpo. But in March 1911, after he accused Adi of stealing and threatened them, several Adi killed him, his traveling companion Dr. John Duncan Gregorsen, and forty-eight of the "native" soldiers accompanying them. In retaliation, in November that year, the British sent another expedition, the Abor Expedition, up the Dihang River Valley, caught and executed those who killed Williamson, and—displaying their usual skill for multi-tasking on military expeditions—discovered multiple birds and moss species as they traveled (Dixon 1914). A smaller group led by Arthur Bentinck even continued upriver but was thwarted by the rains, leading an embittered Bentinck to dismiss Kintup's reports of the falls (Bentinck 1913).

When Williamson was killed in the lower Dihang River Valley, Bailey was traveling between the upper Yangzi and Yarlung Tsangpo Rivers, hoping to enter the gorge from the north. Like Bentinck, his path into the gorge was blocked, but in his case by fighting between the Po and Qing soldiers rather than rains. Locals told him the Po had killed 300 Chinese troops and refused to take him into the gorge (Bailey 1957, 242). Changing directions, he went down the Lohit Valley and was asked to explore the upper Dibang River. During this trip, he determined that the Dibang was not the continuation of the Yarlung Tsangpo and met a group of Tibetan Khampas refugees who agreed to take him into the gorge. With the Khampas' help, in 1913, he traveled along with another British officer, Henry Moreshead (1882–1931), up the gorge, negotiated with Po authorities, and became the first "white man" to discover the gorge and the continuity between the Dibang and the Yarlung Tsangpo Rivers (Bailey 1957, 163). He also discovered that the 1900 flood had originated in a lake outburst on the Po Tsangpo River, which ran through Po territory, as well as significant falls in the steepest part of the gorge near—where Kintup said they would be—but on a minor tributary of the river rather than the main stream (276). After becoming trapped at the side of the roiling river, he had to use a takin track to trek out of the gorge (133) and missed discovering the more significant falls on the main river downstream. Returning to India, he sought out Kintup, who told him he had been misquoted in claiming a large waterfall on the Yarlung Tsangpo itself (287). Bailey and Kintup have tended to share the credit for discovering the gorge and determining the relationship between the Dibang and the Yarlung Tsangpo. The Khampas, Pemaköpa, Adi, and takin are never credited.

The fall of the Qing Empire led to a recalibration of the region's geopolitics as the British and the Tibetans asserted control. In 1913, the British called for a border conference at Shimla with the Tibetans and the Republican Chinese. At the meeting, the British plenipotentiary, Henry McMahon (1862–1949), proposed a new border between British Assam and Tibet that traced the Eastern Himalaya's peaks but cut across the Yarlung Tsangpo River at the bottom of the gorge. The Chinese refused to sign the conference's treaty. The Tibetans signed but ignored it in practice (McGranahan 2003). As Toni Huber's research has shown, the period between 1900 and 1950 showed a marked increase in the Tibetan government's influence on and exploitation of smaller states like the Po and "non-Buddhist, clan-based Tibeto-Burman-speaking communities" in the Eastern Himalaya (2012). This interest coincided with increased turmoil in the plateau's east and the increased movement of refugees into the Hidden Lands (Gamble 2022; Pad bkod 2016; Bailey 1957, 106; Bentnick 1913, 107). These refugees included the Nyingma adherents to whom the Kanam Depa gave land and protection in the Tamang lands in the Yangsangchu River Valley. In 1928, this group lost their protector when the Tibetan government took control of Po, deposing the Kanam Depa, who fled to Sadiya where he quickly died ('Gan 'khur ba 1988; Lazcano 2005). Following this, Sera Monastery, a sizeable Gelug Monastery near Lhasa, was permitted to raise taxes throughout the gorge. From 1938, ignoring the McMahon Line, Sera's representatives began collecting taxes as far down as Karko in Adi territory. This forced the British to send an Assam Rifles regiment back into the valley to threaten them (Huber 2012).

However, by this stage, British interest in the gorge had switched from battles to botanics, as the gorge became a source for one of the era's most lucrative trades, flowers. "Plant hunters" who sought out exotic species that could be cultivated in English gardens had been scouring the Himalaya for nearly a century. One of the most well-known plant hunters was Frank Kingdon Ward, who explored the gorge and commercialized its plants. Some of the flowers he discovered became mainstays of the international botanical trade, including several species of magnolia, rhododendrons, and now-common lilies. During his 1924–25 journey through the region with the Scottish aristocrat John Campbell, Earl of Cawdor (1900–1975), he discovered a new, blue poppy, *Meconopsis betonitificolia*, and a new series of waterfalls in a stretch of the gorge that Bailey and Moreshead had not traversed. He described these adventures in a best-selling book called, *The Riddle of the Tsangpo Gorges* (1925) and returned to the region in 1946 and 1950. Kingdon Ward was primarily interested in plants. He wrote little about the region's wildlife, apart from the takin, whose tracks, he sometimes followed (Kingdon Ward 1926, 221). The only other Western explorers to enter the gorge during this period were naturalists Frank Ludlow (1885–1972) and George Sherrif (1898–1967), who traveled through the district in 1938 and 1945 on either side of World War Two. They did not press far into the gorge but did discover several of its bird species (Ludlow 1940).

Armies and Kayakers

After the Republic of India (1949) and the People's Republic of China (1949) were founded, both post-colonial governments continued the colonial-era claims on the gorge. China occupied Tibet and, once again, its residents headed south through the gorge in search of refuge, this time within the borders of India. India claimed the territory up to the McMahon Line, and China claimed the eastern Himalaya down to the colonial Inner Line. Their conflicting claims led to a short war in 1962, and the gorge's militarization. This militarization transformed the gorge more than any previous human occupation. Both states sought to make the region's people visible through ethnographic taxonomies, and restrict their new subjects' movement and activities (Scott 1998). By curtailing human movement across the border, the governments not only dissected local human communities, sacred sites, trade networks, pastoralists' grazing regions, and hunting tracks (Gamble 2022), but their border patrols and roads also impeded movement in the more-than-human world, for example, wild yaks and takin's movement was hampered by troop positions (Naha et al. 2020).

After the war, China had control of most of the gorge and restricted its access to military and party officials who were sent down to Metok to collectivize its farms and categorize its people into nationalities (Ch. *minzu*). The Tibetan, Memba, and Luoba were classified as "backward" (Ch. *luohou*) and in need of development. Several party officials wrote about their experiences in this isolated space. One author called the region's inhabitants "living fossils" (Ch. *huohuashi*) (Luo 2018).

During a brief opening period in the 1990s and early 2000s, the government allowed a new generation of non-Chinese explorers into the gorge. Imperialism was not an obvious driver for this group. Arguably, they displayed greater respect for local cultures than any explorers who preceded and followed them. But they were still driven by one of imperialism's ghosts, discovery. On the November 8, 1998, Kenneth Storm and Ian Baker discovered a new waterfall, Hidden Falls, within the gorge's "unexplored" region. These explorers and several fellow travelers documented their adventures in the gorge. Botanist Kenneth Cox (2001) republished Kingdon Ward's *Riddle of the Tsangpo Gorge* with appendices from himself and Baker. Baker (2006) authored a book about Pemakö, the hidden land. Their friend, journalist and author Clair Scobie wrote a moving book about discovering a new lily, *Lilium paradoxum*, and her blossoming friendship with a Tibetan nun (2012). Storm later worked with a kayaking expedition led by Scott Lindgren that became the first group to descend the "Everest of Rivers" successfully. Their descent inspired two successful documentaries (Lindren 2009; Sturges 2021) and several books (including Walker 2000; Heller 2004).

"Discovery" and "firsts" are central narratives throughout these works. In his addendum to the republication of *Riddle of the Tsangpo Gorge*, for example, Kenneth Cox argued that he and Cox, rather than a Chinese expedition, were the first to discover the Hidden Falls (2001, 310–312). Somewhat incongruously, he makes this claim two pages after noting that his Mönpa guide, Dungle Phuntsok's grandfather, led "two Ingles," Kingdon Ward and Cawdor, into the same gorge to see the Rainbow Falls decades earlier (2001, 300). The will to discover had proved a persistent imperial ghost.

Engineers and Earth Scientists

Discovery had also become a mainstay of the international scientific project. The leading Chinese scientist who contested Cox and Baker's claim of discovering Hidden Falls was Yang Yichou (1935–2019), and he had reason to be annoyed by their claim. Yang, a Chinese Academy of Sciences (CAS) geomorphologist, had surveyed and documented the gorge since the early 1970s. From his and his academic acolytes' perspective, his work had led to many more discoveries than the Americans could claim, and, unlike the Americans, they had all been conducted with academic rigor.

Yang was a particularly revered scholar of the gorge because his work was interdisciplinary. As the science of the gorge developed, it was often presented within opposing disciplinary boundaries. During Yang's first surveys of the gorge in the early 1970s, he worked with engineers interested in understanding and exploiting the region's hydropower and mineral potentials (Wang 1996). In the 1980s, he worked with engineers, mountaineers, and geographers to map the gorge's upper sections near the village of Pé and its lower sections near Metok. These teams "discovered" the gorge's immense hydropower potential and began advocating for the construction of a large hydropower project within the gorge that would produce three times the power of the Three Gorges Dam ("China to Conduct Feasibility Study on Power Project in Tibet" 2003). For many years, the construction of such a dam was deemed too risky, but it finally became government policy in the fourteenth five-year plan released in 2020.

However, Yang's conception of the gorge was not confined to its hydropower potential. In the early 1990s, he also established the gorge as the globe's deepest terrestrial canyon. He first argued for an international definition of a canyon—its physical characteristics, boundaries, and specific measurements—and then proved that the Yarlung Tsangpo Grand Canyon was larger than the canyon on the Colorado River. Of geopolitical note was Yang's decision to stop measuring the canyon just short of the unofficial border between China and India, even though the gorge continues beyond the border. His claim was published in the official Chinese press on April 17, 1994. The official media hailed it as a great scientific discovery, comparable to the British discovery of Mount Everest's height, and a win for China which now possessed the world's highest and deepest geographical features (Tao 2019).

After establishing the gorge as the Yarlung Tsangpo Grand Canyon, Yang set his sights on a new record. On December 3, 1998, he joined an expedition of scientists, mountain climbers, and journalists who were determined to be "the first humans to travel through the canyon on foot" by focusing their efforts on the "one hundred kilometers of no man's land" in the center of the gorge (Tao 2019, 7). After failing to gain the funds they needed from CAS, they received sponsorship from a tobacco company (Tao 2019, Chapter 3, 5). According to their reports, they did not reach the central part of the gorge that contains the two waterfalls until October 1999 (Tao 2019, Chapter 3, 3), nearly a year after Baker and Storm. But they were carrying out surveys in different parts of the gorge when the two Americans were "discovering" it.

Furthermore, Yang continued to work in the gorge for at least a decade after the Americans returned home. During this next period, he worked across disciplines with biologists intent on documenting the gorge's intense biodiversity. Their studies led to the discovery of a variety of species new to science, including the white-cheeked macaque (Li et al. 2015) and the international recognition of the gorge-cum-canyon as a biodiversity hotspot (Deng et al. 2011). After adopting video surveillance and camera traps in the gorge, the number and density of discovered species increased (Can et al. 2020). Yang and others' work led to the Canyon (not the entire gorge) being declared a Nature Reserve and senior scientists arguing that it should be declared a National Park (Wang et al. 2021). Such a park could severely curtail the gorge's hydropower potential exploitation.

While this interdisciplinary work has been impressive, Yang and his colleagues' attitudes to the region's Indigenous people have been less impressive. Non-local, Han Chinese scientists have driven it, and much like earlier, Western and Qing explorers have downgraded locals' insights and discoveries. Their understanding of "discovery" and practical knowledge has been intrinsically linked to science, which they claim the "backward" locals do not understand (see, for example, Tao 2019, 6).

Conclusions: Who Is Climing Whom?

Despite the variant paradigms, all humans who climbed into the gorge emphasized "discovery"—Adi, Mönpa, Tibetans, British, Qing, Indian, American, and Chinese. All human explorers followed nonhuman critters that entered the gorge before them. They were drawn to their prey but repelled by the gorge's changing weather and impenetrable terrain. All began by cataloging and trying to tame the inhabitants that preceded them. But there have also been differences. Those who entered the gorge in temporally dispersed waves had time to accommodate earlier migrants. They also adapted to the gorge and adapted the gorge to them. By contrast, technology has allowed recent explorers to arrive quickly, remain connected to life outside the gorge, and resist the need to adapt to its climes.

Human transformation of the gorge began in earnest when migrants started cultivating its flat or flattened lands. This led to conflict over this land and the consequent involvement of local polities like the Po Kingdom. The Tibetans' hybrid spiritual-colonialist approach to the gorge drew it into the Tibetan socio-cultural sphere, which led to further transformations and, eventually, Chinese occupation. The British explorations of the gorge and their solving of the "riddle of the Tsangpo Gorge" affected the transformation of the gorge in two ways. First, it put the gorge on the map as a site of geographic and biological interest. Second, it represented yet another instance of the British Empire's combined use of knowledge production and the threat of violence to exercise territorial control. The British imperial rivals, the Qing, and the gorge's subsequent rulers, the Chinese and Indian post-colonial states, enthusiastically embraced the British example while separating themselves from British rule. China's and India's claims to the gorge were based on colonial precedent, and China's administration of it, especially in the past decades, has accelerated discoveries, terraforming, and extraction. It remains to be seen what discoveries and adaptations will come next, but we should not forget that the gorge's ability to transform and regenerate appears to change—or clime—those who stay within it. For example, in the past two decades, conversations about the gorge have shifted from using nuclear weapons to reform it for hydropower purposes to concerted calls to conserve its biodiversity. And these shifts have only occurred on animal and human scales. Every day brings the possibility that tectonic and hydrological shifts will re-transform the region, re-climing it according to its own, *larger-than-human* scale.

Notes

1 The research for this chapter is funded by the Australian Research Council DECRA Fellowship.
2 Po is sometimes referred to as Pomé. The Kingdom of Po (Tib. Spo) covered two main areas, Po tö (Tib. Spo stod), Upper Po and Po mé (Tib. Spo smad), Lower Po.

References

"China to Conduct Feasibility Study on Power Project in Tibet." 2003. *People's Daily*, July 17.
Aiyadurai, Ambika, Navinder J. Singh, and E. J. Milner-Gulland. 2010. "Wildlife Hunting by Indigenous Tribes: A Case Study from Arunachal Pradesh, North-East India." *Oryx* 44 (4): 564–572.
Aiyadurai, Ambika, and Claire Seungeun Lee. 2017. "Living on the Sino-Indian Border." *Asian Ethnology* 76 (2): 367–395.
Bacot, Jacques. 1912. *Le Tibet Revolt. Vers Népémakö, La Terre Promise Des Tibétains*. Paris: Librairie Hachette.
Bailey, Frederick M. 1957. *No Passport to Tibet*. London: Rupert Hart-Davis.
Baker, Ian. 2006. *The Heart of the World: A Journey to Tibet's Lost Paradise*. New York: Penguin.
Bentinck, Arthur. 1913. "The Abor Expedition: Geographical Results." *The Geographical Journal* 41 (2): 97–109.

Blackburn, Stuart. 2008. *Himalayan Tribal Tales: Oral Tradition and Culture in the Apatani Valley*. Leiden: Brill.
Bsod nam bkra shis. 2011. *Gangs ljongs kong po'i lo rgyus dang gnas bshad gyen du 'byur ba'I lhag bsam gangs lha me tog*. Dharamsala: Shes rig dpar khang.
Can, Özgün Emre, Bhupendra Prasad Yadav, Paul J. Johnson, Joanna Ross, Neil D'Cruze, and David W. Macdonald. 2020. "Factors Affecting the Occurrence and Activity of Clouded Leopards, Common Leopards and Leopard Cats in the Himalayas." *Biodiversity and Conservation* 29 (3): 839–851.
Carey, Mark, and Philip Garone. 2014. "Forum Introduction: Climate Change and Environmental History." *Environmental History* 19 (2): 282–293.
Childs, Geoff. 2012. "Trans-Himalayan Migrations as Processes, Not Events: Towards a Theoretical Framework." In *Origins and Migrations in the Extended Eastern Himalayas*, edited by Toni Huber and Stuart Blackburn, 83–106. Leiden: Brill.
Cox, Kenneth, editor. 2001. *Riddle of the Tsangpo Gorges: Retracing the Epic Journey of 1924–25 in South-East Tibet*. New York: Antique Collectors Club Ltd.
Deng, Libin, Chen Duan, Deng Liqun. 2011. "西藏雅鲁藏布大峡谷国家级自然保护区生态评价 [Ecological Evaluation of Yarlung Zangbo Grand Canyon National Nature Reserve in Tibet]." *Scientia Silvae Sinicae* 47(5): 1–6.
Dixon, H. N. 1914. *Report on the Mosses of the Abor Expedition* (12). Calcutta: Superintendent Government Printing House, India.
Dunbar, George. 1917. "Abors and Galongs." *Memoirs of the Asiatic Society of Bengal* 5: 88–200.
Harrison, H. L. and Loring, P. A. 2020. "Seeing beneath Disputes: A Transdisciplinary Framework for Diagnosing Complex Conservation Conflicts." *Biological Conservation* 248: 108670.
Hu, Hai-Ping, Jin-Hua Liu, Jin-Liang Feng, Chang-Sheng Ye, Zhi-Jun Gong, Feng Lü, Feng Chen, et al. 2021. "Geomorphic Processes of a Dammed Palaeo-Lake in the Middle Yarlung Tsangpo River, Tibet." *Science of the Total Environment* 811: 151949.
Gamble, Ruth. 2022. "Surviving Pemakö's Pluriverse: Kunga Tsomo, the Goddess, and the LAC." *Critical Asian Studies* 54 (2): 398–421.
Gamble, Ruth. In Press. "The Mountain's Many Faces: How Geologists Mistook Chomolangma for Everest." In *New Earth Histories*, edited by Alison Bashford, Adam Bobbette, and Emily Kern. Chicago: Chicago University Press.
Gamble, Ruth, John Powers, and Paul Hackett. 2022. "Famines of the Early Little Ice Age (1260–1360): The Impacts of Pre-Modern Climate Change in Central Tibet." *Bulletin of the School of African and Oriental Studies* 85 (2): 1–19.
'Gan 'khur ba U rgyan. 1988. *Spo bo'i lo* rgyus. Lhasa: Bod ljongs mi dmangs dpe skrun khang.
Grothmann, Kerstin. 2012. "Population History and Identity in the Hidden Land of Pemakö." *Journal of Bhutan Studies* 26 (2): 21–52.
Heller, Peter. 2004. *Hell or High Water: Surviving Tibet's Tsangpo River*. Sydney: Allen and Unwin.
Huber, Toni. 2012. "Micro-Migrations of Hill Peoples in Northern Arunachal Pradesh: Rethinking Methodologies and Claims of Origins in Tibet." In *Origins and Migrations in the Extended Eastern Himalayas*, edited by Toni Huber and Stuart Blackburn, 83–106. Leiden: Brill.
Kingdon Ward, Frank. 1926. *The Riddle of the Tsangpo Gorges*. London: Edward Arnold.
Janaki, M. Rohan Pandit, and Rishi K. Sharma. 2021. "The Role of Traditional Belief Systems in Conserving Biological Diversity in the Eastern Himalaya Eco-Region of India." *Human Dimensions of Wildlife* 26 (1): 13–30.
Lang, Karl A., and Katharine W. Huntington. 2014. "Antecedence of the Yarlung–Siang–Brahmaputra River, Eastern Himalaya." *Earth and Planetary Science Letters* 397: 145–158.

Lazcano, Santiago. 2005. "Ethnohistoric Notes on the Ancient Tibetan Kingdom of sPo bo and its Influence on the Eastern Himalayas." *Revue d'Etudes Tibétaines* 7: 41–63.

Li, Baoguo, He, Gang, Guo, Songtao, Hou, Rong, Huang, Kang, Zhang, Pei, Zhang, He, et al. 2020. "Macaques in China: Evolutionary Dispersion and Subsequent Development." *American Journal of Primatology* 82(7): e23142.

Li, Cheng, Chao Zhao, and Peng-Fei Fan. 2015. "White-Cheeked Macaque (Macaca leucogenys): A New Macaque Species from Medog, Southeastern Tibet." *American Journal of Primatology* 77 (7): 753–766.

Lindren, Scott. 2009. *Into the Tsangpo Gorge*. DVD.

Ludlow, Frank. 1940. "Takpo and Kongpo, S.E. Tibet." *Himalayan Journal* 12: 1–16.

Luo Hongzhong. 2018. "雅魯藏布江大峽谷的「活化石：1950年代珞巴族與門巴族留影 [Living Fossils of the Brahmaputra Grand Canyon: Photos of the Lhoba and Monba in the 1950s]". *KK News*, 8 May 2018.

McGranahan, Carole. 2003. "From Simla to Rongbatsa: The British and the 'Modern' Boundaries of Tibet." *The Tibet Journal* 28 (4): 39–60.

Mathes, Klaus-Dieter, and Gabriele Coura. 2021. "Introduction." In *Nonsectarianism (ris med) in 19th- and 20th-Century Eastern Tibet Religious Diffusion and Cross-fertilization beyond the Reach of the Central Tibetan Government*, edited by Klaus-Deiter Mathes and Gabriele Coura, 1–17. Leiden: Brill.

Mosbrugger, Volker, Adrien Favre, Alexandra N. Muellner-Riehl, Martin Päckert, and Andreas Mulch. 2018. "Cenozoic Evolution of Geo-Biodiversity in the Tibeto-Himalayan Region." *Mountains, Climate, and Biodiversity* 429: 448.

Naha, Dipanjan, Suraj Kumar Dash, Abhisek Chettri, Pooja Chaudhary, Gaurav Sonker, Marco Heurich, Gopal Singh Rawat, and Sambandam Sathyakumar. 2020. "Landscape Predictors of Human–Leopard Conflicts within Multi-Use Areas of the Himalayan Region." *Scientific Reports* 10 (1): 1–12.

Nijhawan, Sahil, Iho Mitapo, Jibi Pulu, Chris Carbone, and Justin Marcus Rowcliffe. 2019. "Does Polymorphism Make the Asiatic Golden Cat the Most Adaptable Predator in Eastern Himalayas?" *Ecology* 100 (10): e02768.

Pad bkod 'chi med rig 'dzin. 2016. *Padma bkod yul gyi bkod pa dang byung rabs lo rgyus mdor bsdus*. Dharamsala: Library of Tibetan Works and Archives.

Paterson, Allen. 2000. "The Plant Hunters: Two Hundred Years of Adventure and Discovery around the World." *Studies in the History of Gardens and Designed Landscapes* 20 (3): 258–259.

Peng, Buzhuo, Pu Lijie, Bao Haosheng, and D. L. Higgitt. 1997. "Vertical Zonation of Landscape Characteristics in the Namjagbarwa Massif of Tibet, China." *Mountain Research and Development* 17 (1): 43–48.

Relyea, Scott. 2015. "Conceiving the 'West': Early Twentieth Century Visions of Kham." *Twentieth-Century China* 40 (3): 181–200.

Samuel, Geoffrey. 2020. "Hidden Lands of Tibet in Myth and History." In *Hidden Lands in Himalayan Myth and History*, edited by Frances Garrett, Elizabeth McDougal, and Geoffrey Samuel, 51–59. Leiden: Brill.

Samuel, Geoffrey, and Jamyang Oliphant. 2020. *About Padmasambhava: Historical Narratives and Later Transformations of Guru Rinpoche*. Vienna: Garuda Verlag.

Saikia, Arupjyoti. 2011. *Forests and Ecological History of Assam, 1826–2000*. New Delhi: Oxford University Press.

Scott, James. 1998. *Seeing Like a State: How Certain Schemes to Improve the Human Condition Have Failed*. New Haven: Yale University Press.

Schwieger, Peter. 2002. "A Preliminary Historical Outline of the Royal Dynasty of sPo-bo." In *Tractata Tibetica et Mongolia. Festschrift für Klaus Sagasterzum* 65, edited by K. Kollmar-Paulenz and C. Peters. Wiesbaden-Erbenheim, 209–226. Germany: Harrassowitz Verlag Publishers.

Scobie, Claire. 2012. *Last Seen in Lhasa*. London: Penguin.

Simpson, Thomas. 2021. *The Frontier in British India: Space, Science, and Power in the Nineteenth Century*. Cambridge: Cambridge University Press.

Sperling, Elliot. 1976. "The Chinese Venture in K'am, 1904–1911, and the Role of Chao Erh-feng." *The Tibet Journal* 1 (2): 10–36.

Stewart, Jules. 2006. *Spying for the Raj: The Pundits and the Mapping of the Himalaya*. Cheltenham: The History Press.

Sturges, Rush. 2021. *The River Runner*. Netflix.

Tao, Baoxiang. 2019. *Bainian dili da faxian: Yalu Cangbu Da Xiagu*. Beijing: People's Posts and Telecommunications Publishing House, Kindle edition.

Tibetan Social History Series Editing Team. 1987. *Menbazu shehui lishi diaocha* [Research on the History of the Menba Nationality]. Volumes I and II. Lhasa: Xizang Renmin Chubanshe.

Trungpa, Chogyam. 1985. *Born in Tibet*. Boston: Shambhala Publications.

Tseng, Z. Jack, Xiaoming Wang, Graham J. Slater, Gary T. Takeuchi, Qiang Li, Juan Liu, and Guangpu Xie. 2014. "Himalayan Fossils of the Oldest Known Pantherine Establish Ancient Origin of Big Cats." *Proceedings of the Royal Society B: Biological Sciences* 281 (1774): 20132686.

Tuhiwai Smith, Linda. 1999. *Decolonizing Methodologies: Research and Indigenous Peoples*. London: Zed Books.

Walker, Wick. 2000. *Courting the Diamond Sow: A Whitewater Expedition on Tibet's Forbidden River*. New York: National Geographic.

Wang, Fang, Zhixiang Zhang, Cheng Li, Ge Sun, Xiang Zhao, and Zhi Lu. 2021. "Add Himalaya's Grand Canyon to China's First National Parks." *Nature Correspondance* 13 April 2021.

Wang, Hengyuan. 1996. "对雅鲁藏布江中下游河段水力资源开发工程地质条件的初步认识 [Preliminary Understanding of Geological Engineering Conditions for Hydraulic Resource Development in the Middle and Lower Reaches of the Brahmaputra)." 东北水利水电(Northeast Water Conservancy and Hydropower) 5: 39–42.

Wang, Ping, Dirk Scherler, Jing Liu-Zeng, Jürgen Mey, Jean-Philippe Avouac, Yunda Zhang, and Dingguo Shi. 2014. "Tectonic Control of Yarlung Tsangpo Gorge Revealed by a Buried Canyon in Southern Tibet." *Science* 346 (6212): 978–981.

Wang, Xiaoming, Yang Wang, Qiang Li, Z. Jack Tseng, Gary T. Takeuchi, Tao Deng, Guangpu Xie, et al. 2015. "Cenozoic Vertebrate Evolution and Paleoenvironment in Tibetan Plateau: Progress and Prospects." *Gondwana Research* 27 (4): 1335–1354.

Wang, Yuan, Li Sheng, Liu Wulin, Zhu Xuelin, and Li Bingzhang. 2019. "Coat Pattern Variation and Activity Rhythm of Asiatic Golden Cat (*Catopuma temminckii*) in Yarlung Zangbo Grand Canyon National Nature Reserve of Tibet, China." *Biodiversity Science* 27 (6): 638–647.

Wolfe, Patrick. 2006. "Settler Colonialism and the Elimination of the Native." *Journal of Genocide Research* 8 (4): 387–409.

Xu, Qinqin, Jianqing Ji, Dalai Zhong, Yan Hu, Alan Deino, Jianjun Chen, Jiyao Tu, et al. 2020. "Post-Glacial Entrenchment and Knickpoint Migration of the Yarlung Tsangpo Gorge, Southeastern Tibetan Plateau." *Journal of Asian Earth Sciences* 195: 104337.

Zhang, Peiqi, Xinjun Zhang, Xiaoling Zhang, Xing Gao, Emilia Huerta-Sanchez, and Nicolas Zwyns. 2021. "Denisovans and *Homo sapiens* on the Tibetan Plateau: Dispersals and Adaptations." *Trends in Ecology & Evolution* (Online) https://doi.org/10.1016/j.tree.2021.11.004.

11

OTHER-THAN-HUMAN SUBJECTIVITIES IN A MELTING WORLD

"Climing" as Ontological Disobedience in the Andes[1]

Anders Burman

A Melting World

Looking up from the *barrio* streets of northern La Paz or the noisy city streets of El Alto in the late 1990s you would see the ice- and snowcapped mountain of Chacaltaya at the horizon. To the privileged urbanites of southern La Paz, Chacaltaya was the world's highest ski slope, a glacier where one could indulge in the pleasures of alpine skiing and display wealth and cultural capital in social and athletic practices far beyond the reach of the great majority of the Bolivian population. To many of the both urban and rural Indigenous Aymara people, however, Chacaltaya would be an *achachila*, an ancestral mountain-being dressed in a white poncho, an other-than-human subject with intentionality, social agency, a particular personality, and the capacity to govern the weather. Today, in a warming and melting world, the glacier is gone. The ski lift rusts away. The *achachila* no longer dresses in a white poncho.

A barren landscape of gravel and stone unfolds as high-altitude bodies of ice and snow melt away in the Andes, endangering not only high mountain ecosystems and rural livelihoods, but also water security in urban areas dependent on glacial meltwater (Paerregaard 2018). This partly-new landscape, however, is not divested of life, morality, agency, or subjectivity; rather, it is carved out from within the relations between human social practices on the one hand, and other-than-human agencies and subjectivities on the other. In a context of climate change, these subjectivities appear in a new light. They emerge, as it were, as echoing constituents of past worlds, melting away in the heating face of a capitalist modern-colonial world. Simultaneously though, through incessant acts of ontological disobedience through which other worlds are enacted, they come forward as intentional agents and subjectivities, with the capacity to overthrow the world as we know it and cause *pachakuti*, the Quechua and Aymara term for a profound transformation, radical change, and turn or return of time and space.

DOI: 10.4324/9781003347026-16

Asking "whose reality is allowed to be real in a warming world?" this chapter suggests a conceptual framework for grasping the subjugation of Indigenous lifeworlds, the denial of their ontological weight, and the imposition of dominant ontological schemes: "the coloniality of reality." Moreover, it discusses creative ways of dealing with the ontological dimension of coloniality and delves into concrete practices of "climing" as ontological disobedience. I draw here on Karsten Paerregaard's (this book) and Malene K. Brandshaug's (this book) respective ways of using the word "clime" as not only a noun but a verb, emphasizing practices of worlding (Blaser 2013).

Other-than-Human Subjectivities and Climate Change

Other-than-human subjectivities are ever-present in the Andean landscape; they make the rain fall, the crops grow, and the animals breed; they are fundamental constituents of the relational social landscapes within which humans live their lives, which in turn require humans to know how to interact with and relate to these subjectivities in socially correct ways. At the heart of this social interaction lies ritual practice, as a way to engage in reciprocal relations with other-than-human subjectivities, in a similar way to what Kalzang Dorjee Bhutia (this book) in the context of the Sikkimese Himalayas calls "interdimensional communication," or what Brandshaug (this book) in the context of the Peruvian Andes frames as "emplaced more-than-human engagements." On the one hand, there are everyday practices, such as libations and small offerings of coca leaves, that anyone can carry out. On the other hand, there are complex and lavish ritual practices that require the skills of ceremonial experts or shamans (*amawt'a, yatiri, ch'amakani, qulliri*). As Razzeko Delley and Ambika Aiyaduri (this book) explains: "Shamans [...] are considered as spiritual mediators between humans and nonhumans and between humans and the spiritual world."

Aymara people do not live, however, in a rural Arcadian world of ritual practice or in any antediluvian animist cosmological harmony (Burman 2019, 18). El Alto is one of the largest cities in Bolivia and home to the majority of those Bolivians who identify as Aymara. Still though, in the midst of the everyday concerns of modern urban life, other-than-human persons such as the mountain Chacaltaya are, in one way or another, a patent part of many Aymara people's lifeworlds. What is more, other-than-human persons have come to play a central role in how people experience, relate to, and explain a changing climate. *Achachilanaka* govern the weather, says Aymara shaman Carlos Yujra, and they do so in relation to how humans live their lives (Burman 2017). If humans live morally reproachable lives or if they forget to make offerings, and thereby sever the reciprocal relations with the *achachilanaka*, there are consequences:

> Every year in August all the *achachilanaka* meet around the highest mountains to choose a new *marani* for the year to come. The *marani* is like a

> president and the other *achachilanaka* are like his ministers in a council. They talk about how people have lived their lives, and then they decide who will decide over the weather this year. The weather depends on who is chosen to be *marani*. It can be drought or frost. They can send hail or rain. Some *achachilanaka* are worse than others. It all depends on how the humans have lived their lives, and if we have remembered them.

Agricultural practices in the southern Andes are organized in relation to a yearly cycle of one dry and cold season (approx. April to August), one dry and warm season (approx. September to November), and one rainy and warm season (December to March). The yearly sowing period is from September to December, while most crops are harvested toward the end of the rainy season and the start of the cold season (March to May). During the coldest months of June and July, the earth rests, only to be awakened by way of ritual practices and offerings on August 1 for a new agricultural cycle to begin. The stability and predictability of weather phenomena is of utmost importance for high-altitude agriculture in the Andes (3,000 to 4,500 m above sea level) and farmers tend to dedicate quite a lot of time and effort to reading the signs in the landscape about the weather to come. Indicators for weather forecasting may be anything from where specific bird species build their nests to how sharp certain constellations, such as the Southern Cross, appear in the night sky on specific dates (Tapia Ponce 2012). Yet another indicator is related to the election of the *marani* mentioned above by Carlos Yujra. If clouds gather for several days in August around a specific mountain, it is a sign that this specific mountain has been elected *marani* for the coming year, and people prepare their agricultural activities accordingly. This is so, because each mountain has its own personality and character according to which they will govern the weather. While a mountain such as Illimani tends to generate stable weather throughout the year and produce good pasturage for the livestock, the mountain Qaqajaqi is less reliable and tends to send drought and frost, which may have devastating consequences for the crops. While the mountain Illampu is known for causing a harsh and cold climate, sending snow and frost and strong winds, the mountain Uchumachi may generate torrential rains and flooding. To be a successful farmer in the southern Andes means knowing these other-than-human persons and knowing how to address them and interact with them in ritually correct ways.

I have conducted ethnographic research in the Bolivian Andes for more than twenty years, and increasingly people are telling me about the difficulties they have to predict the weather due to increasing variability in the weather patterns. In 2015, a rural Aymara man in his mid-thirties told me:

> It was different before. There was a time for sowing, a time for harvesting. Not now. Nowadays it comes later in the year in some places, and earlier in other places. There is climate change, and it worries us.

His explanation for the emergence of this new weatherworld involves a reference to a lack of ritual interaction with the other-than-human persons who govern the weather and make the crops grow:

> Our grandparents always said that we always have to ask our Mother Earth for permission to sow, to harvest. But a lot of these things have been forgotten. And it must be so, that this is why these things happen. There has really been a climate change.

Aymara shaman Carlos Yujra adds to this explanation by referring to what he sees as a general degeneration of social life in the communities, which enrages the other-than-human persons of the Aymara social cosmos and makes them distort the weather patterns:

> People think the weather changes just like that, because of nothing. They don't understand that it's our fault. We are unbalanced and that's why the weather is unbalanced. The *uywirinaka* [mountain protectors/shephards] get angry. But no one listens to me when I explain that this is what they call climate change.

Hence, both a lack of ritual interaction with the other-than-human persons who govern the weather, and improper social behavior (including anything from adultery and abortion to theft or not fulfilling social commitments in the community) among humans may cause disruptions in the weather. Apart from extreme weather events such as flooding and drought, many experience an increase in temperature and say that "the sun burns much hotter nowadays," which in turn has tangible implications for high-altitude bodies of ice and snow.[2] Between 1986 and 2014, glaciers across the Bolivian Cordillera Oriental presented an overall areal shrinkage of 43% (Cook et al. 2016, 2399).

When glaciers melt away and ice and snow withdraw from the mountain tops, not only do the hydrology and ecology of the landscape change, but also its chromaticity. Where once there was a white cover of snow and ice, there is now a gray and brown blend of gravel, stone, and sediment. Nevertheless, the *achachilanaka* are still there. They still observe human conduct; they linger on, approachable to humans through ritual practice. As a young woman shaman apprentice, a *soldada*, told me on our way up the slopes of Chacaltaya in 2006:

> The ice melts away, but they [the achachilanaka] don't disappear. They are disappointed with us because we have forgotten about them. That's why they heat up the earth. We have to return to what our grandparents practiced. If not, there will be a terrible disaster, like a *pachakuti* [a radical change, turn or return of time and space].

Thus, other-than-human subjectivities are patent parts of Aymara relational life-worlds within which climate change is lived-through and in which so-called traditional ecological knowledge is generated. Nevertheless, these subjectivities have little or no place in dominant debates on climate change in the Andes, in which, rather, they are belittled as expressions of either curious Indigenous customs or unfortunate misunderstandings, even as traditional ecological knowledge has gained increasing attention by actors such as the International Panel on Climate Change (IPCC) as a key component in climate change adaptation strategies (Nakashima et al. 2012; IPCC 2007; see also Singleton et al. 2021).

The Coloniality of Reality

Coined by Peruvian sociologist Aníbal Quijano (e.g. 1998, 2007), the concept of "coloniality" points to persistent racialized colonial power asymmetries in a world where most formal colonial administrations have been banished as a result of anticolonial liberation struggles in the South during the nineteenth and twentieth centuries. Coloniality is, on the one hand, the underlying logic that informed European colonial expansion and, on the other, its legacy in the present.

Quijano writes of a "coloniality of power" that characterizes Latin American societies, a social, political, and economic organization of society that distributes power and labor in asymmetric ways according to notions of race. The concept has since been used as a tool for increasingly fine-grained analyses of colonial legacies in other areas of societal organization and social life: the "coloniality of knowledge" (Grosfoguel 2013; Mignolo 1999) as a conceptual tool for probing into the epistemic dimensions of colonial domination; the "coloniality of gender" (Lugones 2008) allowing gendered and sexual aspects of colonial power to become visible; the "coloniality of being" (Maldonado-Torres 2007) emphasizing the phenomenological and existential dimensions of being and notions of Self and Other in a world of colonial asymmetries. While these are valuable conceptual tools for grasping the multiscalar and multidimensional nature of colonial domination, they do not fully allow for a thorough understanding of the dynamics at work in a melting world of *achachilanaka*. What is at stake here goes beyond questions of labor and political power, beyond questions of whose knowledge counts, and even beyond questions of whose human subjectivity is given full existential weight. The question that needs to be addressed is: Whose reality is allowed to be real? Elsewhere (Burman 2017), I have suggested the "coloniality of reality" as a conceptual framework for grasping the subjugation of Indigenous lifeworlds, the denial of their ontological weight, and the imposition of dominant ontological schemes.

In the eyes of the Bolivian state, Indigenous notions of other-than-human subjectivities were, at least until the coming to power of Evo Morales in 2006, a radical alterity, an existential, epistemological, and ontological otherness; a primitive alterity doomed to extinction. Nevertheless, due to the fact that Indigenous people constitute a majority of the Bolivian population, it was an alterity the state had to deal with, not least in its educational practices.

During the last century, five principal strategies have been used by the Bolivian state to deal with radical difference, sometimes overlapping, sometimes coalescing, often in seemingly contradictory ways (Burman 2019, 22). Firstly, during the early twentieth century, Indigenous alterity was segregated from dominant society within the so-called *educación indigenal* (Indigenous education) (Choque and Quisbert 2006). It was not suppressed per se, but rather maintained as difference, as a rural subaltern alterity that could be controlled to secure a constant supply of labor. Secondly, in the 1940s, when Indigenous Others used the *educación indigenal* for their own purposes and emerged as knowledgeable subjects in their own right and, to a certain extent, on their own terms, they were repressed and the *educación indigenal* was closed down. Thirdly, as an integral part of the national revolution of 1952, an educational reform was launched and would prove to be instrumental in the production of a new national subjectivity in which "the Indian problem" would be solved by conditionally assimilating Indigenous people into the new Bolivian mestizo nation as *campesinos* (peasants) (Ströbele Gregor 1994, 108). Fourthly, after four decades of attempts at assimilating Indigenous difference into dominant society, a peculiar combination of neoliberal socioeconomic policies and official state multiculturalism emerged in the 1990s. This would involve the recognition of Indigenous alterity, but only in terms of a politically defanged celebration of Indigeneity as a smokescreen behind which a neoliberal assault on Indigenous collective rights was carried out (Burman 2019, 2020; Postero 2007). Lastly, since the coming to power of Evo Morales and the MAS (*Movimiento al Socialismo*; Movement toward Socialism) in 2006, yet another strategy of the Bolivian state to deal with radical difference has been crafted (Canessa 2007; Postero 2017). While sharing certain characters with the neoliberal multiculturalism of the 1990s, the current strategy involves the fashioning of a distinctive form of state Indigeneity used as a legitimizing foundational trope of the new nation-building process in which Bolivia is transformed from a unitary republic with formally recognized ethnic and cultural diversity to a plurinational state in which Indigenous peoples are recognized as nations.

Interestingly, current debates and politics regarding climate change adaptation and mitigation manifest an assortment of all the above strategies for dealing with Indigenous alterity: recognition, subalternization, repression, assimilation, and celebration. While certain government representatives celebrate Indigeneity and have introduced other-than-human subjectivities into standard state rhetoric and protocol, national policies of climate change adaptation and mitigation leave little room for "worlds otherwise." Certainly, there is the "Law of Mother Earth and Integral Development for Living Well (Ley Marco de la Madre Tierra y Desarrollo Integral para Vivir Bien)," passed in congress in 2012. However, its general formulations about *Madre Tierra* as a living being and its endorsement of the concept *vivir bien* ("living well") are quite remote from the notions and concrete practices of Indigenous communities. The legislative text is rather an example of a romanticized and esoteric version of Indigeneity that has been instrumental to the government in presenting itself as an Indigenous government to gain political legitimacy (Burman 2020). The following quotes from Article 6 in the "Law of Mother Earth and Integral Development for Living Well" are a case in point:

> *Vivir Bien*, is to dance in gratitude for Mother Earth and in celebration with the community and in harmony among people in which the spiritual and energetic dimension is manifested.[3]
>
> *Vivir bien*, is to communicate and to know how to speak. (...) Everything we say is written in the hearts and in the genetic memory of the beings of Mother Earth.[4]

While the law argues for "considering the experiences and the wisdom of the Indigenous native peasant nations"[5] (Article 17), rather than reflecting notions and practices of Indigenous communities it displays a specific form of governmentalized Indigeneity and the esotericism of a small group of people working in the Ministry of Foreign Affairs and the Vice Ministry of Decolonization who have had great influence over governmental discourse and protocol, and also over the formulations present in this legislative text. A young Aymara shaman apprentice (*soldado*) talked about the people behind these formulations in the following way:

> They are *pajpaqus* [charlatans], mixing the knowledge of our grandparents with New Age ideas of harmony and energies and whatnot. It has very little to do with how people in the rural communities see the world.

Opposing the government's environmental politics is a growing urban environmental/climate movement that points to the contradictions between, on the one hand, state rhetoric of decolonization, *vivir bien*, and Mother Earth, and, on the other, developmentalist real politics of industrialization, extractivism, and the expansion of the agricultural frontier further into the Amazon rainforest. Similar to the government, however, the movement uses a rhetoric sprayed with certain decontextualized Indigenous concepts. Moreover, the movement, constituted primarily by upper and middle-class urban actors, tends to present itself as a movement that works *with* and *for* Indigenous peoples. In the current Bolivian political context, Indigeneity bestows legitimacy and just as the government uses the trope of Indigeneity to claim legitimacy, the movement is dependent on the legitimacy that Indigenous people can offer. Still though, having scrutinized the movement's rhetoric and having participated in a number of its seminars, workshops, and rallies over the years, I have identified discursive patterns in which Indigenous peoples are romanticized as "guardians of nature" while simultaneously being marginalized as political actors and portrayed as ignorant (Burman 2017, 2021). The following quote by an urban middle-class environmentalist at an activist meeting is a case in point:

> People in the villages have no idea about climate change. They live in the deepest ignorance regarding these issues. How is it possible that they don't have the slightest idea about climate change and the environment?[6]

Unsurprisingly, there is a certain hesitance among many Indigenous people toward the environmental movement, as reflected in the words of an Aymara leader whom I interviewed at a workshop organized by the environmental movement in the wealthier parts of the city of La Paz in 2015:

> I don't understand why they invite us, when they talk all the time and don't listen to us. What do they know? Do they by any chance work the land? I don't think so. And I don't know what they mean by "civil society.". I don't even know if they include us in it. And still, if they do, I don't want to be included on their terms.

Thus, Indigenous people find themselves in the crossfire between the state and the environmental movement, both sides celebrating their respective versions of Indigeneity and using Indigenous concepts and symbols in instrumental ways. However, nowhere do we find Indigenous notions of climate change taken seriously, neither by the movement, nor by the government. Instead, we are faced with "a global, and mainly, scientific metanarrative" replacing "popular conceptualizations and discourses of climate, and its manifestations through local weather" (Endfield 2011, 161; see also de Wit & Haines 2020; Johnson, Parsons & Fisher 2021), though sprinkled with selected features of Indigeneity.

This is, I argue, an example of the coloniality of reality, that is, the denial of the ontological weight of Indigenous lifeworlds, the imposition of dominant ontological notions, and the subalternization of worlds that do not fit within the straightjacket of the one-world World. Confronting this coloniality of reality, however, Indigenous Aymara shamans and their young apprentices explore ontological openings within the cracks and fissures of dominant reality to make room for worlds otherwise, insisting on the ontological weight of a reality that is not allowed to be. Let us return to the mountain Chacaltaya.

Climing as Ontological Disobedience

It was close to midnight as we began the ascent. A group of four Aymara shamans and their apprentices, friends, and family members making our way up the snow-clad slopes of the mountain Chacaltaya. At more than 5,000 m above sea level during the cold and dry season, it was bitterly cold. The year was 2006, and there were still a few remnants of the once impressive glacier, but mostly there was snow that would melt away again once the warm season started in September. No skiers go there anymore, only a few tourists and occasionally a company similar to ours, on a ritual mission. Our backs were yoked with firewood, cane alcohol, beer, sweet wine, and, most importantly, the meticulously prepared bundles with offerings for the *achachila* that we had prepared in the old ski hut.

The preparation of the offerings (*luqtanaka*) had begun as the sun set behind the cordillera and we entered the more or less abandoned ski hut to escape from the bitter cold. We were sitting in a circle, women to the left and men to the right, sharing coca leaves among us and selecting the freshest undamaged green leaves to make small bundles, called *k'intus*, as offerings for the *achachilanaka*. Coca leaves are at the core of Aymara sociality and protocol, and no favor should be asked without offering coca. The sharing and ingestion of coca leaves, an act called *akhulliku*, implies a mutual recognition of subjectivity and social personhood among the parties involved, be they human or other-than-human.

Holding a bottle of cane alcohol in his right hand, the eldest of the four shamans spoke: "Achachila Chacaltaya, we come to visit you. We ask for your permission to prepare the offerings!" He sprinkled alcohol in the four corners of the room and then over the two paper sheets that were placed on colorful textiles in the middle of the circle. In a low voice, he mentioned the names of other important mountain beings: "Illimani, Mururata, Qaqajaqi, Sajama, Janq'u Uma, Illampu, Pachjiri! *Licenciamampi!* (With your permission!)" From a woven bag slung over his shoulder, the shaman took out the elements for the offerings to be prepared on the paper sheets: white sugar blocks and llama yarn for the male forces of the cosmos, colorful sugar blocks and llama yarn for the female forces of the cosmos; the fragrant herbs *juyra q'uwa* and *k'ili q'uwa*, and the incense and the flowers; and finally, two desiccated llama fetuses that were handed over to two young apprentices together with lumps of llama fat. The apprentices, a man in his early twenties and a woman in her late twenties, knew what to do. No instructions were needed. They started rubbing the llama fat onto the fetuses, after which they covered them from hoof to head with llama yarn, while speaking softly but intensely to the mountain beings. The young man sat to my left and I saw tears filling his eyes as he raised the fetus to his lips and whispered: "Achachila, I'm returning. With your permission and wholeheartedly. Please receive what we prepare for you."

Meanwhile, the shaman kneeled among us and started distributing the elements among the participants who one by one spoke and breathed into the sugar blocks, the herbs, and the incense, whereupon we put them on the paper sheets in the middle of the circle while continuously speaking in low voices to the mountain beings, the older participants mainly in the Aymara language, the younger in Spanish mixed with key concepts of ritual oratory in Aymara. After little more than an hour, the two apprentices put the adorned fetuses—reinvigorated by fat and dressed up in wool—on top of the prepared offerings, upon which all the participants placed their small bundles of selected coca leaves. The offerings were now completed. Two exquisite meals for the mountain beings; two offerings bursting with symbolism, emotions, wishes, and respect; two bundles insisting on the presence of other-than-human subjectivities—mountain beings, or, in Marisol de la Cadena's (2015) words, "Earth beings"—in a melting world.

On the summit a few hours later, the sweat on our faces turning to ice in the cold wind, we arranged the firewood and put the two bundles on top of it. The shaman poured cane alcohol over it all, invoking *achachilanaka* far and near, and set a match to it. The fire burned fervently, a good sign, a sign that the *achachila* was hungry and pleased with our offerings. As the bundles turned to smoke and ash, we sat down in the snow and shared coca leaves, cane alcohol, wine, and beer among us and with the mountain. Holding a beer bottle in his hand, the young man who had prepared one of the llama fetuses smiled and looked at me: "You see? The *achachila* is happy, happy that we hadn't forgotten about him!"

While having lost most of his impressive white poncho, Chacaltaya is thus invoked together with a number of other other-than-human subjectivities. Their names pronounced through the wind as libations are poured and the snowfall intensifies, they are summoned and made present there among the rocks. Their subjectivity is confirmed, their personhood, their agentive presence in the landscape. The realness of their world is substantiated, engraved in memories and bodies, by words, sensory experience, bodily movements, and emotions. These are practices through which other worlds are enacted, allowing for other-than-human subjectivities to emerge from within the cracks and fissures of dominant reality. These are acts of ontological disobedience, carried out, not in isolation from a dominant world, but in relation to it, in spite of it, in defiance of it (Blaser 2013; Burman 2016). Moreover, these are insistent feats that lead to a process of *climing* the climate change debate in the Andes, a process that is not *only* about reconnecting climate to specific places, regions, cultures, and ecologies (see Carey et al. 2014) or to collectively lived experiences in "a pluralistic world with many climes and cultures" (Fleming 2010, 7, 268). Nor does it imply reducing "clime" to the prevailing climate and "the terrestrial environment, vegetation, etc., that goes with that" (Lamb 1995, 8) in a specific region. It also entails, I would suggest, expanding the ontological horizon of climate change, recognizing how spaces therein are carved out for other-than-human subjectivities.

Using the concept "clime" to discuss Andean understandings of weather, place, and subjectivity is not about assuming a local clime as particularity in relation to a global climate as universality (Smyer Yü, this book). Nor does it imply denying the causal links behind climate change as identified by climatology; it is rather the result of an insight into how climate change, both as physical reality and narrative, becomes intertwined with the relational fields in which human and other-than-human subjectivities come to be and in which people live their lives in what Delley and Aiyadurai (this book) call "a set of animist climes." Thus, climatological chains of causality are knitted into people's own ontologically informed lifeworlds, where personhood is not synonymous with humanness and agentive subjectivity is not limited to humans. As Paerregaard writes in this book: "geological time and climatic space are woven into social relations and configurations" and the landscape and the climate constitute "a universe of beings that are more-than-human and yet approachable to humans."

Climing, in this sense, is a process of ontological disobedience, a process of defying the coloniality of reality, which in turn is a fundamentally political act that aims at lending ontological weight to a reality that is not allowed to be. Ontological disobedience, then, is about exploiting ontological gaps (Sullivan, 2013, 2017) in order to generate existential possibility; the generation of the possibility for other worlds to be disclosed and for other subjectivities to emerge. Climing, as concrete practices of ontological disobedience, is not, however, a practice of closing-in on one's specific locality, nor an attempt to hermetically seal one's ontologically informed lifeworld in relation to other ways of knowing and being. Rather, it involves interrelating in and with a melting world in ways that are known, in ways that render radical change meaningful in a profoundly existential sense and that opens up a space for critiquing the mechanisms of existential and ontological negation, and also, as I have argued elsewhere (Burman 2017), the global structures of fossil-fueled capitalism. In ritual practice, the mountains make themselves present not only as geological formations, not only as mineral deposits, not only as rock and gravel, but as other-than-human subjectivities; they emerge at the melting frontier of partially connected worlds, as constituents of a reality on the negated side of the colonial difference, a reality that is not allowed to be real.

After a night of ritual engagement with the mountain and a mere two hours of sleep, we sit among the boulders on the slopes of Chacaltaya as the first rays of light appear at the horizon. "Look" says the young woman who had prepared one of the llama fetuses the night before, and points toward the spectacular mountainous landscape surrounding us. Then she continues:

> People think it's lost, that it's something of the past. White people say we're wrong, the Church says we're wrong, the ecologists think we're ignorant. But we live this reality every time we climb a mountain, or every time we prepare a *luqta*. Our *achachilanaka* are here, and they will continue to be here, even if their glaciers disappear.

Notes

1 The essay builds on research funded by the Swedish Research Council (VR) and Riksbankens Jubileumsfond (RJ). Parts of the ethnographic data presented here, including some quotes from research participants, have been analyzed in Burman (2017, 2019, and 2022), but then primarily within other analytical frameworks and in relation to other scholarly debates.
2 As early as 2011, the average temperature in the region had increased 1.5° Celsius compared to preindustrial times (Hoffman and Requena 2012: 36).
3 "Vivir Bien, es danzar en gratitud a la Madre Tierra y en celebración de la comunidad y armonía entre las personas donde se manifiesta la dimensión espiritual y energética."
4 "Vivir Bien, es comunicarse y saber hablar. (...) Todo lo que hablemos se escribe en los corazones y en la memoria genética de los seres de la Madre Tierra."
5 "...considerando las experiencias y la sabiduría de las naciones indígena originario campesinas..."
6 A mere euphemism for "Indigenous people."

References

Blaser, Mario. 2013. "Ontological Conflicts and the Stories of Peoples In Spite of Europe: Towards a Conversation on Political Ontology." *Current Anthropology* 54(5): 547–568.

Burman, Anders. 2016. "*Damnés* Realities and Ontological Disobedience: Notes on the Coloniality of Reality in Higher Education in the Bolivian Andes and Beyond." In R. Grosfoguel, E. Rosen Velasquez, & R. D. Hernandez (eds.) *Decolonizing the Westernized University: Interventions in Philosophy of Education from Within and Without*. Lanham, MD: Rowman & Littlefield/Lexington Books, 71–94.

Burman, Anders. 2017. "The Political Ontology of Climate Change: Moral Meteorology, Climate Justice, and the Coloniality of Reality in the Bolivian Andes." *Journal of Political Ecology* 24: 921–938.

Burman, Anders. 2019. "Against the Episteme of Domination and the Coloniality of Reality: Andean Formations of Subversive Subjectivities, Dissident Knowledges and Rebel Realities." In R. Aman & T. Ireland (eds.) *Educational Alternatives in Latin America: New Modes of Counter Hegemonic Learning*. London: Palgrave MacMillian, 17–38.

Burman, Anders. 2020. "Black Hole Indigeneity: The Explosion and Implosion of Radical Difference as Resistance and Power in Andean Bolivia." *Journal of Political Power* 14(2): 179–200.

Burman, Anders. 2021. "A Taste for Ecology: Class, Coloniality and the Rise of a Bolivian Environmental Movement." *Latin American and Caribbean Ethnic Studies*. https://doi.org/10.1080/17442222.2021.1918841

Burman, Anders. 2022. "Indigenous Knowledge and the Coloniality of Reality: Climate Change Otherwise in the Bolivian Andes." In M. Roué, D. Nakashima, & I. Krupnik (eds.) *Resilience through Knowledge Co-production: Indigenous Knowledge, Science and Global Environmental Change*. Cambridge: Cambridge University Press, 261–274.

Canessa, Andrew. 2007. "A Postcolonial Turn: Social and Political Change In The New Indigenous Order Of Bolivia." *Urban Anthropology* 36(3): 145–159.

Carey, Mark, Garone, Philip, Howkins, Adrian, Endfield, Georgina, Culver, Lawrence, White, Sam, Johnson, Sherry, and Fleming, James Rodger (eds.) 2014. "Forum: Climate Change and Environmental History." *Environmental History* 281–364.

Choque, Roberto and Quisbert, Cristina. 2006. *Educación Indigenal en Bolivia: Un siglo de ensayos educativos y resistencias patronales*. La Paz: IBIS.

Cook, Simon J., Kougkoulos, Ioannis, Edwards, Laura A., Dortch, Jason, and Hoffmann, Dirk. 2016. Glacier change and glacial lake outburst flood risk in the Bolivian Andes. *The Cryosphere* 10: 2399–2413.

de Wit, Sara and Haines, Sophie. 2020. Climate change reception studies in anthropology. *Wiley Interdisciplinary Reviews: Climate Change* 13(1): e742.

Endfield, Georgina. 2011. Reculturing and particularizing climate discourses: Weather, identity, and the work of Gordon Manley. *Osiris* 26(1): 142–162.

Fleming, James Rodger. 2010. *Fixing the Sky: The Checkered History of Weather and Climate Control*. New York: Columbia University Press.

Grosfoguel, Ramón. 2013. "The Structure of Knowledge in Westernized Universities Epistemic Racism/Sexism and the Four Genocides/Epistemicides of the Long 16th Century." *Human Architecture* 10(1): 73–90.

Hoffman, Dirk and Requena, Cecilia. 2012. *Bolivia en un mundo 4 grados más caliente: Escenarios sociopolíticos ante el cambio climático para los años 2030 y 2060 en el altiplano norte*. La Paz: Fundación PIEB, Instituto Boliviano de la Montaña.

IPCC. 2007. IPCC Fourth Assessment Report. Working Group II: Impacts, Adaptation and Vulnerability. 20.8.2: "Participatory Processes in Research and Practice." https://archive.ipcc.ch/publications_and_data/ar4/wg2/en/ch20s20-8-2.html#box-20-1

Johnson, Danielle, Meg Parsons, and Karen Fisher. 2021. Engaging indigenous perspectives on health, wellbeing and climate change. A new research agenda for holistic climate action in Aotearoa and beyond. *Local Environment* 26(4): 477–503.

Lamb, Hubert. 1995. *Climate, History and the Modern World*. 2nd edition. London & New York: Routledge.
Lugones, Maria. 2008. "Colonialidad y género: hacia un feminismo descolonial." In W. Mignolo (ed.) *Genero y descolonialidad*. Buenos Aires: Ediciones del Signo, pp. 13–54.
Maldonado-Torres, Nelson. 2007. "On the Coloniality of Being: Contributions to the Development of a Concept." *Cultural Studies* 21(2, 3): 240–270.
Mignolo, Walter. 1999. "I Am Where I think: Epistemology and the Colonial Difference." *Journal of Latin American Cultural Studies: Travesia* 8(2): 235–245.
Nakashima, Douglas, Galloway McLean, Kirsty, Thulstrup, Hans, Ramos Castillo, Ameyali, and Rubis, Jennifer. 2012. *Weathering Uncertainty: Traditional Knowledge for Climate Change Assessment and Adaptation*. Paris & Durban: UNESCO & UNU.
Paerregaard, Karsten. 2018. "The Climate-Development Nexus. Using Climate Voices to Prepare Adaptation Initiatives in the Peruvian Andes." *Climate and Development* 10(4):360–368.
Postero, Nancy. 2007. *Now We Are Citizens: Indigenous Politics in Postmulticultural Bolivia*. Stanford: Stanford University Press.
Postero, Nancy. 2017. *The Indigenous State: Race, Politics, and Performance in Plurinational Bolivia*. Oakland: University of California Press.
Quijano, Aníbal. 1998. "La colonialidad del poder y la experiencia cultural latinoamericana." In R. Briceño-Leon and H. R. Sonntag (eds.) *Pueblo, época y desarrollo: la sociología de América Latina*. Caracas: Nueva Sociedad, 27–38.
Quijano, Aníbal. 2007. "Coloniality and Modernity/Rationality." *Cultural Studies* 21 (2, 3): 168–178.
Singleton, Benedict., Gillette Boyd. Maris, Burman, Anders, and Green, Carina. 2021. "Towards Productive Complicity: Applying 'Traditional Ecological Knowledge' in Environmental Science." *The Anthropocene Review*. https://doi.org/10.1177/20530196211057026
Ströbele-Gregor, Juliana. 1994. From Indio to Mestizo…to Indio: New Indianist movements in Bolivia. *Latin American Perspectives* 21(2): 106–123.
Sullivan, Sian. 2013. "Nature on the Move III: (Re)Countenancing an Animate Nature." *New Proposals: Journal of Marxism and Interdisciplinary Enquiry* 6(1, 2): 50–71.
Sullivan, Sian. 2017. "What's Ontology Got to Do with It? On Nature and Knowledge in a Political Ecology of 'The Green Economy'". *Journal of Political Ecology* 24: 217–242.
Tapia Ponce, Nelson. 2012. *Indicadores del tiempo y la predicción climática: Estrategias* agroecológicas campesinas para la adaptación al cambio climático en la puna cochabambina. La Paz: PIEB.

12

COLLAPSING ELEPHANT CLIME SINCE THE LITTLE ICE AGE

Climatic Refugees, Animal Zomia, and Elephant Modernity in Yunnan

Dan Smyer Yü

In March 2020, a family of fifteen elephants departed from Xishuangbanna National Nature Reserve. After walking northward for 500 km, they reached the outskirts of Kunming, the capital of Yunnan Province, in June 2021. In August, they returned to the nature reserve, led by food-incentives spread along the truck-corralled routes organized by a government response team of wildlife specialists, administrative staff, and law enforcement officers. Thanks to the human intervention, their return journey was thirteen months shorter than their trip to Kunming from Xishuangbanna, and was spared physically violent incidents. Consisting of twelve adults and three juveniles, this family of elephants is known as "the Short-Trunk Clan" named after one of the adult members' injury-shortened proboscis. Their entry into densely populated towns and agricultural fields soon made national headlines and hit social media with millions of daily followers across China. In mid-October 2021, the popular charisma of the Short-Trunk Clan made it to the inaugural session of the UN Biodiversity Conference (Cop15) in a six-minute film produced by Yunnan's provincial government and TV station. *Elephants' Journey in Yunnan* features the clan's departure from and safe return to the nature reserve as a successful case of human-elephant conflict management (Feng 2021). Meanwhile, questions are being asked by many scientists, scholars, and elephant fans: what prompted the elephants to leave their home range in Xishuangbanna National Nature Reserve? Why were the elephants headed north instead of south into Myanmar or Laos as they are known to travel? Is the nature reserve becoming less habitable for the elephants? Is elephant extinction or domestication inevitable in China?

This chapter is intended to address these questions by seeking more enlightened stories of Asian elephants beyond Yunnan and before the Anthropocene—in a larger geographical region and a longer historical period. By correlating findings from climate science, history, geography, and folklore, it tells three

DOI: 10.4324/9781003347026-17

interrelated stories of the elephants under the rubrics of "climatic refugees," "animal Zomia," and "elephant modernity." This story of the ancestors of the Short-Trunk Clan is centered on the Little Ice Age and the subsequent human-induced environmental changes of the last 800 years. This chapter thus sketches out a climatic-geographical history of the elephants that is laden with meteorological and environmental changes revealed by a historical detour into the Mongolian Steppe and northern China. This historical-geographical approach is a practice of what Emily O'Gorman and Andrea Gaynor call "more-than-human histories" (2020) and of what Chris Philo and Chris Wilbert (2000, 4), Julie Urbanik (2012, 3), and Henry Buller (2015, 423) call "new animal geographies." "More-than-human-histories" call for relational understandings of the earth whose past and present are "co-constituted by multiple organisms" (O'Gorman and Gaynor 2020, 3, 6). "New animal geographies" focuses attentiveness on networked human-animal relations grounded in concrete places (Philo and Wilbert 2005, 5), on "a making visible of animals within our human geography" in ethical terms (Buller 2015, 422), and on animals and humans as "experiential beings" lodged in power relations (Urbanik 2012, 42). This chapter is thus written with combined approaches of more-than-human histories and geographies centered on a historical climatic understanding of the fragmented, shrinking, and human-managed habitats of Asian elephants over time. Like the members of the Short-Trunk Clan who made a circular journey, the historical turn is meant to return to Yunnan with the stories of the clan's ancestors who once upon a time lived in a grand territory further north.

Climate in the Shapes and Colors of Clime

Before unfolding the stories of the Short-Trunk Clan's ancestors over 800 years ago, I feel it is pertinent to inform my readers of how I use the word *clime* as the primary medium of my storytelling and conceptual discussions. The currently renewed interest in clime appears in the works of James Fleming (2010) and Mark Carey and Philip Garone (2014). While clime is tangentially emphatic in Fleming's remark on the local-global nexus of climate in "a pluralistic world with many climes and cultures" (2010, 7, 268), it is given fuller thought by Carey and Garone to underscore "the powerful link between weather and place," "to see weather as regions (and vice versa)," and to recognize climate as "a cultural and scientific construction" (2014, 284–286). Clime, in their works, is thus intended for place-based approaches and understandings of climate histories and climate change in the form of weather and its associative impacts on human and environmental health. This powerful concept of clime has not yet been applied in individual case studies of climate change. While my peer authors showcase their conceptualizations of clime in non-animal cases, this chapter takes a few experimental steps to illustrate a changing elephant clime over time.

My use of clime inherits the place-grounded understanding of climate change from Fleming, Carey, and Garone; however, I give more emphasis to the terrestrial nature of clime given the historical fact that the elephants, under the pressures of climate change and human imperial expansions, migrated from the north to the south on foot. My application of clime is thus intended to capture the terrestrial meanings of climate change. To be noted, I do not treat clime as local and regional (Carey and Garone 2014, 284) and climate as global (Fleming 2010, 7). Instead, I regard a clime as both a recipient of and a contributor to global climate change. From the perspective of earth science, I view a clime as an inherent part of the Critical Zone or the earth's life-giving surface "between the top of the tree canopy and the base of weathered rock" (Dybas 2013, 1). Clime is thus both local and global. It is local because of its specific geology, ecology, and geomorphology that shape human and nonhuman dwelling places. It is global because, as a part and parcel of the earth's Critical Zone, it is an integral part of the global climate system.

Throughout the chapter, clime is taken for granted for its synonymity with place, nation, landscape, territory, and region, all of which are embodied with climate change in the physical world but without their own etymological pointers toward climate. In this regard, clime is interchangeably used with climate in the terrestrial sense. My use of clime minimizes the commonly recognized disconnection between abstract climatological knowledge and lived experiences of climate change (Shepherd and Lloyd 2021, 2; Bristow and Ford 2016, 6), particularly in this case of a weathered and storied elephant clime. As the saying goes, "Climate is everywhere, yet nowhere" (Haberman 2021, 1; Bristow and Ford 2016, 6); the practical value of the clime perspective in this chapter, then, lies in its efficaciousness in assisting us, my readers and myself, to see the omnipresence of climate from the firmament to the terrestrial worlds of humans and elephants, to recognize how changing climes in history topple empires and initiate replacements, to correlate and illustrate the similar historical migratory patterns of humans and elephants as climatic refugees, and to identify both past and present anthropogenic forces of change as lived in the once-expansive nation of the Short-Trunk Clan's ancestors. It is my interdisciplinary conviction that this perceptual gravity, shifting from climate change to clime change, will be generative of new insights and conceptual advancements.

The Little Ice Age and Climatic Refugees

Four thousand years ago, the contiguous clime of Asian elephants comprised an area of over nine million square kilometers (Barua et al. 2013, 76), reaching an area archeologically confirmed to be as far north as the Yellow River region and as far east as the lower reaches of the Yangtze River in China, as far west as Western Asia, and as far south as the Indonesian islands (Zhang 2011, 43; Sukumar 2006, 1). This expansive elephant clime was once an "elephant nation" (Smyer Yü 2021, 19; Aiyadurai and Banerjee 2021, 86), an expansive natal habitat interwoven with the nations of other megafauna such as rhinoceroses, tigers, leopards, and humans. Its national constitution wasn't a piece of paper declaring the rights and privileges of a single people to be provided for and defended; instead, it was a co-constitution of the earth in which the ecologically niched lifestyles of the elephants and their companion species were coexistent in shared or interconnected habitats, pathways, and corridors that allowed for foraging, travel, and reproduction. The elephant had very few predators until the human population grew proportionally larger than any other megafauna and until climate change began to transform the elephant nation ecologically to an inhabitable clime. Within the reach of human memories, the southward migration of the Short-Trunk Clan's ancient ancestors began with the forces of change arising concurrently from human activity and climate shifting.

The thirteenth century marked the beginning of the last push of the elephants into Yunnan and Southeast Asia from their already shrunken clime south of the Yangtze River. It was a combined force from the Little Ice Age and the mass migrations of humans from the Mongolian Steppe and northern China into southern China and Southeast Asia. The shared clime of the elephants and humans became colder and dryer, with far less provision. Both elephants and humans were victims of climate change; however, inter-human wars and increased populations that squeezed into the south from the north soon became a systemic lethal force. Combined with the force of the Little Ice Age, it afforded humans to take possession of the remaining elephant clime and convert it into agricultural fields or sites of natural resources that were reserved and extracted for human use only. This lethal force of change to the elephant clime specifically refers to the anthropogenic force starting from the time of the Mongol Empire to its territorial successors the Ming Dynasty, the Manchu Empire, and the modern Chinese republics. The Mongols' climate-motivated migration and imperial conquest laid the territorial and political foundations for the historical anthropogenic transformations of the elephant clime, as these dynastic and republican successors built their sovereign geographies on the territory of the Mongol Empire or the Yuan Dynasty, which geographically overlapped the clime of the elephants in East and Southeast Asia.

The emergence of the Little Ice Age coincided with the reign of Chinggis Khan (1206–1228). As a climate event taking place in the northern hemisphere, it led to an average 2 °C temperature drop, a reduction in rainfall, and an extension of the snowline to lower latitudes. The timeline of this climate change in the northern part of the planet varies from the 1200s or the 1300s to the mid-1800s (Matthews and Briffa 2005, 17; Berglund 2003, 10; Fagan 2000, 18; Mann 2002, 1). Currently, there are two divergent climatological stories concerning how the Little Ice Age and the Mongol Conquest were causally correlated. In one story, dendroclimatologists and historians whose work orients around Central Mongolia and Eastern Europe, recently found evidence that the Little Ice Age had little negative environmental impact on the Mongols' Eurasian campaigns. Based on their tree-ring reading, a fifteen-year interval of warmer temperatures and mild weather in northern Mongolia afforded the Mongol army lush grasslands to raise enough horses and gather other resources for their westward conquest into Eastern Europe (Pederson et al. 2014, 4375). In the other story, by correlating the climatic data of the Little Ice Age with court records and local histories, climate scientists and scholars with a Mongolia and China orientation emphasize the temperature drop and prolonged cold and dry weather in the 1200s as the primary environmental causes of the Mongols' migration southward (Zhu 1973; Wang 1996; Zhang 2006, 460).

The difference between the two stories is that the former reinforces the existing historiography of the Mongols as well-prepared empire-builders under favorable climatic conditions, while the latter deems the Mongols as "climatic refugees" (Smyer Yü 2021, 11; Li 2020, 157) based on evidence of the climatically conditioned agro-nomadic conflicts along the Great Wall over a longer period of time between the 200s BCE and the 1200s CE, and based on the historically observable mass, southward human migrations, including that of the Mongols, during the Little Ice Age. The point here is to demonstrate that there are regional and local variations beneath a clear-cut, averaging model of global climate change whose value lies in its informative but abstracted assessments of climatic trends, not in locally and regionally specific cases whose varied geo-conditions, e.g., geological, geomorphological, and ecological, are part and parcel of their climes in terms of how they each embody global climate change. Both research outcomes thus speak of the same historical climate change in two geographical and eco-geological variations.

In the case of Asian elephants, both migrating humans and elephants were climatic refugees; however, the latter soon became the animal victims and subjects of the former. The human climatic refugees were not merely migrants seeking new shelter and provision, but were also the builders of empires, dynasties, and modern republics. Elephant-human relations shifted from co-existence to the imperial humans' domination and their marginalization of the elephants and their indigenous human neighbors. Political domination in this case engendered anthropogenic forces of change, which were lethal to the species-wide survival of the elephants.

The elephants first encountered by the Mongols were recorded as *zhan-xiang* or war elephants trained by local peoples such as the Dai and the Burmese to battle the advancing Mongol troops in the hills of Southwest China and Southeast Asia. To recount, in the mid-1200s, the Dai people in Xishuangbanna unsuccessfully organized a large battalion of war elephants to battle the invasion of the Mongol troops led by Khungjil Khan, the fifth son of Chinggis Khan. The Mongols captured over 200 war elephants (Zhang and Xiao 2016, 153). Later, in March 1388, the Ming troops used their firearms to overcome a Dai battalion of eighty war elephants and 40,000 soldiers led by Chief Silunfa. The better-armed troops killed 30,000 indigenous fighters and captured over forty elephants (Xia 1418, 2860). In April of the same year, the chiefdom surrendered to the Ming Dynasty. Emperor Hong Wu demanded "five hundred elephants, thirty thousand buffaloes, and three hundred mahouts" as tributes to the Ming court (Xia 1418, 2868). Throughout the reigns of the Yuan, Ming, and Qing Dynasties, the imperial palaces in Beijing had a designated quarter for elephants and mahouts sent from Southwest China and Southeast Asia as tributes. The Ming Dynasty marked the disappearance of elephants in China except in Yunnan (He 1984, 50; Wen et al. 1979, 54).

Converging with human imperial expansion, the domino effects of the Little Ice Age from the Mongolian Steppe and the heart of the Chinese agricultural plains to the highlands of Southwest China and Southeast Asia inevitably manifested themselves as environmental changes concurrently induced by the earth as well as by humans. The clime of the Short-Trunk Clan's ancestors was inescapably transformed into one that was human-dominated. Climate change in this case was a multifaceted clime change in the ecological experiences of both elephants and humans, and in their uneven interspecies encounters. The climatic force of the Little Ice Age caused by the Earth may have been indiscriminate to everyone and everything in the northern hemisphere; however, in the sense of the terrestrially-understood, horizontal interspecies relations, it was incarnated in a variety of common—as well as species-specific—ecological survival stories involving humans in every respect. The common experience was what I describe as an "inter-climing experience." By converting "clime" into a verb as Karsten Paerregaard does in his Andean case in this book, the historical picture I see is that the elephants and the northern humans moved away from their home-climes into the remaining elephant climes in southern China and Southeast Asia. By moving southward into more hospitable environments, the elephants and the northern humans were looking for fecund lands and adopting new modes of subsistence in new environments. The inter-climing experience was then an inter-regional migration from one clime to another, cramming more people (human and elephant) into more habitable but shrinking climes. Everyone on the move was a climatic refugee.

The climing elephants joined their compatriots in southern regions, currently Fujian, Guangdong, Guangxi, and Yunnan. While the elephants were regaining their strength by foraging their familiar foods, for example, bamboo, wild banana, and tiger grass, the Chinese farmers from the north renewed their farming in newly accessed arable lands adjacent to or appropriated from the elephant habitats in the south. By relocating their political center from the Mongolian Steppe to Beijing and expanding their territorial claims, the Mongols forcefully combined the Steppe, much of Central Asia, the agricultural plains of the Chinese, the high plateau of the Tibetans, and part of Southeast Asia into a single empire. Both humans and elephants inter-climed within this imperial geography for the next 800 years. All of them were seeking refuge for survival and renewal; however, the elephants were not only climatic refugees. They were also fleeing human encroachment on their habitats. Given the fact that the climate in southwest China has stayed relatively the same since the 1200s (Li and Cao 2019, 7), the elephants in Guangdong and Guangxi could have survived to this day if their elephant-friendly environments had not been imperialized under a single political system created by the Mongols and sustained by its dynastic and modern successors. The ecological conditions required for the survival of the elephants were soon eroded by human farming and resource-extraction activities. Those elephants who escaped the ecological devastation of the Little Ice Age would then go on to experience endangerment from their human counterparts.

Anthropogenic Forces of the Mongol, Ming, and Manchu Dynasties

When the Ming Dynasty succeeded the Mongol-led Yuan Dynasty in 1368, the ancestors of the Short-Trunk Clan were only found in Guangxi and Yunnan. Over 3,000 elephants lived in southern Guangxi (He 1984, 50; Liu 2011, 55), a clime with "flowers blooming throughout four seasons and winters without snow" (Zhou and Wang 1992 [1756], 41) and with abundant water and vegetation. The elephants and the indigenous peoples such as the Zhuang and the Yao had lived peacefully in close proximity since before the arrival of the Mongol and the Ming troops and before the migrant (human) populations from the north became overwhelmingly large and ecologically invasive. The places where the elephants lived most comfortably were those where the land was fertile. As the Ming Dynasty brought its subjects from the north into Guangxi, the intensive agricultural use of the land soon became the norm (Elvin 2004, 241). In Nanning township, between 1503 and 1531 CE, land put to agricultural use nearly doubled from 5,691 hectares to 10,247 hectares. Logging of the forested hills became more and more frequent. Between 1556 and 1561 during the reign of Emperor Jiajing, the annual quota for the local people to log hardwoods increased from 16,460 to 51,060 trees (Liu 2011, 66). In Hengzhou Township, under pressure from the Ming court, the Zhuang and Yao people replaced the natural vegetation on hilly slopes and in narrow valleys with upland rice. Both migrant and indigenous peoples also began to raise large herds of water buffalo ranging from several hundred to over a thousand head per household. As humans and their livestock were expanding into and overtaking the elephant clime, the elephants also began to raid the crops (Liu 2011, 66).

While the elephants attempted to evade the human empire and concomitant agricultural expansion by diving deeper into the forested hills or by occasionally striking back at human crops, many of them were captured or killed by the Ming troops through what is historically known as the Elephant Training Forts (*xunxiang wei*). Established in the Shiwanshan, Nanning, and Ningming areas where the elephant population was concentrated, the forts, serving as military posts, were commissioned by the Ming court to solidify Ming rule in non-Han territory and to capture and train elephants for labor and as tributes (He 1984, 50). Elephants sent to the court served as a material symbol of the submission of those indigenous societies to the Ming Dynasty. Aided by military efforts, this imperial tribute custom lasted well into the Qing Dynasty.

Ming Emperor Hongwu (1328–1398) received 192 elephants as gifts from Guangxi, Champa (in current Vietnam), and Kmir (currently Cambodia) between 1386 and 1388. With the elephants from Yunnan included, Emperor Hongwu received over 3,000 elephants as tributes. Emperor Yongle (1360–1424) was another ruler known for collecting elephants. In 1408, general Zhang Fu brought 112 elephants back to the emperor from the Guangxi-Vietnam borderland region (Yang 2010, 38) as the spoils of his military campaigns in indigenous lands. The estimated, total number of elephants sent from the Ming Dynasty's indigenous subjects was between 1,140 and 1,339 (Li and Cai 2019, 7). The ending of the Ming Dynasty by the Manchus was tragic to the elephants who were imperial subjects living in captivity in Beijing, a cold, human urban clime. In 1644, the last year of the crumbling dynasty, the elephants were "roaring and crying with pouring tears" because of hunger, unattended to by the court caretakers during a time of war (Yang 2010, 42). Finally, by the mid-1800s, elephants were no longer seen in Guangxi.

The survivors of the onslaughts of human imperial expansion and concurrent agricultural use of the elephant clime headed southwest into the current clime of the Short-Trunk Clan in Xishuangbanna, Yunnan. Geologically located in the southeastern end of the Himalayan-Tibetan Plateau, Xishuangbanna became the last refuge of the elephants, a clime where the elephants had an abundance of water and food, and where they neighbored with elephant-friendly indigenous human groups such as the Dai, the Bulang, and the Jinuo. In their studies, He Zhaoxiong (1984, 49) and Li Chun and Cao Dafan (2019) find that from the sixth to the twentieth century, the southward migration of the elephants from the Yellow River region to Yunnan was made at a speed of average of one arcdegree (111 km) per hundred years, and was due to climate change. After the Mongols fully conquered the homelands of the Han, Tibetans, and smaller ethnic societies in southwest China, the speed of elephant migration further south increased to four arcdegrees (444 km) every hundred years. This rapid change started especially after the Ming Dynasty, which was systematically moving agricultural populations into southern regions within the Mongols' previously defined imperial geography. Both the Ming and Qing Dynasties expedited the total disappearance of elephants in Guangxi Province through warfare, the replacement of indirect rule with direct rule, deforestation, agricultural land use, and mining. Thus, needless to say, the credit for the additional three-arcdegree per century anthropogenic reduction in the elephant clime starting in the 1200s and continuing to the early 1900s go to the Mongols, the Ming, and the Manchus. All this occurred before more anthropogenic changes were wrought during the modern republics, between 1911 and the present.

Animal Zomia

The histories of the Mongol Yuan, the Ming, and the Manchu Qing Dynasties, then, were a story of the shrinking geography of the elephant clime under the pressure of anthropogenic forces. This particular geography could be seen as a new animal geography with the focus on "the complex entanglings of human-animal relations" (Philo and Wilbert 2000, 4) in spatial terms. It could also be a more-than-human geography that "pertains to an antithetical but symbiotic world of humans and nonhumans grounded in land and environmental flows" (Smyer Yü 2021, 16) and that "entails critical reflection on the values shaping past and current responses to competing claims to life" (O'Gorman and Gaynor 2020, 712). These more-than-human perspectives are centered on the multinatural ontologies of human-dominated biogeographies (Lorimer 2012, 3). With the interconnection of animal geographies and human geographies considered, the recognition of the ecological, bodily, and moral intertwinement of animality and humanity (Viveiros de Castro 1998, 469, 470) is becoming an interdisciplinary norm in multispecies studies. From this perspective, I see the geography of the last dwindling elephant clime in Southwest China between the 1200s and the 1800s as an "animal Zomia" inclusive of its derivative phrases like "elephant Zomia," "more-than-human Zomia," and "Zomian clime." These coinages point to the shared experience of being encroached upon and marginalized by both natural and human forces of change.

Initially coined by Willem van Schendel from the Tibeto-Burman term "zomi" meaning "remote-people" and referring to the highlands of Southeast Asia and much of the Himalayan region (Van Schendel 2002, 653, 654), the political use of Zomia flourishes in James Scott's *The Art of Not Being Governed* (2009). Delimited mostly to the Southeast Asian Massif inclusive of Southwest China (Michaud 2006), the mountainous geography of Scott's Zomia is understood as "a zone of refuge or 'shatter zone'," which was a sanctuary for peoples who were "fleeing the oppressions of state-making projects in the valleys—slavery, conscription, taxes, corvée labor, epidemics, and warfare" (Scott 2009, ix, 7) in the span of 2,000 years with an emphasis on the 1200s—1800s, a period coinciding with the Little Ice Age. If I include the climatic factor and the animal refugees in the human-centered geography of Scott's Zomia, it is a more-than-human Zomia in the multispecies sense. Thus, I see the historical Xishuangbanna as an animal Zomia in which the elephants were animal Zomians fleeing the dispossessing forces of the Little Ice Age and imperial humans. No less than indigenous human Zomians in the mountains of the eastern Himalayas, the elephants were hard-hit refugees and endangered victims on the run from humans' "state-making projects" (Scott 2009, 22), territorial advancements, and agricultural practices. This geographical and political overlap of human and animal Zomians is evident when the histories of the Little Ice Age, Asian elephants, and human imperial expansions are pieced together.

My qualification of the elephants as animal Zomians is meant to show the presence of humans in animal geographies and vice versa. In other words, the habitats of both elephants and humans were (are) in the same clime, but the former had been severely endangered since the Mongols founded the Yuan Dynasty, a continental imperial system that afforded humans to encroach on elephant habitats in a systematic and organized fashion for the subsequent 800 years. These imperial humans transformed their climatic victimhood into their triumphant domination of the homelands of both humans and animals in faraway places. Thus, seen from the more-than-human perspective, Scott's Zomia is also an animal Zomia, one in which the elephants were both climatic and political refugees.

To look more carefully at its geology, ecology, and multispecies worlds, Scott's human-centered Zomia is a vast montane ecological refugia, a biodiversity hotspot hosting diverse populations of humans, animals, and plants. To be specific, the north-south direction of the Hengduan Mountain Range, which geologically and ecologically hosts Xishuangbanna, allows the annual monsoon to bring plentiful precipitation and warm air from the Indian Ocean into its riverine valleys and the vertically spread-out climing/climbing ecological zones (Stokstad 2020, 493; Ebersbach et al. 2017, 7). The great latitudinal range and the high-altitudinally rugged terrain of the mountains respectively provide different species with options to escape unfavorable climatic conditions within the mountain region and afford plenty of microclimatic refugia for the continuity of biodiversity (Muellner-Riehl 2019, 9). Zomia in this eco-geological and climatic sense is thus a zone of biodiversity and a more-than-human clime or refuge accommodating runaway humans and animals.

After the end of the Qing Dynasty, the Zomian clime of the Short-Trunk Clan's ancestors in Yunnan shrank further and quickly became a human-managed geography of multispecies in the modern era. The elephants eventually stopped being animal Zomians as did their indigenous human Zomians. As elephants took on the new role of flagship species assigned by humans to showcase the achievements of their nature conservation efforts, their Zomian style of survival was transformed from evasion of human dominance to overt confrontation as well as negotiation once they had nowhere else to go.

Elephant Modernity

My northward detour into the historical clime of the Short-Trunk Clan's ancestors is meant to show that the popular questions at the opening of this chapter, concerning the clan's year-long departure from their human-designated habitats in Xishuangbanna, are reflexively grounded on an assumption of the nature reserve as the "home" of the elephants without the awareness of the 60% of the total population roaming outside the reserve. The contemporary human perceptual field of the elephants in Yunnan reflects what Daniel Münster and Ursula Münster call "ecological modernity" (2012) and what Susan Nance calls "animal modernity" (2015). The former suggests the loss of Asian elephant habitats to human agrarian development. The latter highlights the paradox of humans' "unbridled destruction of wild animals and their habitat" on the one hand and the preservation of the selected species as "nature pets" (Nance 2015, 6) on the other. The stories of the Short-Trunk Clan and its early modern ancestors resonate with both perspectives except for the marked absence of climatic approaches in their discussions. However, when factoring in the impacts of historical climate change and modern anthropogenic transformations of the elephant clime, I see a picture of "elephant modernity" that is ecologically and politically specific to the Short-Trunk Clan and its relatives in Xishuangbanna.

In the field of new animal geographies, Latour's proclamation "we have never been modern" (1993) is widely referenced to battle the nature-culture dualism which "provides an impoverished analytical repertoire" (Lorimer 2007, 913). The debates on modernity, whether animal or human, are an unending maze with very little chance of finding an exit. While I see multiple intellectual hindrances from the systemic nature-culture dichotomy, I also recognize a clear ontology of this dichotomy embodied in the physical construction of nature reserves across the world that continue to reinforce the perception of the nature-culture divide. Not to be disoriented in the maze of the modern–anti-modern debates, modernity in the case of the Short-Trunk Clan refers to a series of physical and perceptual changes to a specific elephant clime, which were initiated by a particular historical climate change 800 years ago and intensified by subsequent human activities particularly since the sixteenth century. Elephant modernity entails qualitative transformations of a multispecies clime. Humans have subsequently turned this shared clime into an entirely anthropogenic clime that includes isolated nature reserves designated to wild animals and plants. Elephant modernity, therefore, means a string of events on different scales: the collapsing of an elephant nation from a historically vast clime to a few contemporary isolated nature parks, the loss of the transitional habitats maintained by elephants and their indigenous human counterparts, the increase in the endangered elephant population with human intervention, the dependency of the elephants on the sustenance provided by humans, and the tourist commercialization of elephant charisma. Overall, this particular elephant modernity is materialized in the modernization of the elephant lifeways and habitat on human terms.

The Xishuangbanna National Nature Reserve, a point of pride in China's nature conservation endeavor, was officially established in 1986. Not a continuous, integrally single area, the reserve consists of five separated sub-reserves with a total size of 242,510 hectares. It is not only the last refuge of elephants, but also hosts research stations and tourist sites. In addition to existing county and provincial roads, the Kunming-Bangkok highway and the Kunming-Vientiane highspeed railway cut across the reserve to facilitate both domestic and international flows of traveling humans and goods. The goal of the nature reserve in practice is concurrently to preserve endangered species and to be a revenue-generating site (Yunnan 2013, 37; Yunnan 2015, 15). Since its incorporation, it has been successful in increasing its elephant population from 79 in 1958 and 101 in 1979, to 241 in 1984, and 300 in 2021 (Zhang and Xiao 2016, 8; Li and Cao 2019, 10). At the same time, its nature-tourism programs steadily bring in annual revenue to supplement the state budget for operating the reserve and to bring job opportunities and higher incomes to local people.

From the state-corporate perspective, Yunnan's nature conservation program is a win-win case for both humans and nonhumans; however, it has been subjected to many painful lessons. One of the political and environmental characteristics of modern China is that everything and everyone within the territory of China is the subject of the state, including the land, the plants, the animals, and of course the people. While each indigenous ethnolinguistic group has been identified as an ethnic minority, so too, the animals have been identified and counted; the elephants' numbers and their status in the modernization course of the new human nation-state were debated too. In the process, local resources were not meant merely for local people but have also been collected for national use, with the end result being that the carrying capacity of the local environment has thus been overstretched. This is where elephant modernity commenced, with a new set of elephant-human and human-earth relations detrimental to the elephants and their relative peaceful co-existence with local human societies.

The Short-Trunk Clan was among the 60% of the total population that had become vagabonds outside the nature reserve. The emotionally stirring images of the clan members were charismatic enough for the state to cover the 6.8 million-yuan human property losses they left behind (Zhao 2021) after their safe return to the nature reserve. Both the public and scientific researchers are still pondering the paradox of the shrinking elephant clime and the increasing elephant population in Yunnan. My understanding of this paradoxical elephant modernity resonates with Lorimer's "nonhuman charisma" and with Nance's discussion of the fate of Jumbo, a nineteenth-century orphaned elephant from Sudan whose parents were killed by hunters and who spent most of his life in the London Zoo and as a circus performer in the USA. Both perspectives pinpoint the anthropogenic forces that alter the fate of elephants in human worlds. Relating to our case study, Lorimer's etymological evocation of charisma as "the gift of grace" (Lorimer 2007, 915) with a Weberian emphasis on the innate quality of given nonhuman beings helps us recognize the charisma of the Short-Trunk Clan and its relatives as an agentive charisma. It enables the elephants to express their gift of grace as a saving grace that protects their lives by winning the hearts of their growing number of human fans in spite of the shrinking elephant clime and the increasing human-elephant conflict.

The internationally-marketed "Wild Elephant Valley" inside the reserve is simultaneously a scientific field station for observing elephants and a popular tourist destination as a national eco-tourism model zone. It also hosts the Elephant Training School and the Asian Elephant Breeding and Rescue Center. The former serves as part of the eco-tourism package, allowing for controlled human-elephant contact with entertainment themes. The latter has been credited for rescuing thirty-two injured wild elephants and the birth of ten calves since its founding in 2008 (Xiong 2019). As these facilities are owned and operated by Yunnan Investment Group, a multi-billion-yuan for-profit government organization, the elephants who recovered from their injuries and who were born in the breeding center are trained to perform as "representatives of their kind" (Nance 2015, 18) in front of ecotourists. The modern fate of the elephants living inside the Elephant Training School in Yunnan resembles that of Jumbo whose life was saved by humans and yet who had to repay his human owners with the proceeds earned from his emotional bond with Queen Victoria and the one hundred thousand children who pleaded for his life (Sutherland 2014, 93, 99) as an "animal celebrity" in the London Zoo (Nance 2015, 84). Undoubtedly, the elephants of Yunnan are charismatic megafauna that have bonded with millions of their fans throughout China. Meanwhile, their freedom, and their very existence, is in human hands.

The human caretakers of the rescued elephants are all men; nicknamed by the elephant fans as "Elephant Daddies," they teach the elephants how to communicate with humans and to behave like humans during their educational-entertainment performances in front of a human audience. While humans commercialize elephant charisma, the elephants earn the preservation of their species with their innate gift of grace under human "parental" guardianship.

In responding to the public question "Is elephant extinction or domestication inevitable in China?" I do not see a forthcoming extinction of the elephants in Yunnan. Both the innate charisma and the human-felt, commercially-valued charisma of the elephants are their saving grace. Both do not necessarily redeem their inborn freedom in the ecological and ethical sense but sustain the continuity of their species under human custody, here connoting both captivity and guardianship. This leads me to see their current and near-future status as what I call "free-range domestication," an oxymoron but one that realistically frames the paradox of elephant modernity in Yunnan. It suggests that the elephants have the freedom of movement limited to five isolated habitats and receive human assistance for food, medical care, and reproduction rights. All these are already taking place steadily except for the likely full-fencing of the nature reserve as a fortress conservation site, which is a matter of time.

Collapsing Climes

In summary, the history of the Little Ice Age and its aftermath gives us an 800-year-long panorama of the irreversible changes to elephant habitats, human societies, and their shared provision from the earth. The terrestrial pattern of this climate change is a process of what I call "collapsing climes" over time. The collapsing process started from the Mongolian Steppe and northern China and continued to southern China. It entailed multiple environmental and political changes in a more-than-human sense. By "collapsing," I mean:

- Climate change and human agricultural advancement squeezed the elephants out of their vast clime in the north. Its inhabitability turned the northern elephants into climatic, environmental, and political refugees migrating southward into the warmer climes in Southwest China and Southeast Asia;
- The Mongol conquest impresses many world historians as a continental empire building process (Onon 2001); however, from my climatic perspective, the conquest seen in the context of the elephant case study was rather an organized, armed mass migration of humans from the Mongolian nomadic

clime into the agricultural and the semi-agricultural climes in China and Southeast Asia. The entire regional nomadic empire in Inner Asia literally collapsed into the Chinese agricultural clime as a new short-lived empire known as the Yuan Dynasty that lasted only ninety-seven years (1271–1368) but that left behind profound environmental and political effects on everyone wherever the Mongols and their imperial and modern republican successors made contact;

- After the demise of the Mongol Yuan Dynasty, the Little Ice Age continued to progress well into the mid-1800s. Environmentally and politically, more northern human populations took refuge in southwest China. The climes in the south inevitably had to shoulder more responsibility to feed additional mouths and to supply the new empires and new republics with natural resources. The increased human demographic density marked the advent of the systemic anthropogenic impacts on the remaining elephant climes in southwest China;
- The foothills of the eastern Himalayas became new but short-lived micro-climes that served as refuges for elephants and indigenous human communities. As the integrally continuous elephant clime was shattered into isolated habitats and human-managed nature parks in the modern era, the elephant nation was decimated; however, its survivors continue to reproduce in captivity or in corralled wilderness patches under human "parental care."

Climate change in this case is synonymous with clime change concretized in the terrestrial worlds of humans and nonhumans. Brought about by the Little Ice Age, the average 2 °C temperature drop in more-than-human worlds of the northern hemisphere turned out to be a continental force compelling many human and nonhuman societies to flee from their ancestral lands into the ancestral lands of others. The new, successive imperial political systems on the scale of continental empires played a destructively creative role in unifying and standardizing the human extraction of natural resources and in transforming nonhuman worlds into a single human world at the expense of diminishing nonhuman nations. Over half of this chapter addresses the human-induced aftermath of the clime change that had little effect in southwest China, and yet that had a profound impact in that region due to the climatic refugees arriving from the north. The clime change then shapeshifted into imperial conquests, state-sanctioned resource acquisitions, and the political struggles of indigenous humans and nonhumans. In essence, the clime change evolved into human-with-human and human-with-animal conflicts. Its consequences are shown in the fate of the Asian elephants in Yunnan and are matters of "geoethics" (Urbanik 2012, 39) and animal-human "relational ethics" (Whatmore 2002, 146) in which the continuity of the elephants as a megafauna species is at the mercy of human "experts" and their institutions. The elephants' innate gift of grace and their persistent skirmishes with human farmers, tourists, scientists, and administrators are not winning back their lost nation but are compelling the human world to enter deeper ethical self-reflections on how we have treated nonhuman beings and how we can rebuild equitable and ethical elephant-human relations found in the co-constituted histories of elephants and their indigenous human neighbors. Clime change and anthropogenic consequences are thus concurrently demanding us to preserve more nonhuman habitats and to repair our currently ailing moral clime toward a sustainable multispecies future.

References

Aiyadurai, Ambika and Sayan Banerjee. 2021. "Understanding Borderlands through Elephant Corridors in the Yunnan–Myanmar–Bengal Landscape." In *Yunnan-Burma-Bengal Corridor Geographies: Protean Edging of Habitats and Empires* edited by Dan Smyer Yü and Karin Dean. London/Delhi: Routledge, 85–104.

Barua, Maan, Shonil A. Bhagwat, and Sushrut Jadhav. 2013. "The Hidden Dimensions of Human-Wildlife Conflict: Health Impacts, Opportunity and Transaction Costs." *Biological Conservation* 157: 309–316.

Berglund, Björn E. 2003. "Human Impact and Climate Changes—Synchronous Events and a Causal Link?" *Quaternary International* 105: 7–12.

Bristow, Tom and Thomas H. Ford. 2016. "Climates of History, Cultures of Climate." In *A Cultural History of Climate Change* edited by Tom Bristow and Thomas H. Ford. London: Routledge, 1–14.

Buller, Henry J. 2015. "Animal Geographies III: Ethics." *Progress in Human Geography* 40(3): 422–430.

Carey, Mark and Philip Garone. 2014. "Forum Introduction" 284–295. In Mark Carey, Philip Garone, Adrian Howkins, Georgina Endfield, Lawrence Culver, Sherry Johnson, Sam White, and James Roger Fleming "Forum: Climate Change and Environmental History." *Environmental History* 19(2): 281–364.

Dybas, Cheryl Lyn. 2013. *Discoveries in the Critical Zone: Where Life Meets Rock*. Alexandria, Virginia: National Science Foundation.

Ebersbach, J., J. Schnitzler, A. Favre, and A. N. Muellner-Riehl. 2017. "Evolutionary Radiations in the Species-Rich Mountain Genus *Saxifraga L*." *BMC Evolutionary Biology* 17: 1–13.

Elvin, Mark. 2004. *The Retreat of the Elephants: An Environmental History of China*. New Haven: Yale University Press

Fagan, Brian. 2000. *The Little Ice Age: How Climate Made History 1300–1850*. New York: Basic Books.

Feng, Yin, Director. 2021. *Elephants' Journey in Yunnan* ("象"往云南), a documentary film designated to the UN Biodiversity conference (COP15) inaugural session on October 13, 2021, http://yn.people.com.cn/n2/2021/1013/c378439-34953366.html.

Fleming, James Roger. 2010. *Fixing the Sky: The Checkered History of Weather and Climate Control*. New York: Columbia University Press.

Haberman, David L. 2021. "Introduction: Multiple Perspectives on an Increasingly Uncertain World." In *Understanding Climate Change Through Religious Lifeworlds* edited by David L. Haberman. Bloomington: Indiana University Press, 1–20.

He, Zhaoxiong. 1984. "Southward Migration of Wild Elephants and Deforestation in China" (中国野象南移与森林破坏). *Journal of Ecological Studies* 《生态学杂志》 58(5): 48–51.

Latour, Bruno. 1993. *We Have Never Been Modern*. Translated by Catherine Porter. Cambridge: Harvard University Press.

Li, Cun and Cao Dafan. 2019. "Revisiting and Reflecting on the Conservation History of Asian Elephants in China" (中国亚洲象保护历史回顾与思考). *Forestry Construction* 《林业建设》210(6): 6–10.

Li, Tana. 2020. "The Mongol Yuan Dynasty and the Climate, 1260–1360." In *The Crisis of the 14th Century: Teleconnections between Environmental and Social Change?* edited by Martin Bauch and Gerrit Jasper Schenk. Berlin: De Gruyter, 153–168.

Liu, Xiangxue. 2011. "An Examination of the Elephant Training Fort in the Ming Dynasty" (明代驯象卫考论). *Historical Research* (历史研究) (1): 51–67.

Lorimer, Jamie. 2007. "Nonhuman Charisma." *Environment and Planning D: Society and Space* 25: 911–932.

Lorimer, Jamie. 2012. "Multinatural geographies for the Anthropocene." *Progress in Human Geography*, published online 28 February 2012, http://phg.sagepub.com/content/early/2012/02/27/0309132511435352.

Mann, Michael. 2002. "The Little Ice Age." In *Encyclopedia of Global Environmental Change* edited by Michael MacCracken and John Perry., vol. I. Chichester: John Wiley & Sons, Ltd.

Matthews, John and Keith R. Briffa. 2005. "The Little Ice Age: Re-Evaluation of an Evolving Concept." *Geografiska Annaler* 87 A 1: 17–36.

Michaud, Jean. 2006. *Historical Dictionary of the Peoples of the Southeast Asian Massif.* Toronto: The Scarecrow Press.

Muellner-Riehl, Alexandra. 2019. "Mountains as Evolutionary Arenas: Patterns, Emerging Approaches, Paradigm Shifts, and Their Implications for Plant Phylogeographic Research in the Tibeto-Himalayan Region." *Frontiers in Plant Science* 10: Article 195, 1–18. https://doi.org/10.3389/fpls.2019.00195.

Münster, Daniel and Ursula Münster. 2012. "Human-Animal Conflicts in Kerala: Elephants and Ecological Modernity on the Agrarian Frontier in South India." In *Fields and Forests: Ethnographic Perspectives on Environmental Globalization* edited by Daniel Münster, Ursula Münster, and Stefan Dorondel, *RCC Perspectives*, no. 5, 41–49.

Nance, Susan. 2015. *Animal Modernity: Jumbo the Elephant and the Human Dilemma.* New York: Palgrave.

O'Gorman, Emily and Andrea Gaynor. 2020. "More-Than-Human Histories," *Environmental History* 00: 1–25.

Onon, Urgunge, translator. 2001. *The Secret History of the Mongols: the Life and Times of Chinggis Khan.* London: RoutledgeCurzon.

Pederson, Neil, Amy E. Hessl, Nachin Baatarbileg, Kevin J. Anchukaitis, and Nicola Di Cosmo. 2014. "Pluvials, Droughts, the Mongol Empire, and Modern Mongolia." *PNAS* 111(2): 4375–4379.

Philo, Chris and Chris Wilbert, eds. 2000. *Animal Spaces, Beastly Places: New Geographies of Human-Animal Relations.* New York: Routledge.

Scott, James C. 2009. *The Art of Not Being Governed: An Anarchist History of Upland Southeast Asia.* New Haven: Yale University Press.

Shepherd, Theodore G. and Elisabeth A. Lloyd. 2021. "Meaningful Climate Science." Forthcoming from *Climatic Change.*

Smyer Yü, Dan. 2021. "Perpendicular Geospatiality of Corridors and Borderlands: An Introduction." In *Yunnan-Burma-Bengal Corridor Geographies: Protean Edging of Habitats and Empires* edited by Dan Smyer Yü and Karin Dean. London/Delhi: Routledge, 1–26.

Stokstad, Erik. 2020. "Mountains and Monsoons Created Tibetan Biodiversity." *Science* 369(6503): 493.

Sukumar, R. 2006. "A Brief Review of the Status, Distribution and Biology of Wild Asian Elephants." *International Zoo Year Book* 40(1): 1–8.

Sutherland, John. 2014. *Jumbo: The Unauthorized Biography of a Victorian Sensation*. London: Aurum Press.

Urbanik, Julie. 2012. *Placing Animals: An Introduction to the Geography of Human-Animal Relations*. New York: Rowman & Littlefield Publishers.

Van Schendel, Willem. 2002. "Geographies of Knowing, Geographies of Ignorance: Jumping Scale in Southeast Asia." *Environment and Planning D: Society and Space* 20: 647–668.

Viveiros de Castro, Eduardo. 1998. "Cosmological Deixis and Amerindian Perspectivism." *Journal of the Royal anthropological Institute* (N.S.) 4: 469–488.

Wang, Zheng. 1996. "The Impacts of Climate on the Society of China During Historical Times." *Acta Ecologica Sinica* 51(4), 329–339.

Wen, Huanran, Jiang Yingliang, He Yeheng, and Gao Yaotin. 1979. "A Preliminary Study of Chinese Elephants in Different Historical Periods" (历史时期中国野象的初步研究). *Thought Frontier* 3(6): 43–57.

Whatmore, Sarah. 2002. *Hybrid Geographies: Natures, Cultures, Spaces*. London: SAGE.

Xia, Yuanji. 1418. *Records of Ming Taizu*《明太祖实录》. 189: 190. https://ctext.org/wiki.pl?if=gb&res=555283&remap=gb.

Xiong, Yingqi. 2019. "26 'Elephant Daddies' Help Asian Elephants Return Home in Yunnan" (云南26位"象爸爸"帮助亚洲象回家). *Beijing Youth Daily*《北京青年报》Sept. 5, 2019.

Yang, Ho-Ji. 2010. "Textual Research on Elephants in the Imperial Palace during the Ming Dynasty" (明廷養象考). *Bulletin of Association for the History of Science* (中華科技史學會學刊) (14): 35–44. http://www.sciencehistory.url.tw/?page_id=968

Yunnan Provincial Multispecies Protection Joint Conference. 2013. "Yunnan Provincial Strategies and Actionable Plans for Multispecies Protection," the Government of Yunnan Province, April 2013.

Yunnan Provincial Academy of Environmental Science. 2015. "The Evaluation of Xishuangbanna Dai Autonomous Prefecture's Ecological Construction and Environmental Protection during the Eleventh Five-Year Plan." March 2015.

Zhang, David, C. Y. Jim, George C-S Lin, Yuan-Qing He, James J. Wang, and Harry F. Lee. 2006. "Climatic Change, War and Dynastic Cycles in China over the Last Millennium." *Climatic Change* 76: 459–477.

Zhang, Jun and Xiao Chunmei. 2016. "Elephant Survival and the Sociocultural Changes in Xishuangbanna since the Ming and Qing Dynasties" (明清以来西双版纳社会文化变迁与象的生存). In *Social Governance and Development in Yunnan's Borderlands* (云南边境地区社会治理与社会发展) edited by Xiong Shuxin, Yan Xiong and Han Quanfang. Beijing: Social Science Press.

Zhang, Li. 2011. "Current Status of Asian Elephants in China." *Gajah* 35: 43–6.

Zhao, Peiran. 2021. "Do the Wild Elephants Know the Way Home? Following the Northward Footsteps of the Fifteen Wild Elephants in Yunnan." *Xinhua News*, May 29, 2021, http://www.xinhuanet.com/2021-05/29/c_1127508405.htm.

Zhou, Shuoxun and Wang Jiaxian. 1992 [1756]. *Chong Zhen Annals of Lianzhou Prefecture, vol.2. Climate* (崇祯《廉州府志》卷2《气候 》). In *Rare Collections of Chinese Annals in Japan: Chong Zhen Annals of Lianzhou Prefecture and Yong Zheng Annals of Lingshan County* (日本藏中国罕见地方集刊：崇祯廉州府志和雍正灵山县志). Beijing: Bibliographic Literature Press (书目文献出版社).

Zhu, Kezhen. 1973. "A Preliminary Research on Climate Chang of China in the Last Five Thousand Years." *Scientia Sinica* (in Chinese) 3(2): 168–189.

CONCLUSION

Multilateral Clime Studies

Jelle J.P. Wouters

As expounded by Dan Smyer Yü in the introduction and further substantiated by our authors, this book extrapolates the notion of clime from Fleming (2010) and Carey and Garone (2014). It deepens, thickens, and enlivens it into a multifaceted opening for reorienting, reimagining, and reconnecting with anthropocenically impacted places, species (humans included), ecosystems, and relations. This opening for rethinking climate as clime adds terrestrial, affective, relational, and more-than-human dimensions to clime/climate studies. Contributors to this book variously located these openings in heterogeneous worlding practices, storied landscapes, emplaced imagination, and pluriversal communities of life, along with their co-constitutive earth and elemental entanglements. While climate and clime can appear as synonymous, and are at times used interchangeably in this book's chapters, clime studies draws distinctly earth-gravitated perspectives. Simultaneously, clime is synonymous with "place"—places, that is, embodied with climate history, agency, and patterns, and thus places as recipients, grinders, makers, and changers of climate.

Our explicitly multipolar approach, in turn, emerges in full recognition of the intimate connections between latitudinal polar regions, altitudinal poles, global climate systems and change, and the fecundity and fate of the "indigenous earth" (Smyer Yü, Introduction), as a living, pre-human, and self-regulating entity. In centering on geological, climatological, and watery (solid, vaporous, and liquid) perspectives of the living earth as a multispecies planet, multipolar clime studies offsets purely anthropocentric conceptions of space and time. Thus, our approach fronts a multipolar-earth vision better attuned to grappling with our anthropocenic earth and global climate crises that transcend political and ethnic boundaries.

In this terrestrial vision, we rely on climate science for meteorological models, averages, projections, simulations, and technical discourses on the condition of the indigenous-turned-anthropocenic earth. This is an authoritative knowledge

DOI: 10.4324/9781003347026-18

field that has long shaped global politics, policies, and public opinions. Simultaneously, we observe that climate science has created a fairly monolithic, largely inaccessible, and often detached kind of knowledge that is frequently self-referential in the sense that it abstracts, quantifies, and models its own variables that subsequently take on a reality of their own (see Hulme 2009). These understandings of the physics of global climate systems, and the scientific capacity to measure and model them, too easily misses the deeper, diverse, and more-than-human meanings, experiences, and implications of actually lived climate change; dimensions that are more readily available to committed ethnography, "lively ethography" (Van Dooren and Rose 2016), and historical ecology.

Identifying the need for convergence, this book operationalizes the clime approach to mutually complement and bridge climate science with the social sciences, humanities, policymakers, and the public. In its form and substance, multipolar clime studies initiates and builds interdisciplinary, interscientific, and inter-epistemological approaches to researching, knowing, and response-ability of climate change. Our empathic attempts variously in bridge-building, line-drawing, dot-connecting, and co-creating—what I will elaborate as "multilateral clime studies" further below—is by foregrounding "place" as an intertwinement of climate/clime, nature/earth, human/more-than-human, and historically situated cultural ontologies.

It is by offering lived and lively climes that we converse with climate science, thus clearing the ground for mutually nourishing inhabitations of the notions of clime and climate. The upshot of this is that clime studies variously translates, transmits, refracts, and enriches climate science by adding more grounded, holistic, more-than-human, and relational modes of knowing and living with climate change. This culminating fusion between climate and clime generates more wholesome and inclusive knowledge(s) that we insist our anthropocenic earth demands.

Thinking Like a Clime

The chapters of this book reveal that a clime is not one thing, but many things at once—it is a versatile, open-ended, and more-than-human approach useful for relating localized, pluralized, and multiple kinds of knowledge (science-making included), capable of conceiving place-based and more-than-human histories, experiences, and meanings of climate (change). In broad strokes, climes emerge as humans find elevational and ecological niches within which they live with others, human and not, organic and inorganic, earthly and elemental. Thus conceived, climes emerge in togetherness, tangles, and intertwinements that are nevertheless everywhere embedded in distinctive, yet connected cultural ontologies. Everywhere, too, it is in this process of co-constituting, co-evolving, and co-becoming that climes congeal the formerly parted domains between climate science and the lived earth, the biotic and abiotic, human and nonhuman, nature and culture, and the material and ideational.

In recognizing multibeing co-becomings, this book reaffirms ancient knowledge of the mutual embodiment of seeming opposites, such as that there are no individual or autonomous wholes in this world, and that it never makes sense to make a clear-cut distinction between what is natural and what is not. Authors experience and express the nonhuman world, the earth included, in the active voice (Plumwood 2009), while they find that "fleshier, livelier stories make us care more" (Van Dooren 2014, 10). In this way, climes allow for humans to emerge as ecological beings deeply entangled with places, climates, and nonhuman others, and as geological beings deeply knotted with the indigenous earth. And so it is in the Andes, Arctic, and Himalayas as detailed in this book, but equally everywhere else.

To apprehend climate change in all its diversity and difference, clime studies critically evaluates climate change as both the one and the many—the Anthropocene (Crutzen 2006), Capitalocene (Moore 2016), Necrocene (McBrien 2016), Plantationocene (Tsing 2015), Chthulucene (Haraway 2016)—and the relations between them, such as the socio-cultural, ethico-moral, spiritual-religious, political, and economic dimensions of life in specific places. It is through these multiplicities that clime studies diversifies and stretches engagements with climate change, as well as connects climate science with diversely lived everyday lives, in all their complexities, internal dialectics, and totalities.

Thus, clime-thinking and multipolar clime studies is an innovation-driven scholarship that interlaces climate science variously with extant and emergent scholarship that relates traditional ecological knowledge, ecological and political ontology frames, relational imagination, indigenous epistemologies, ecologies of practice, more-than-human becomings, anthropologies of life, multispecies landscapes and conversations, and pluriversal worldings (Viveiros de Castro 1998; de la Cadena 2015; De La Cadena and Blaser 2018; Blaser 2009, 2016; Latour 2004; Kohn 2013; Escobar 2018; Haraway 2008; Tsing 2015; Van Dooren 2019; Chao 2022). Contributors, indeed, variously illuminate how clime/climing facilitates thinking and theorizing with and from the animatedness, agency, relationalities, and intra-active co-becomings of places, glaciers, mountains, weather, plants, species, elements, water, deities, earth beings, and other supernatural beings. Our authors transmit these, as do the varied communities they work amid, as living and sentient beings with their own histories and perspectives, and with the ability to experience and communicate in their own ways. The climes so revealed are bustling with life and relations and are always in a state of co-becoming, particularly in the Anthropocene context that now strains and rearranges many of these relations of life.

While it is impossible to translate the diversely embodied entanglements and experiences of clime and clime change into a singularly concrete clime-theory, it *is* possible to identify some of the main currents and implications of thinking in, through, with, and like climes. Inexorably, climes are metabolic, ecological, and affective, bringing into focus not only the biogeochemical landscapes, but also their ethos, consciousness, and soul, as can be apprehended ethnographically, for example *sila* in the Arctic, *Tirukuna* (earth beings) in the Andes, *eno* in the

Eastern Himalayas (Andersen; Brandshaug; Paerregaard; Delley and Aiyadurai, this book). Inexorably, too, climes always constitute more-than-human histories, geographies, and agencies, for example, multispecies migrations, elephant climes, beyond-human subjectivities (Gamble; Smyer Yü; Burman, this book). In their shape and substance, climes reveal and enact through subjectivity and experience, relate affective bonds and affordances of a living earth as a multispecies planet, and rotate around living local histories, indigenous epistemologies, and creative imaginations. Crucial, in all of this, is that climes consist of the multiple lived relationships that people are nourished by in particular places—for it is solely by virtue of these relationships, and of these assemblages and entanglements, that climes acquire meaning, often arrayed through relational, storied, and multibeing imaginations.

In their relational properties, climes relate both as noun and verb, with the former focusing on the particularities of emplaced clime-worlds, and the latter detailing the ways in which peoples socialize, engage, and interact with the earth, water, weather, mountains, and other physical systems and flows, in full appreciation of their agency, animatedness, and sentience (Paerregaard; Smyer Yü, this book). In their shape and shape-shifting, climes are further locked, adjectivally, to terrestrial spheres, i.e. riverine, glacial, mountain, alluvial, gorge, and monsoon climes. Unlike climate science models and predictions that float in academic journals and travel in policy documents, climes are places, they are terrestrial and are as much a part of us as we are a part of them, and senses of climes intricately partake of both. Biogeochemically, climes recognize how human life is produced and sustained by the nourishing circuit of the biosphere, lithosphere, atmosphere, cryosphere, and hydrosphere; a continuous life-giving and life-receiving exchange between life and matter. Phenomenologically, climes recognize both how humans think in terms of places embodied with geological, ecological, and climate (change) history and how these "clime-worlds" operate in human thought-worlds, both consciously and unconsciously. Stated otherwise: a clime relates a terrestrial sense and sensibility of the connection between what it is that constitutes a particular clime/climate-place and how that clime/climate-place is biogeochemically, affectively, and cognitively experienced, acted upon, and inhabited by human beings, amid many other beings.

As most chapters relate ethnographically, climes are distinctly alive and emerge out of diverse kinds of relationalities, or modes of life-enabling entwinements of biotic beings and the physical environment. These relationalities are locally abstracted into norm-worlds, expressed in many vocabularies and voices, and they embed cosmologies and ontologies, or theories of reality, including, in this book, a more-than-human sociality of *nuna* (landscape), cosmopolitical ecologies, and cosmic and communicative waterways (Iqbal; Andersen; Brandshaug; Bhutia, this book). The act of making and narrating climes provides a crucial way of engaging directly with climate and the ways it is changing. After all, how people experience, think, and talk about climes informs their actions and responses to climate change.

In much of the multipolar arc forefronted in this book, clime change is experienced and explained as a breakdown of relations and norm-worlds, leading to social, ecological, and aesthetic ruin and despoliation. The resultant "re-climing" (Paerregaard, this book), as other chapters, too, illuminate, is a resilient, creative (albeit often confusing, always unsettling) process and discourse of adaptation to changing conditions and the reimagining of self and others. Amid the variously experienced climate crises, these social processes of re-climing have now become integral to the contemporary human condition.

Earth Architecture, Migrating Waters, and Teleconnection Geographies

The multipolar earth is at once a geological vision, a climatological configuration, and a physical reality to visualize and rethink climate- and environmental studies. It accentuates more-than-human histories and geographies as expansions of the exclusively anthropogenic nation-states, ethnic groups, and academic area-studies that have long structured and limited our earth vision. This limitation extends from the deeply epistemological entrenchment of, what we might call "methodological humanism" that analyzes socio-spatial forms and scales in exclusively human terms and units at the expense of other-than-human imaginations, histories, and geographies. Clime-thinking explodes into such distinctly human-centric imaginations.

Gamble (this book) details multispecies migration into the Yarlung Tsangpo Gorge. Plants, animals, birds, and humans arrived at different times, co-adapted to the gorge's ecological niches, and turned it into a "multispecies pluriverse." Smyer Yü (this book) historically traces the Asian elephant clime and transmits it as an "elephant nation" variously displaced and dispossessed by climate change and human expansion. Besides the elephant clime, it is equally possible to think of the yak-clime, crane-clime, vulture-clime, leopard-clime, reindeer-clime, caribou-clime, or alpaca-clime. These animal climes express norm-worlds and social contracts, albeit not written in modern-Western constitutional language, but as ecologies of practices grounded in co-dwelling and co-becoming with the earth and companion species in shared ecological niches, habitats, and corridors. Such an intellectual exercise in animalizing, avianizing, and vegetalizing territories (See Banerjee and Wouters 2022) both offer an affective critique of anthropogenic borders that block, divide, and frustrate clime-commons, and bring more-than-human climes into purview.

If animal climes make us think otherwise about histories and geographies, so do elemental bodies such as glaciers, mountains, rivers, and ice-sheets that many indigenous peoples acknowledge as sentient, agential, and animate—a view scientific scholarship is beginning to reinscribe through integral Gaia perspectives (Lovelock 2000) and the animist critical zone (Smyer Yü 2020). Niche scholarship knows that glaciers are layered selves who smell, hear, morally evaluate, act, and respond (Cruikshank 2005); that, like humans, ice-sheets are gendered

(Hewitt 1989); that forests think (Kohn 2013); that rivers have memory (Lyons 2018); that mountain bodies cultivate kinship (Pommaret 1994); that rocks listen (Povinelli 1995); that lakes are sentient and imbued with spiritual essence (Phuntsho 2013); and that fog and mist are interdimensional communicators and connections (Bhutia, this book). Their life-endowed and life-enabling and sustaining presences are integral to locally and relationally scripted clime-worlds. In incorporating the animatedness of earth bodies and systems, and in centering on earth architecture, multipolar clime studies fosters a sense of deep horizontal fraternity.

While the Himalayas, the Andes, and the Arctic are not contiguous as a physical space, from the vantage point of earth architecture they—together with other highlands, mountains, and the Antarctic—connect as planetary climate makers and changers. They unite as the earth's storehouses, affective stewards, and circulatory nodes of water, the largest nonhuman presence in the world and integral to scientific and indigenous understandings of climate change. These landscapes are in fact so enmeshed with water that they can be seen and storied as "terrestrial oceans" (Smyer Yü 2021, 9), and as the towers, tanks, tubes, and taps of the global hydrological cycle.

Unlike other known planets, which are thought to be mostly dry, the indigenous earth is a watery earth. Certainly so from a terrestrial vantage point as over 70% of the earth surface is covered by water, as well as from a bodily vantage point as between 50% and 65% of human substance is water. Over two-thirds of the earth's freshwater is stored in latitudinal and altitudinal highlands. These are the earth's headwaters from which flow life-giving, life-sharing, life-sustaining (and in its absences or excesses, life-taking) forces that irrigate natural environs, multispecies civilizations, and spiritual-religious worldviews; forces that ultimately mediate the fecundity and reproduction of the terrestrial earth and its many earthlings. These highlands attained their affective affordances through the metaphysics of Deep Time; a geological epoch that encompasses both prehuman and human eras, an almost mystical world in which everything was possible, from mountain-building and glacier-formation to the birth of rivers. If the Anthropocene, the turning of humans into geological agents, has breached the walls between human and natural history (Chakrabarty 2009), climate studies must also breach the wall between human temporality and earth temporality. To breach these temporalities is to recognize Deep Time as an earth archive, to be aware that we live, walk, and breathe Deep Time, that we must therefore think with Deep Time, and that, in its final evaluation, we are Deep Time.

While climatology's strength lies in abstracting from analyses and averages of usually thirty-year periods (Shepherd and Truong, this book), clime studies foregrounds a much longer pattern and memory, cognizant of the fact that the planetary history of the present (and the future) are the products of geology and climate that can only be approached within the expanded time of the earth's biography. Through "Deep Climing" (Paerregaard, this book), clime studies recognizes humans also as geological beings, in the sense that the places we live in and with, including the rhythms of other-than-humans, are thickened by other times (Irvine 2020). Deep Climing, as a worlding practice, enlivens Deep Time, as well as helps to visualize a Deep Future in which climatic conditions will be progressively affected by a host of human variables. Deep Time's corollaries: "deep space," "deep freeze," "deep seas," "deep ice," and "deep winds" further reveal the ongoing history of the geological present, forever shaping and sculpting climes.

In its metaphysics, Deep Time created relationships of ecological interdependence. It nourished positive, relational aspects of biotic and abiotic activities through the making of varied climes in the course of geological history. It is ultimately Deep Time that enabled the currently life-sustaining "internatural bonds" between the Bay of Bengal, Meghalayan mountain ranges, and the Bangladeshi *haor* (wetlands) that Iqbal (this book) details. Here the world's rainiest place, the southern edge of the Meghalaya plateau, and the world's wettest place, the Bangladeshi lowland *haor* co-create each other through evapotranspiration and trans-regional drainage regimes, nourishing lively biotic communities of fishes, wildlife, and humans in vastly different, but co-constitutive, elevational and ecological niches. Such Deep Time, inter-clime, relational bonds also reveal when humans are left outside the equation.

Amrith and Smyer Yü (this book) show this by tracing the dense interlacing of Himalayan and monsoon climes in South Asia. Consider also the Indian alluvial climes. They exist only in relation and with reference to, even as part of, Himalayan climes that channel water, silt, and sediment downstream. Van Schendel's insight that "Bangladesh is the Himalayas, flattened out" (Van Schendel 2020, 3; also Amrith and Smyer Yü, this book) cements this point. Reciprocally, monsoon climes shape the melt and accumulation of glacier climes (Fugger et al. 2021).

Besides unfolding in Deep Time and through relational bonds, multipolar climes are also always on the move, geologically, geomorphologically, and glacially. They are ever in flux, in a perpetual state of "being as co-becoming." The condition and circulation of global water in particular have always changed, since the inception of the indigenous earth. When water changes, as for instance during the Little Ice Age, this results in migrations of human, animal, and plant polities, as well as in climate-induced political coups (McMichael 2017). However, it is the current anthropocenic climate change that fundamentally alters the global water cycle so that, in comparison to Holocene climes, water is migrating, both voluntarily, but excessively and forcibly by humans.

At the heart of current anthropocentric visions of development and sustainable futures is powerful geo-engineering that drains, diverts, dams, encloses, tunnels, and floods, in human efforts to harness for ourselves the more-than-human life-giving forces of integrated water bodies and systems. As a case in point: integral to China's vision of "ecological civilization" is the "discovery" of the Yarlung Tsangpo Gorge as a Deep Time treasury of mineral wealth and its unparalleled water storage and hydropower potential (Gamble, this book). The damming and diversion of the Yarlung Tsangpo River—the highest major river in the world, and the upper stream of the Brahmaputra River in India, and the Yamuna River in Bangladesh—was adopted in China's 14th five-year plan, released in 2020. It envisages its diversion toward the deserts of Xinjiang, to make these places viable for anthropogenic expansion. In this way, China is rehearsing already-existing water diversion activities in Yunnan, intent on draining lakes and rivers in the Yunnan-Guizhou plateau to feed central drought-prone areas (Persio 2017; Rippa et al. 2020). Similar geo-engineering is also underway in the Andes, where canals and dams are constructed as part of mega-irrigation projects that displace large volumes of water to irrigate drier areas, "to make the desert green" for human use (Brandshaug, this book). In China, but equally in the Andes, the groundwater tables and earth currents mingling with cryogenic and hydrological ecosystems in subsurface terrains, are now met by anthropogenic intrusions, giving and taking away soil, sediments, and water through redistribution and management, and manipulating the biosphere from below (Wouters and Heinese 2022: 26).

Inherent to these anthropogenic alterations of watery climes is the blurring of the multiple temporalities these climes invariably inhabit: Deep Time, divine time, anthropogenic time, and—in the context of accelerated climate change—troubled time. Global temperature rise is melting Himalayan and Andean glaciers and is now discharging waters that may have been stored in solid form there since the Pleistocene. In the Arctic (but also elsewhere) capitalist yields are sought in the commercialization, indeed erasure, of Deep Time (Andersen, this book). Here ecological destruction is the extraction from climes originating in Deep Time to serve the needs of present time, leading to crises of ecological reproduction that are particularly hard felt in the cryospheric worlds that are integral to the multipolar arc. From an exclusively human vantage point, this contributes to water being in all the wrong places, whether it is glacial melt, sea-level rise, increased rates of evaporation, overflows of inland waters, or erratic and excessive precipitation.

The multipolar arc forefronted in this book is indispensable to global water and is disproportionally affected by anthropocenic climate change. Its polar regions, poles, glaciers, and other water-forms unite as nodal points in intricate global "teleconnection geographies." Loosely affiliated with "process geographies" in the social sciences (Appadurai 2000, 7) and "Actor-Network-Theory" in Science and Technology Studies (Latour 2005) that replace borders and delimited societies with "relations" and "interactions" as dominant metaphors of social connectedness, the notion of teleconnection is a spatial reality based on earth architecture and climatological connectivity. In climate science, teleconnections refer to a linkage or correlation between meteorological, climatological, or other environmental phenomena that occur a long distance apart, creating climate variability and change (Nigam 2003). Teleconnections compel us to think about currents, anomalies, flows, fractals, streams, waves, scales, and other forces that join noncontiguous areas in a range of positive feedback loops. Multipolar mountain bodies, glaciers, and ice-sheets figure as nodal points in now emergent telecommunication maps, so that a change or fluctuation in the multipolar arc creates varied ripple-effects that effectuate climate variability and change elsewhere, both near and far. In forefronting the multipolar arc and through following and thinking with teleconnections, clime studies can complement climate science models' recognized bias toward "nearest neighbor interactions, both in space and time" (Shepherd and Truong, this book), to also apprehend connections across currently dominant and human-centered spatial and scalar fixes.

Take albedo (Latin, *albus,* or whiteness), which is the proportion of the incident light that is reflected by the earth's surface. The whiteness of ice and snow has a cooling effect for the surface of the earth, as it reflects light back to the atmosphere. When more snow and ice-melt, the chromaticity of the landscape changes, revealing more dark surfaces. Surfaces without ice and snow again absorb more heat from the sunlight, and this, through a self-reinforcing effect, raises surface temperatures (Coakley 2013, 1914). In addition to cooling the earth, the multipolar arc also cleans the earth through absorbing industrial dust and pesticides, sometimes blown to them many decades ago and from thousands of miles away, and once there they are incapacitated into snow and ice. The great, global melt is now releasing decades of accumulated pollution back into ecosystems, from where it again travels through waters and winds. The currently weakening albedo so teleconnects both to earth warming and increased pollution that is experienced globally.

Such teleconnections are many and varied. Fresh research indicates how the Arctic region teleconnects with the Indian monsoon, several thousands of miles away. Accelerated ice-melt leads—through disturbances in land-ocean-atmospheric processes—to both variability and intensification of the Indian monsoon, the liquid lifeline of over one billion people. As the Indian monsoon intensifies, more heat dissipates. This extra heat finds its way back to the Arctic, further accelerating its melt in a positive feedback loop (Chatterjee et al. 2021). Scholarship also relates teleconnections between the retreat and advance of Andean and New Zealand glaciers (Fitzharris et al. 2007). Several other such teleconnections are now emergent, with many of them involving and radiating from the multipolar arc, especially in relation to the global water cycle. It is these—albeit often still poorly understood—teleconnections that the science of climate tipping points inhabits, and the local experiences and understandings of which clime studies presents through its multipolar terrestrial vision.

The more we become aware of teleconnection geographies, especially in relation to watery multipolar climes, the more we realize how earth systems and anthropocenic climate change resist the human-centered spatializations and scalars of both climate science and social/critical theory. In repudiation of this, multipolar clime studies equips itself with the potential to relate both locally constituted clime-worlds and long-distance, large-scale, and noncontiguous climatic relationships that readily defy both human-centric spatial imaginations and any straightforward local-global binaries. In the wider context of climate studies, some critical issues include how our multipolar-earth vision, with its emphasis on hydrological stewardship and teleconnection geographies, relates to other approaches and methodologies of understanding climate change, what the mediations are between these and overarching climate studies, and what specific contributions multipolar clime knowledges can make to ongoing climate studies. Further advances in multipolar clime studies will need to engage and untangle these issues.

Multipolarity, Indigenous Climes, and Decolonizing Climate Science

Our multipolar terrestrial approach is derived from earlier bipolar—Arctic, Antarctic—and tri-polar—with the Himalayas included as the "Third Pole"—earth visions. Concurrently, "multipolarity" is also a metaphor for global power circulation and diversion that evolved from a post-World War II Cold War bipolar order, followed by a US unipolar world, to a currently unfolding multipolar political world. Here, the central sites of contestation are not only political, military, and economic, even as the GDP-driven rise of India, China, Brazil, and other emergent economic powerhouses keep earth-ravaging Capitalism, in upper case, securely on planetary terrain. Increasingly also, contestations are over questions of what knowledges are worth knowing, what worlds are worth worlding, what lives are worth living, and what ethics and justice are worthy of an Anthropocene world in which humans, other-than-humans, and the indigenous earth are so deeply enmeshed and vulnerable that violence and dispossession laid upon the one inevitably harms the other.

Amid multipolar political emergences, the US's temporary withdrawal from the Paris Climate Agreement, and its general moral economy of inaction, has weakened the US-led politically integrative potential of the capitalist recuperation of climate change, that is, the Green New Deal, growth models, emissions trading, and the manufactured consent of the acceptability of 2° of global warming. Simultaneously, the Intergovernmental Panel on Climate Change (IPCC), in its sixth global assessment (2022), finally acknowledged colonialism as a driver of climate change. The central dispatch here is that decolonizing climate change, and the dominant knowledge production thereof, must now be epistemologically and institutionally arranged. Multipolar clime studies is multiply resonant with these intellectual-political transformations.

Our integral contribution to these progressive movements is the offering of lively and innovative explorations of multispecies "indigenous climes" (Smyer Yü, Introduction). Chapters move beyond important but ultimately shallow notions of colonially-implicated political indigeneity by also framing indigeneity in terms of the affective, ecological, social, and cultural relations humans share with other humans, the living earth, and nonhuman co-dwellers, both past and present, both actual and potential. While distinctly diverse, indigenous climes have been comparatively efficacious in enacting more-than-human norm-worlds and affective earth-relations. At the same time, indigenous climes now emerge as particularly vulnerable to grievances of climate change, which they disproportionally receive. In the multipolar arc many indigenous climes are currently in varied stages of anthropogenic alteration, ruin, despoliation, and necessary "re-climing" (Paerregaard, this book).

The focus on "indigenous climes" in this book allies with the vocal rise of indigenous voices, knowledges, and alliances that are catalyzing a thickly braided critique against climate change mediations that continue to find capitalist value-production in extraction, depletion, and extinction. Our groundbreaking accounts of indigenous climes acknowledge that indigenous ways of knowing and being in the world will generate unprecedented attention on a global scale with their coherent alternatives to the colonial-capitalistic logic and its extractive and exhaust-expelling ethos. At another level, which is the level this book primarily addresses, indigenous climes also both critique and complement the largely distanced, disembodied, and secular science models. This book shows how the emplaced experiences and knowledge practices of indigenous climes quickly overwhelm the received limits of climate science knowledge and associated lexicon and set off an avalanche of other perspectives.

To be sure, the indigenous critique of climate science models, variously voiced, is usually not a nativist rejection of climate science, but rather is a call for more collaborative and capacious knowledge generation, including that of indigenous-driven science (Snively and Corsiglia 1998) and the indigenization of science (Tsosie and Claw 2020). The indigenous grievance is not that climate science is not true, but that it is often regarded as the single truth, to the exclusion of other ways of knowing. In fact, a case can be made that in its truth-effects the mathematical and meteorological models that climate science produces are more real than the climate change they represent. This is especially the case considering these models' far-reaching political and policy implications. Because of this and other truth-effects, it is not obvious that climate science is less real than more concrete and actually lived indigenous climes. The indigenous critique, emergent from indigenous climes, is that this climate modeling has been the nearly-exclusive driver of the unfolding narrative of climate change. Relatedly, and variously, it critiques climate science's elevated status as the dominant apparatus and taken-for-granted technique of knowing climate change, as the main filter of prevailing political narratives, and as the fuel of normative philosophy that remains highly anthropocentric in nature. Indigenous knowledges have been deemed inconsequential in these debates. Considering further that the two dominant traditions within climate science, the descriptive and the explanatory, are both deeply entangled with Empire (Shepherd and Truong, this book), many indigenous communities are now rightly doubtful about the potential of scientific climate knowledge and associated Euro-American-led theories and policies to ever free themselves from racial-imperial associations.

For all its invaluable knowledge production, climate science does not inhabit actual glaciers, mountaintops, or water bodies, as indigenous climes do. Climate science predominantly resides in laboratories, super-computers, and advanced software, the most superior versions of which are located in the Global North, even as climate scientists now progressively also engage in fieldwork. As late as 2010, moreover, the climate historian Edward (2013, 171) could remark that "the elite world of global climate simulation still includes no members from South America, Central America, Africa, the Middle East, or South Asia." While the chapters of this book do not necessarily assert this point, Edward's emphasis on an "elite world" of climate scientists intuitively counter-positions their climate knowledge and worldviews with those subalternized, those who—as subaltern studies and other post-colonial matrixes convincingly reveal through their focus on exploitation, exclusion, abjection (but also resilience and autonomy)—have different ontological roots, think and speak in different idioms, and yield different norm-worlds, norm-worlds that have been marginalized and silenced. This necessitates an additional dialogue about climate science and scientists in which we ask who they are, where they live, what their expected professional citational practices are, and how prestige and precarity are allocated among them.

The general point of all this is that the history and generative knowledge of climate science should not be construed as an isolated odyssey of value-neutral knowledge production, but is simultaneously, equally, and inseparably also an exercise in political economy, a Foucauldian regime of power/knowledge, a product of Enlightenment modernity, and ultimately also, what Mignolo (1995) calls an "epistemology of coloniality" that reveals, and is resisted, as a "coloniality of reality" to which the indigenous response is "ontological disobedience" (Burman, this book). If the emergent multipolar world order and the recognized need to decolonizing climate change demand us to think of climate history, climate science, and climate change policies as *also* a space mapped by knowledge/power, attentiveness to indigenous climes opens up this space by questioning the implicit occlusion and sequestering of indigenous epistemologies and worlding practices. What the notion of indigenous climes does, then, is to give pause to the often technical and universal abstractions of climate science knowledge to also consider climate change from different epistemic and place-specific grounds.

By plugging indigenous climes into the equation of climate science, this book expands the lexicons and horizons of conventional climate science. Our focus on indigenous climes reveals that to the degree that the planetary life-system, as large as it is diverse, is facing anthropogenic alterations (or rather, altercations), very different encounters, experiences, and responses have resulted from it. For one thing, even as climate change is generally spoken of in the universalizing abstractions of physics, chapters by Paerregaard, Burman, Brandshaug, and Delley and Aiyadurai relate that many indigenous communities also regard climate change as a religious problem (of gods and deities evacuating human abodes) and foretelling (of death, judgment, and the end of the world) (See also Haberman 2021). This leads to ritual responses, such as offerings to earth summits and water bodies (Burman, Delley and Aiyadurai, this book). More broadly, it prompts religious wakes and funerary moments that mediate mourning and explore the ethics of species extinction, glacial death, and geological annihilation (Van Dooren 2019; Schmidt 2021).

However, rather than universally mourned, in its warming effects, climate change incidentally offers opportunities to highland humans, plants, and animals to expand ranges of pasturing and growth (Paerregaard, this book). In the Arctic, in turn, climate change is appropriated into a new capitalist frontier, as in commercialized Greenlandic freshwaters rendered available through anthropocenic ice-melt and leveraged for capitalist production-consumption (Andersen, this book). For yet another, clime change inspires new experiments with governmentality and biopolitics, of risk-management, technocratic totalitarianism, and geo-engineering (Gamble, this book), but also cascades experiments with new legal grammars that explicitly recognize nonhumans as rights-bearing subjects and as explicit victims of climate change.

Brandshaug (this book) shows ethnographically that how "to live well together" in the Peruvian Andes is contingent on "water climing"; a committed communion, that is, between Yanque humans, Pachamama (Earth Mother), *apu* (mountain beings), and Tata Mallku (Father Water), amid other autochthonous earth beings. "Water climing" recognizes the sentience of water, both as a cyclical circuit and in its many bodily manifestations (springs, lakes, streams, glaciers) who are named persons "with feelings, thoughts, and will." Also in the Peruvian Andes, Paerregaard details "social climing," a worlding practice in which humans speak through "gifting," in the form of ritual interaction and offerings, and earth beings respond by making earth affordances available. Clime change is experienced as a breakdown of more-than-human communion, and results, in Brandshaug's ethnography, in the common human exclamation of "No hay agua (there is no water)."

Water's refusal to nourish human life is so engaged not primarily in secular terms, as an imprint of climate change, but as "cosmopolitics through and through." To restore communion with water, two Peruvian municipalities extended legal rights to water bodies, thus recognizing them as rights-bearing persons. This move is part of a wider trend, emergent from indigenous climes, that restores personhood and legal subjectivity to nonhumans and ecosystems, and recognizes them as being capable of being harmed and deserving of redress, of which examples include Whanganui River in New Zealand (Salmond 2014), the Gangotri and Yamunotri glaciers in India (Alley 2019), the constitutionally incorporated "natural contract" of Ecuadorians with Pachamama (Fitz-Henry 2012), and the "Mother Earth Law" in Bolivia (Tănăsescu 2022). These represent unevenly-successful attempts to restore more-than-human norm-worlds and relations, and to translate cosmopolitics in a legal grammar intelligible to humans, so that more-than-human "multilateralism" (more in the next section) can be better facilitated.

This ancient and now re-emergent more-than-human legal order, a more-than-human demos, also encompasses supernatural entities. Burman's (this book) Andean ethnography can be translated as a long process of multilateralism between the Aymara people and Chacaltaya, an ancestral mountain-being (*achachila*), "dressed in a white poncho," who is known to have intentionality, social agency, personality, and who is invested with the capacity to govern weather and water. Chacaltaya's poncho is now melting away in the face of a "capitalist modern-colonial world." Refusing to reinscribe themselves with the explicit "coloniality" of this reality, Aymara humans attempt to reestablish communion with Chacaltaya through incessant ritual acts "through which other worlds are enacted," worlds that carry the capacity "to overthrow the world as we know it" More-than-human climes are also revealed in the Himalaya, where Delley and Aiyadurai show how human sustenance and sustainability are continuously mediated by engagement between Idu Mishmi humans and beyond-human spirits who dwell in water bodies, as well as in other parts of a densely inhabited eco-spiritual landscape.

These chapters show that clime change is a key part of the human condition and experience. Their lively explorations builds bridges with the natural sciences by contributing stories that need to be told about more-than-human communities of life; stories that prompt discussions about knowledge, ethics, and justice. These stories are intellections from peoples who are intimately affected by climate change; they experience and see the things that are being represented in climate science models. If climate science, then, is challenged by a loss of meaning, multipolar clime studies instills it with meaning generated by actually existing peoples, places, and patterns of being.

Multilateral Clime Studies: The Way Forward

In diversely exploring and demonstrating the "translatability, interchangeability, and complementarity between the notions of clime and climate" (Smyer Yü, Introduction), multipolar clime studies builds pathways for more critical, collaborative, creative, curious, caring, and capacious ways of knowing, thinking, and acting in relation to clime/climate change. Our authors engage in interdisciplinary, epistemological, and ethical wayfaring in the Anthropocene context. Everyone's openness renders disciplinary comforts uncomfortable, transforms presumed interdisciplinary incommensurabilities and incompatibilities into connections, conversations, and complementarities. In the upshot, it affirms a spirit of collaboration and cross-pollination across indigenous climes knowledges, sciences, and humanities. The lively explorations that result now enable mediations of the multi-polarization that characterizes the Anthropocene, our presently unfolding epoch in which the "new," or anthropocenically impacted earth, is reconsidering its principally life-enabling properties.

Anthropogenic polarizations fold and unfold between scientific and indigenous knowledge practices, between a range of invented dichotomies, or what Latour (1993) called the "Great Divide": of nature versus culture, humans versus nonhumans, modern versus nonmodern politics, anthropogenic versus indigenous earth, and between the unevenly shared causes and consequences of climate change. This multi-polarization was heralded by the growth, across the planet, of colonial-capitalism, and, in parallel, by the sciences and humanities that long relied epistemologically on dichotomies, divergences, and bifurcations. Multipolar clime studies repudiates this multi-polarization by enlivening the multi-pronged, multi-nodal, multi-facetedness, multi-sidedness, and more-than-human dimensions and totalities of clime/climate change. Through lively explorations, this book's chapters contribute and reaffirm ancient and newly emergent knowledges of the mutual embodiment of the above anthropogenic opposites. This final section extrapolates this mediation as inherent to the programmatic turn of multipolar clime studies and captures it as, variously, "multilateral clime studies," "multilaterality of clime studies," "multi-nodality of clime studies," and "multimodality of clime studies."

The envisaged multilaterality of clime studies communicates with Chakrabarty's (2009, 2021) injunction that anthropogenic climate change must make us revise the fundamental assumptions of history, modernity, and globalization across disciplines. Chakrabarty's revision is crucial, but also complicated by the observation that there is no single way of knowing and experiencing climate change or studying its effects. What this book shows is that its consequences are felt differently, and across domains, spanning the economy and the sacred, the substantive and the metaphysical, the human and the beyond human. We discern in climate change's polymorphy and multiplicities an emphatic call for communication across lines of disciplinary cleavage, making it possible to mutually complement knowledges, to engage in hybrid collaborations, and to diminish divergences in a shared responsibility to engage clime/climate change in more holistic terms.

Multilateralism, as employed here, wrests the term free from its modern association with nation-states and international relations. Instead, it attaches it to "geopolitics," in its original sense with "geo" signifying the integral earth, Gaia (Smyer Yü, Introduction). It is with this accommodating geo/*Gaia* that human ancestors negotiated original, animist social contracts that made them accountable not just to other humans and later to governments, but especially also to nonhuman beings—to animals, plants, waters, forests, and the earth itself (Banerjee and Wouters 2022). Several chapters implicitly foreground this original, life-giving social contract and its continual reproduction—through ritual, offerings, orature, and mythical storytelling: knowledge practices that reach deep into the cultural imagination—that enliven, memorialize, and visualize more-than-human constitutions and demoi. In this constitutive and inalienable context, the multilaterality of clime studies impels earth, epistemological, and ethical connections to multilateralism's formal geopolitical usage.

When thus adapted and tailored to clime and clime change, multilaterality, multimodality, and multi-nodality invite different histories, knowledges, and subjectivities, both human and nonhuman, and different experiences and epistemologies to be represented and rendered equally present in climate change knowledge and policy-mediations. It asks of researchers, across disciplines, to overcome currently stark epistemic antinomies through cultivating a common interest in climes as places embodied with climate histories, and to be an emphatic and reliable partner to each other's knowledges, methodologies, and perspectives. In its strong commitment to multimodal coalitions, the multilaterality of clime studies is practiced not to artificially neutralize or ignore divergences, to streamline heterogeneous worlding practices, or to put all beings, human and not, in common agreement. Instead, it pursues the mutual translatability, complementarity, nourishment, and ultimately the co-becoming and co-creation between disciplines and epistemologies, the interlacing of models and meanings, meteorological and terrestrial perspectives, observations, and ontologies.

Multilateral clime studies expects the natural sciences to be receptive to knowledges and concerns currently outside its epistemological scaffolding. Simultaneously, it expects social scientists and humanists to avoid too generalized a conception of climate science and its scientists, or to primarily use climate science dialectically by turning it into a ready-made strawman against which to frame alternative understandings. This highly modernist tradition of the strawman-ing of disciplines and knowledge-fields, pitting one against the other, has been short-circuiting potentialities of intellectual co-creation and is ultimately reductionist. Genuine interdisciplinary attentiveness and curiosity enables translatability, convergence, and complementarity between disciplines.

As a case in point, Shepherd, a physicist and climate scientist, and Truong, an anthropologist and development practitioner (this book) co-inhabit traditions, trajectories, and new departures in their respective fields. They find convergence in climate science's storylining approach, of which Shepherd is an advocate (Shepherd et al. 2018, see also Shepherd and Lloyd 2021). Storylining climes coalesces "contingency" in climate science with "conditionality" and enlivening ethnography in anthropology, and merge these into "storying climes." Shepherd and Truong's innovative bricolage, or braiding, is expressed in their Himalayan and Arctic case studies, and results not only in enriched knowledge, but in "the realization of response-ability and responsibility" of this knowledge as well.

Shepherd and Truong are at the frontline of an emergent trend in climate studies in which vanguard scientists and philosophers are addressing divergences, discords, dichotomies, and ways toward efficacious interdisciplinary understandings (Barad 2007; Hulme 2009, 2016; Bristow and Ford 2016; Kimmerer 2014; Victor 2015). Singh (this book) is also part of this search for multilateral convergence and complementarity. She confronts her self-perceived failure in teaching climate change solely through the prism of physics. Adopting radical openness and humility in the Arctic, where she learned from Inupiaq town officials, elders, whalers, and a multitude of scientists, she developed a multi-pronged approach to clime pedagogy consisting of: (1) the scientific-technological, (2) the transdisciplinary, (3) the epistemological, and (4) psychosocial action dimensions of climate change, and resulting into more engaged, affective ways of teaching and learning back in her classroom.

If Shepherd and Singh represent climate scientists open to thinking across disciplines and epistemologies, social and human scientists must now also symbiotically associate themselves with earth and climate sciences to expand the scope of their research. Such efforts potentially open up new and productive research fields that lie beyond the grasp of any single discipline: earth anthropology, gender and glaciology, social geology, animal ontologies, vitalist volcanology, political seismology, plant philosophy, among other innovative possibilities. Besides garnering a truly interdisciplinary community vested in intellectual co-creation, multilateral clime studies further relates a collaborative and creative search for life-sustaining intelligence that is cognizant of more-than-human integrity and justice. The indigenous climes that inhabit this book detail experiential and affective worlding practices that offer viable paths for dwelling and being in the world premised upon a holistic, relational, and more-than-human imagination (Paerregaard, Brandshaug, Burman, Delley, and Aiyadurai, this book). Thus, multilateral clime studies is integral to decolonizing climate change by opening up space for thinking and living *with* the terrestrial earth as a multispecies planet in order to do better justice to the multiply lived realities and experiences of climate change.

Walking the multilateral path requires collaborating, unlearning, and relearning, and raises demanding questions of interpretation, possibility, and representation (of the translatability, for one, between the voices of humans and those of glaciers, rivers, ice-sheets, the hydrosphere, and the beyond-human more generally). These are questions the environmental humanities are pursuing with vigor and with notable success (e.g. Cruikshank 2006; Krupnik et al. 2010; Smyer Yü 2016, 2020; Chao and Enari 2021). Ultimately, the multilateral path will itself reveal the way guided by cross-pollinating, collective, hybridized, and inclusive forms of knowledge generation that cut across and connect epistemological, ethical, and political domains.

Amid our indigenous-turned-anthropocenic earth, this book sows life-enabling and sustaining seeds, and waters the saplings of multipolar and multilateral clime studies and collaborations. Our cross-disciplinary chapters emphatically relate how this results in much fuller and richer understandings of clime/climate change, and how such enriched understandings can be productively applied to mediate the multi-polarizations of the Anthropocene—epistemic, ethical, or otherwise—with genuine multilateralism, pluralism, and pluriverses.

References

Alley, Kelly, D. 2019. "River Goddesses, Personhood and Rights of Nature: Implications for Spiritual Ecology." *Religions* 10: 501. https://doi.org/10.3390/rel10090502

Appadurai, Arjun. 2000. *Globalization and Area Studies: The Future of a False Opposition.* Amsterdam: Centre for Asian Studies Amsterdam.

Banerjee, Milinda and Jelle J.P. Wouters. 2022. *Subaltern Studies 2.0: Being against the Anthropocene.* Chicago: Chicago University Press/Prickly Paradigm Press.

Barad, Karen. 2007. *Meeting the Universe Halfway: Quantum Physics and the Entanglement of Matter and Meaning.* Durham: Duke University Press.

Blaser, Mario. 2009. "Political Ontology." *Cultural Studies* 23(5, 6): 873–896.

Blaser, Mario. 2016. "Is Another Cosmopolitics Possible?" *Cultural Anthropology* 31(4): 545–570.

Bristow, Tom and Thomas H. Ford. 2016. "Climates of history, cultures of climate." In *A Cultural History of Climate Change*, edited by Tom Bristow and Thomas H. Ford. London: Routledge, 1–14.

Carey, Mark and Philip Garone. 2014. "Forum Introduction," 284–295. In "Forum: Climate Change and Environmental History," edited by Mark Carey, Philip Garone, Adrian Howkins, Georgina Endfield, Lawrence Culver, Sherry Johnson, Sam White, and James Roger Fleming, *Environmental History*: 281–364.

Chakrabarty, Dipesh. 2009. "The Climate of History: Four Theses." *Critical Inquiry* 35(2): 197–222.

Chakrabarty, Dipesh. 2021. *The Climate of History in A Planetary Age*. Chicago: University of Chicago Press.

Chatterjee, Sourav, et al. 2021. "A Possible Relation between Arctic Sea Ice and Late Season Indian Summer Monsoon Rainfall Extremes." *NPJ: Climate and Atmospheric Science* 4. https://doi.org/10.1038/s41612-021-00191-w

Chao, Sophie and Dion Enari. 2021. "Decolonising Climate Change: A Call for Beyond-Human Imaginaries and Knowledge Generation." *E-Tropic: Electronic Journal of Studies in the Tropics* 20(2): 32–54.

Chao, Sophie. 2022. *In the Shadow of the Palms: More-Than-Human Becomings in West Papua*. Durham: Duke University Press.

Coakley, J.A. 2003. "Reflectance and Albedo, Surface." *Encyclopedia of Atmospheric Sciences* 1914–1923. https://www.sciencedirect.com/science/article/pii/B0122270908000695

Cruikshank, Julie. 2006. *Do Glaciers Listen? Local Knowledge, Colonial Encounters, and Social Imagination*. Vancouver: University of British Columbia Press.

Crutzen, P.J. (2006). The "Anthropocene". In: Ehlers, E., Krafft, T. (eds) Earth System Science in the Anthropocene. Springer, Berlin, Heidelberg. https://doi.org/10.1007/3-540-26590-2_3

De La Cadena, Marisol. 2015. *Earth Beings: Ecologies of Practice across Andean Worlds*. Durham: Duke University Press.

De la Cadena, Marisol and Mario Blaser (eds.). 2018. *A World of Many Worlds*. Durham: Duke University Press.

Edward, Paul. 2013. *A Vast Machine: Computer Models, Climate Data, and the Politics of Global Warming (Infrastructures)*. Cambridge, MA: MIT Press.

Escobar, Arturo. 2018. *Designs for the Pluriverse: Radical Interdependence, Autonomy, and the Making of Worlds*. Durham: Duke University Press.

Fitzharris, Blair, et al. 2007. "Teleconnections between Andean and New Zealand Glaciers." *Global and Planetary Change* 59(1–4): 1590174. https://doi.org/10.1016/j.gloplacha.2006.11.022

Fitz-Henry, Erin. 2012. "The Natural Contract: From Lévi-Strauss to the Ecuadorian Constitutional Court." *Oceania* 82: 264–277.

Fleming, James Roger. 2010. *Fixing the Sky: The Checkered History of Weather and Climate Control*. New York: Columbia University Press.

Fugger, Stefan, et al. 2021. "Understanding Monsoon Controls on the Energy and Balance of Himalayan Glaciers." *The Cryosphere*. https://doi.org/10.5194/tc-2021-97

Haberman, David (ed.). 2021. *Understanding Climate Change Through Religious Lifeworlds*. Bloomington: Indiana University Press.

Haraway, Donna, J. 2008. *When Species Meet*. Minneapolis: University of Minnesota Press.

Haraway, Donna J. 2016. *Staying with the Trouble: Making Kin in the Chthulucene*. Durham: Duke University Press.

Hewitt, Flarida. 1989. "Woman's Work, Woman's Place: The Gendered Life-World of a High Mountain Community in Northern Pakistan." *Mountain Research and Development* 9(4): 335–352.
Hulme, Mike. 2009. *Why We Disagree About Climate Change: Understanding Controversy, Inaction and Opportunity.* Cambridge: Cambridge University Press.
Hulme, Mike. 2016. *Weathered: Cultures of Climate.* London: Sage.
IPCC. 2022. *Climate Change 2022: Impacts, Adaptation, and Vulnerability. Contribution of Working Group II to the Sixth Assessment Report of the Intergovernmental Panel on Climate Change.* Cambridge: Cambridge University Press.
Irvine, Richard, D.G. 2020. *An Anthropology of Deep Time: Geological Temporality and Social Life.* Cambridge: Cambridge University Press.
Kimmerer, Robin, W. 2014. *Braiding Sweetgrass: Indigenous Wisdom, Scientific Knowledge, and the Teachings of Plants.* Minneapolis: Milkweed Press.
Kohn, Eduardo. 2013. *How Forests Think: Towards an Anthropology Beyond the Human.* Berkeley: University of California Press.
Krupnik, Igor, Claudio Aporta, Shari Gearheard, Gita J. Laidler, and Lene Kielsen Holm (eds.). 2010. *SIKU: Knowing Our Ice. Documenting Inuit Sea Ice Knowledge and Use.* New York: Springer.
Latour, Bruno. 1993. *We Have Never Been Modern.* Cambridge, MA: Harvard University Press.
Latour, Bruno. 2004. 'Whose Comsos? Which Cosmopolitics? A Commentary on Ulrich Beck's Peace Proposal.' Common Knowledge, 10(3) 450-462
Latour, Bruno. 2005. *Reassembling the Social: An Introduction to Actor-Network Theory.* Oxford: Oxford University Press.
Lovelock, James. 2000. *Gaia: A New Look at Life on Earth.* Oxford: Oxford University Press.
Lyons, K. 2018. "Rivers have Memory: The (Im)Possibility of Floods and Histories of Urban De-and-Reconstruction in the Andean-Amazonian Foothills." *City and Society Forum.* https://doi.org/10.1111/ciso.12191
McBrien, Justine. 2016. "Accumulating Extinction: Planetary Catastrophism in the Necrocene." In *Anthropocene or Capitalocene? Nature, History, and the Crisis of Capitalism,* edited by Jason W. Moore. Los Angeles, CA: PM Press/Kairos.
McMichael, Anthony. 2017. *Climate Change and the Health of Nations; Famines, Fevers, and the Fate of Populations.* Oxford: Oxford University Press.
Mignolo, Walter D. 1995. *The Darker Side of the Renaissance: Literacy, Territoriality, and Colonization.* Ann Arbor: University of Michigan Press.
Moore, Jason (ed.). 2016. *Anthropocene or Capitalocene? Nature, History, and the Crisis of Capitalism.* Los Angeles, CA: PM Press/Kairos.
Nigam, S. 2003. "Teleconnections." In *Encyclopedia of Atmospheric Sciences,* edited by Fuqing Zhang and John C. Pyle, Cambridge, MA: Academic Press, 2243–2269.
Persio, L. 2017, October 30. China Plans World's Longest Tunnel to Turn Desert into California. Newsweek. Retrieved from https://www.newsweek.com/china-plans-worlds-longest-tunnel-turn-desert-california-696326
Phuntsho, Karma. 2013. *The History of Bhutan.* New Delhi: Random House.
Plumwood, Val. 2009. "Nature in the Active Voice." *Australian Humanities Review* 46: 113–129.
Pommaret, Francoise. 1994. "On Local and Mountain Deities in Bhutan." "Le culte des montagnes sacrées dans l'aire tibétaine et tibéto-birmane," 1994, Paris, France, 39–56. halshs-00717941f

Povinelli, Elizabeth A. 1995. "Do Rocks Listen? The Cultural Politics of Apprehending Australian Aboriginal Labour." *American Anthropologist* 97(3): 505–518.

Rippa, Allesandro, Galen Murton, and Matthaus Rests 2020. "Building Highland Asia in the Twenty-First Century." *Studies in Global Asia* 6(2): 83–111.

Salmond, Anna. 2014. "Tears of Rangi: Water, power, and people in New Zealand." *Hau: Journal of Ethnographic Theory* 4(3): 285–309.

Schmidt, Jemery J. 2021. "Glacial Deaths, Glacial Extinction." *Environmental Humanities* 13(2): 281–300.

Shepherd, Theodore, G., et al. 2018. "Storylines: An Alternative Approach to Representing Uncertainty in Physical Aspects of Climate Change." *Climate Change* 151(3): 555–571.

Shepherd, Theodore G. and Elisabeth A. Lloyd. 2021. "Meaningful climate science." *Climatic Change* 169: 17. https://doi.org/10.1007/s10584-021-03246-2.

Smyer Yü, Dan. 2016. *Mindscaping the Landscape of Tibet: Place, Memorability, Ecoaestethics.* New York: De Gruyter.

Smyer Yü, Dan. 2020. "The Critical Zone as a Planetary Animist Sphere: Ethographing an Affective Consciousness of the Earth." *JSRNC* 14(2): 271–290. http://doi.org/10.1558/jsrnc.39680

Smyer Yü, Dan. 2021. "Situating environmental humanities in the New Himalayas: An introduction." In *Environmental Humanities in the New Himalayas Symbiotic Indigeneity, Commoning, Sustainability*, edited by Dan Smyer Yü and Erik de Maaker. London: Routledge, 1–24.

Snively, Gloria and John Corsiglia. 2001. 'Discovering Indigenous Science: Implications for Science Education.' *Science Education* 85(1): 6-34.

Tănăsescu, Mihnea. 2022. *Understanding the Rights of Nature: A Critical Introduction.* (New Ecology, 6). Bielefeld: transcript Verlag. https://doi.org/10.14361/9783839454312.

Tsing, Anna Lowenhaupt. 2015. *The Mushroom at the End of the World.* Princeton: Princeton University Press.

Tsosie, Krystal. S. and Katrina G. Claw. 2020. 'Indigenizing Science and Reasserting Indigeneity in Research. *Human Biology* 91(3): 137-140. Van Dooren, Thom. 2014. *Flight Ways: Life and Loss at the Edge of Extinction.* New York: Columbia University Press.

Van Dooren, Thom. 2019. *The Wake of Crows: Living and Dying in Shared Worlds.* New York: Columbia University Press.

Van Dooren, Thom and Rose, Deborah, B. 2016. "Lively Ethography: Storying Animist Worlds." *Environmental Humanities* 8(1): 77–94.Van Schendel, Willem. 2020. *A History of Bangladesh.* Cambridge: Cambridge University Press.

Victor, David. 2015. "Climate Change: Embed the Social Sciences in Climate Policy." *Nature* 520: 27–29.

Viveiros de Castro, Eduardo. 1998. "Cosmological Deixis and Amerindian Perspectivism." *The Journal of the Royal Anthropological Institute* 4(3): 469–488.

Wouters, Jelle J.P. and Michael T. Heneise. 2022. "Introduction: Highland Asia as a World-Region." In *The Routledge Handbook of Highland Asia,* edited by Jelle J.P. Wouters and Michael T. Heneise. London and New York: Routledge.

INDEX

For Product Safety Concerns and Information please contact our EU
representative GPSR@taylorandfrancis.com
Taylor & Francis Verlag GmbH, Kaufingerstraße 24, 80331 München, Germany

www.ingramcontent.com/pod-product-compliance
Ingram Content Group UK Ltd.
Pitfield, Milton Keynes, MK11 3LW, UK
UKHW020144200926
475702UK00005B/912